BASIC PROGRAMS FOR PRODUCTION AND OPERATIONS MANAGEMENT

Pricha Pantumsinchai
Decision Systems, Inc.

M. Zia Hassan
Illinois Institute of Technology

Ishwar D. Gupta
Harris Trust and Savings Bank

Prentice-Hall, Inc., Englewood Cliffs, NJ 07632

Pantumsinchai, Pricha, (date)
  BASIC programs for production and operations management.

  Includes bibliographies and index.
  1. Production management—Computer programs.
2. Basic (Computer program language) I. Hassan, M. Zia,
(date). II. Gupta, Ishwar, (date).
III. Title. IV. Title: B.A.S.I.C. programs for
production and operations management.
TS155.P257 1982   658.4′0028′5425   82-23040
ISBN 0-13-066266-6

Editorial/production supervision and
  interior design: Susan Adkins
Cover design: Diane Saxe
Manufacturing buyer: Ed O'Dougherty

© 1983 by Prentice-Hall, Inc., Englewood Cliffs, New Jersey 07632

All rights reserved. No part of this book may be
reproduced, in any form or by any means,
without permission in writing from the publisher.

Printed in the United States of America

10 9 8 7 6 5 4 3 2 1

ISBN 0-13-066266-6

Prentice-Hall International, Inc., **London**
Prentice-Hall of Australia Pty. Limited, **Sydney**
Editora Prentice-Hall do Brasil, Ltda., **Rio de Janeiro**
Prentice-Hall Canada Inc., **Toronto**
Prentice-Hall of India Private Limited, **New Delhi**
Prentice-Hall of Japan, Inc., **Tokyo**
Prentice-Hall of Southeast Asia Pte. Ltd., **Singapore**
Whitehall Books Limited, **Wellington, New Zealand**

*This book is dedicated
to our families
and
to you*

CONTENTS

PREFACE		ix
CHAPTER 1	Introduction	1
CHAPTER 2	Capital Budgeting and Financing	10
CHAPTER 3	Forecasting	41
CHAPTER 4	Linear Programming Applications	107
CHAPTER 5	Project Management (CPM/PERT)	159
CHAPTER 6	Work Measurement	201
CHAPTER 7	Layout of Facilities	232
CHAPTER 8	Inventory Analysis	257
CHAPTER 9	Statistical Quality Control	314
CHAPTER 10	Waiting Line Analysis	400
INDEX		431

PREFACE

The techniques of production and operations management are widely accepted as a means of dealing with the complex operational problems facing operating organizations. These problems usually require computers to solve them. However, although students, analysts, and managers have access to computers of all sizes, the availability of software in operations management on these machines is currently very limited.

Our aim in writing this book is to provide a set of easy-to-use BASIC computer programs in operations management for students and professionals in business, industrial engineering, management science, and operations research. The book can be used as a supplement to the myriad textbooks in these areas.

For the students who are learning these techniques for the first time, the use of the computer programs to solve realistic problems provides an experience that will foster the understanding of, and encourage insight into, the power and the limitations of the techniques.

For the professionals, we believe that some of the programs in this book will be useful for analyzing problems that arise in day-to-day operations as well as those that require careful planning. These programs can easily be adapted to run on large main-frame computers, minicomputers, and microcomputers.

As far as possible, we have tried to include some short descriptions of the techniques used in the programs. The users should review any relevant material from other sources. We strongly suggest that the users familiarize themselves with the commands contained in the example terminal sessions before attempting to use the programs. These example terminal sessions are designed to illustrate the input requirements and the essence of the output capability of each program, and its operating characteristics.

Our efforts in writing this book span over two years during which a number of our Ph.D. students have provided assistance. First and foremost, we thank Mr. Toemchai Bunnak for assisting in the development of the programs in the quality control chapter and for helping with many aspects of typing the manuscript. We also thank Mr. M.V. Sarman for his assistance in the development of the program ASSMBLY, and Mr. S. Palaniswami for helping with the proofreadings.

We are indebted to our colleagues who have read various chapters and have made valuable comments and suggestions. They are Dr. Martha Evens, Dr. James E. Hall, Dr. Thomas W. Knowles, Dr. Spencer B. Smith, Dr. Nick T. Thomopoulos, Dr. V.K. Venkataraman, and Dr. Reino V. Warren. Thanks also to Prof. Daniel G. Shimshak of the University of Massachusetts and Prof. Richard Discenza of the University of Colorado who reviewed the manuscript. Last but not least, we are indebted to Ms. Patricia Davis who has typed many drafts of the various chapters.

The Authors

CHAPTER 1

INTRODUCTION

I. Organization of Chapters

II. Design Philosophy and Program Characteristics

III. Program Structure

IV. BASIC Language Used

V. Experience on One Microcomputer

VI. About This Book

This book is a compendium of twenty-three computer programs written in BASIC language dealing with production and operations management. It covers a large variety of aspects of operations within an organization, ranging from planning functions such as financial analysis, forecasting, and project management to detailed operations such as work measurement, inventory control, and quality control. It contains over seventy distinct mathematical models and techniques that have been widely adopted in industry today. The book is primarily written to fill the gap in the ever-increasing software demands for mini-micro computers in these areas.

The book contains almost ten thousand lines of source codes written in the most elementary form of BASIC language, thus making the programs highly portable. Care has been taken during the development of these programs so that the effort required to implement them on different computers, large or small, is minimized.

A brief description of the programs follows.

```
     Program            Description
-----------------------------------------------------------------
 1.  CAPBUD             Capital Budgeting and Investment Decisions
 2.  LEASE              Lease versus Purchase Analysis
 3.  8CURVS             Fitting eight curves to X,Y Data
 4.  FORCST             Forecasting and Time Series Analysis
 5.  LP                 Linear Programming via Simplex Method
 6.  AGG                Aggregate Planning Model using LP
 7.  TRANSP             Transportation Algorithm
 8.  CPM                Critical Path Method
 9.  PERT               Program Evaluation and Review Technique
10.  TIME.STY           Time Study Analysis
11.  WORK.SMPLG         Work Sampling Analysis
12.  PLAYOUT            Process Layout Design
13.  ASSMBLY            Assembly Line Planning
14.  PBREAK             Price Breaks in Inventory Control
15.  EXCHNGE            Exchange Curves for Inventory Planning
16.  INVCOST            Inventory Cost Models
17.  LOTSZ.MRP          MRP Lot-Sizing Techniques

18.  XBARR              $\bar{X}$ and R Statistical Control Charts
19.  PUC                P, U, and C Statistical Control Charts
20.  LBL.AOQL           Design of Single-Sampling Plans for
                          Attributes based on AOQL
21.  AQL.LTFD           Design of Single-Sampling Plans for
                          Attributes based on AQL and LTFD
22.  LBL.SDV            LBL Single-Sampling Plans for Variables
23.  QUEUE              Single- and Multiple-Server Queueing Models
```

I. Organization of Chapters

Each chapter contains a brief discussion of the underlying theory of the subject and the assumptions used in each computer program. It identifies all of the necessary inputs to each program and describes the outputs from the program. There is at least one complete example terminal session for each program in order to demonstrate its usage. The program source listing follows the example terminal sessions. At the end of each chapter, the dimensional specification of each program is discussed so that possible expansion of the program to handle larger amounts of data can be made with ease. Each chapter concludes with a list of references relevant to the materials contained in the chapter.

II. Design Philosophy and Program Characteristics

Inasmuch as it is impossible to write a computer program that will satisfy everyone's needs and wants, we have attempted to develop the programs so that modifications to the programs can be done with relative ease. This is accomplished by organizing a similar structure for all programs. We have also chosen to develop most of the programs with a command-driven characteristic. This means that the user will be giving commands to instruct the computer to perform the desired tasks. The advantage of having a command-driven structure is that additional commands can be added to a program to perform extra functions rather easily. This is contrary to the query-driven programming approach in which it may be quite difficult to modify the program once its structure is built.

Developing computer programs that are strictly command-driven is by far the most powerful approach because it permits great flexibility in implementing various options within a command. However, this type of programs is more difficult to develop because the command structures can be quite complicated. Also, they may not be as friendly to use as query-driven programs because they may require a lot of familiarity with the syntax of the command structure itself.

Strictly query-driven computer programs, although friendly to use, may lack the flexibility that the command-driven type has in terms of the user being able to choose among the possible functions that a computer program offers. This is especially true when a program requires answers to a large number of questions before it can determine precisely what the user wishes to do.

We have, therefore, adopted a combination of both approaches in designing our programs. With the exception of the programs LP and AGG, which are strictly query-driven, all other programs in this book have one level of command structure. In other words, each command will perform its function if all of the input data are readily available in the computer memory. If there is additional information required by the command, the program will prompt for the necessary information with one or more queries. Each command consists of one word which attempts to describe its function as clearly as is practical. For example, the HELP command will display all available commands, the RESTART command will start the execution of the program without having to recompile it again, the EXIT command will terminate the program execution.

All computer programs are designed to run interactively with no requirements of extra input/output devices. All input data are either prompted for from the computer terminal or entered through the use of DATA statements in BASIC. The use of DATA statements can serve as a method of storing data as part of the program. But it means that the user has to recompile the program every time the data are altered. The task of converting some programs to accept input data from the DATA statements should be quite trivial. The following section on program structure will discuss it further.

On some computer systems, especially the larger ones, facilities are provided to create a file in which all input data and commands can be stored in free format. This is called a "control file." The program execution can be initiated by activating the control file using some appropriate system command. This is especially helpful when the amount of data to be entered through the terminal becomes prohibitive and provisions are not made within the program to edit the data. Other alternatives would be to modify the program to read input data from the DATA statements or to modify the program so that it is capable of reading input data from a file such as a disk file.

All outputs from these programs are designed to appear on an 80-column output device such as a standard CRT terminal or a paper terminal that can print 80 characters or more per line.

III. Program Structure

All computer programs in this book have been written using a specific structure. The structure of the source listing can be broken down into six major segments as follows.

1. Dimensional Specifications. All declarations of array sizes are made in this first segment, and every array is explicitly declared. If the dimensions of the arrays need to be redefined, the section on dimensional specifications at the end of each chapter describes the necessary procedure.

2. Brief Introduction to the Program Usage. This

segment contains descriptive information about the program and its input requirements. The programs LP, TIME.STY, WORK.SMPLG, and ASSMBLY are the only exceptions in that the input requirements are not explicitly stated. This is because the input data for these four programs are somewhat more involved and we require the user to enter them using DATA statements. In this case, the programs will inform the user to obtain the detail of proper input sequence from examples contained in this book.

3. Prompting of Input Data. This segment exists only if the program requires its input data interactively from the terminal. If the user wishes to modify the program to read the input data from DATA statements, then this section should be modified. Perhaps the simplest approach is to modify the relevant INPUT statements to the READ statements leaving the prompting queries unmodified.

4. Command Verification Segment. All programs, except LP and AGG, contain the HELP command and the EXIT command. Each command will either set up some flag variables or call a subroutine using the GOSUB statement. If additional commands are desirable, this section is easily modified.

5. Subroutines. The bulk of the computation associated with a command including any output reports is done through subroutines. If the output from a program requires fixed formatting (e.g., a table), there is one subroutine that contains all of the format statements. This is called the initialization routine. It is usually located at the end of the source listing. This is extremely convenient to modify because all format statements are grouped in one place. If there is any incompatibility between the user's BASIC interpreter and the one that we use, this should be the first place to look for the differences.

6. DATA Statements. As mentioned above, the programs LP, TIME.STY, WORK.SMPLG, and ASSMBLY are the only four programs that require the input data to be entered via DATA statements. They are usually designed to follow the END statement, which is always the last statement in our source listings. Except for the program LP, any DATA statement that appears in the source listing is necessarily part of the program and should not be modified. This is especially crucial in the quality control chapter programs (i.e., XBARR, LBL.AOQL, and AQL.LTFD), which store some statistical tables using the DATA statements.

IV. BASIC Language Used

All programs have been developed using the BASICV✦REV18.1 Interpreter on the PRIME 400 computer at Illinois Institute of Technology, Chicago. The language, although very powerful, is somewhat limited in terms of naming a variable. In other words, a variable name or an array name cannot be more than two characters in length. Most microcomputers nowadays can accept longer variable names. This works, in fact, to our advantage in that the portability of these programs is enhanced.

In addition, we have limited ourselves only to certain types of BASICV statements which we feel are fairly universal. The following characteristics are inherent in our programs:

1. No distinction is made between an integer variable and a real variable.

2. Array Variables. At most, arrays of two dimensions are used.

3. Constants. Both numerical and string constants are utilized -- e.g., -1, 1.E30 refer to numerical constants and "HELP" is a string constant.

4. Arithmetic Operators:

```
+     addition
-     subtraction or negation
*     multiplication
/     division
**    exponentiation
```

5. Relational Operators:

```
=     equality
<>    inequality
<     less than
>     greater than
<=    less than or equal to
>=    greater than or equal to
```

6. Only one type of assignment statement is used, that is,

VARIABLE = EXPRESSION

where EXPRESSION can be a constant, or a combination of arithmetic operations of numerical variables and numerical constants, or a string variable.

7. Two types of IF-THEN statements are used:

IF logical expression THEN line number

If the logical expression is true, the program branches to the specified line number.

        IF logical expression THEN statement

If the logical expression is true, the program executes the statement.

   8. FOR-NEXT Loop for Iterative Operations. Two standard types are used:

        FOR variable=value1 TO value2 STEP value3
         . . .
        NEXT variable

        FOR variable=value1 TO value2
         . . .
        NEXT variable

   9. Input Statements. When the input data are prompted for from the terminal, the following statements are used:

        INPUT   variable
        INPUT   variable1,variable2,...,etc.

When the input data are entered through the use of DATA statements, the BASIC statements used are:

        READ   variable
        READ   variable1,variable2,...,etc.

   10. Output Statements. Outputs are always routed to the computer terminal using the following statements:

        PRINT
        PRINT variable
        PRINT variable1,variable2,...
        PRINT USING String Variable,variable1,variable2,...

The String Variable - e.g., F$(1) - always contains the necessary format. For example, if F$(1)="-####.##", this means that the variable to be printed using this format can be positive or negative and the number of decimal places is two. If the variable is nonnegative the minus sign is suppressed by BASICV. If the variable is negative, the minus sign will appear in the exact position as shown in the format -i.e., the minus sign will not be adjusted to the right as it would naturally be. Thus some of the example terminal session outputs from our programs may appear a bit strange at first glance when the formatted numbers are negative.

In addition, the use of the free format PRINT statement results in printing numbers with up to fourteen significant digits. This is inherent in the interpreter that we use. Other interpreters may produce fewer significant digits.

11. Other standard BASIC statements used are

```
REM     (for remarks or comments)
GOSUB line number
GOTO line number
RETURN
END
```

12. Intrinsic or Built-in Functions Used. Each function must have an argument that can be an arithmetic expression or a numerical constant.

```
ABS(argument)   - Takes absolute value of the argument.
EXP(argument)   - Returns exponential of the argument.
INT(argument)   - Truncates the numeric value of
                  the argument to integer. Note
                  that on some BASIC interpreters,
                  this function means rounding down.
                  Hence, if the resulting argument
                  is -1.7, rounding down will result
                  in -2 instead of -1 for truncation
                  of the number.
LOG(argument)   - Takes natural logarithm.
RND(argument)   - Generates a random number between
                  0 and 1. The argument used in all
                  of our relevant programs is zero.
SQR(argument)   - Takes square root of the argument.
TAB(argument)   - Moves to the indicated print
                  position specified by the
                  argument.
```

V. Experience on One Microcomputer

We have converted these programs to run on an IBM Personal Computer in order to assess the running times of these programs. The IBM-PC we used had 64K of random-access memory and one diskette drive.

The first step we did was to transfer the source listings from the PRIME-400 minicomputer to the IBM-PC through the telephone line using a communication program, which operates on the IBM-PC. The second step was to convert the programs to make them compatible with the BASIC interpreter under PC-DOS. In the conversion process, we found the differences in the "PRINT USING" statement, the print-format convention, the exponentiation, and some built-in functions (i.e., RND and INT functions).

When we ran the examples that are given in this book, we obtained the same results. However, for those examples that require simulations using random numbers, the numerical results were different due to a different random-number sequence we obtained from the IBM-PC. The response times for most examples were quite good. There were a number of exceptions where computational requirements were very intense. These were the program FORCST when the optimization option was utilized, the programs PERT and MRP when the simulation features were activated, and the program AGG when the number of planning horizon was large. For these cases, it may not yet be attractive to use microcomputers. In fact, it took roughly 38 minutes to solve the example of the AGG program, which is a linear programming formulation with 72 variables and 37 constraints. The same problem was solved in less than two minutes on the PRIME-400.

There is growing belief that microcomputers will become faster and more powerful in the future, and we are witnessing such happenings. Hence there is no doubt that microcomputers are going to have a significant impact on how production and operations management problems are solved in the future.

VI. About This Manuscript

Ninety-five percent of this manuscript was prepared on the PRIME-400 computer, except for some pages that require Greek symbols for the equations. The example terminal sessions, the program listings, and most of the descriptive materials were generated on a hardcopy computer terminal in order to reduce possible error.

CHAPTER 2

CAPITAL BUDGETING AND FINANCING

I. Capital Budgeting

    The Program CAPBUD
   - Input
   - Output
   - Example
   - CAPBUD Example Terminal Session
   - CAPBUD Program Listing

II. Financing

    The Program LEASE
   - Input
   - Output
   - Example
   - LEASE Example Terminal Session
   - LEASE Program Listing

III. Dimensional Specifications

IV. References

Capital budgeting is concerned with establishing the acceptability of a capital investment project. This can be achieved using criteria such as rate of return, net present value, or payback period. If the problem involves multiple projects, the profitability index (PI) of a project can be computed and used as the basis for determining its relative ranking.

Financing a project, on the other hand, is only considered after its acceptability has been established. Our discussion will be confined to the problem of asset acquisition in which the asset will be financed through leasing or establishing a debt.

In this chapter, two computer programs are provided to aid in solving these problems. They are CAPBUD and LEASE. CAPBUD deals with the capital budgeting problem, whereas LEASE deals with the financing aspects.

I. Capital Budgeting

The problem of capital budgeting involves the outlay of funds with the expectation of receiving inflow of funds in the future. The following procedure can be used in a capital budgeting analysis.

    1. Estimate the economic life of the investment in years.

    2. Estimate the operating cash flows for each year of the economic life using the following steps:

        a. Estimate the current total cash outlay and the working capital required for the project.
        b. Select the appropriate depreciation method and estimate the salvage value, if applicable. The most commonly used depreciation techniques are straight-line depreciation, sum-of-years-digits depreciation, and double-declining depreciation.
        c. Estimate the cash outflow for each year of the economic life.
        d. Estimate the cash inflow for each year of the economic life.
        e. Determine the appropriate corporate marginal income tax rate.

    3. Calculate the Internal Rate of Return (IRR), the Net Present Value (NPV), or the Payback Period of the investment in order to establish its acceptability.

The IRR of an investment is the discount rate that equates the present value (PV) of outlay of funds relating to the investment to the PV of all inflow of funds. Let n be the economic life, and R_i, $i=1,\ldots,n$ be the net cash flow (cash inflow minus cash outflow) for period i. Also, let r be the discount rate. The IRR can then be determined by solving for r in the following equation:

$$C = \sum_{i=1}^{n} R_i / (1+r)^i$$

where C = total initial investment required including any working capital.

The NPV of the investment can be calculated by selecting an appropriate discount rate or interest rate, k:

$$NPV = -C + \sum_{i=1}^{n} R_i / (1+k)^i$$

The Payback Period represents the number of years required to recover the initial investment (C). This method does not use discounted cash flows.

4. Rank the various projects using the profitability index, PI, which is defined as the ratio of the PV of future net cash flows to the initial investment:

$$PI = \left(\sum_{i=1}^{n} R_i / (1+r)^i \right) / C$$

5. Determine whether or not the project is worthy of the investment based on the analysis and other relevant factors.

The Program CAPBUD

CAPBUD has been designed such that the above analysis can be performed easily once the initial data are entered. The program can be used to determine the net cash flow if the revenue and expense are known separately. The program is command-driven, and commands are provided to alter the discount rate and the depreciation method. The depreciation methods implemented in this program are straight-line depreciation (SLDEP), sum-of-years-digits depreciation (SYDDEP), and double-declining depreciation (DDDEP). The default depreciation method selected by the program is the double-declining balance method.

INPUT
-----
1. The name of the investment
2. Life of the investment in whole period
3. Discount rate in percentage
4. The initial cash investment
5. If the net cash flows by periods are unknown:
   - the working capital amount
   - the corporate tax rate in percentage
   - revenues by periods
   - total expenses by periods
   - the salvage value of the investment
6. If the net cash flows by periods are known:
   - the net cash flows by periods

OUTPUT

There are two output generating commands: the SUMMARY and the DETAIL commands. The SUMMARY command produces a short report consisting of the IRR, the payback period, the PI, and the NPV. The PI and the NPV are based on the selected discount rate. If the net cash flows are unknown and the user is using the program to determine them, then the IRR depends on the depreciation method selected. If the net cash flows are known, this implies that the user has preselected a depreciation technique when determining the net cash flows. Hence the commands for selecting the depreciation methods have no significance in this instance. The DETAIL command, on the other hand, generates detailed output of depreciation, tax amounts, cash flows by periods, and the PV of the net cash flows by periods where applicable.

EXAMPLE

ABC company is considering investing in a punch press with a cash outlay of $20,000 for an operation that also requires $3000 in working capital. The economic life of the press is estimated to be 5 years after which its salvage value is $2000. The company marginal corporate tax rate is 50 percent. The applicable annual discount rate is 9 percent. The revenues generated and the total expenses by year are:

Year	1	2	3	4	5
Revenue	12000	15000	15000	18000	14000
Total expense	4000	5000	6000	7000	3000

The company wishes to assess the economic worthiness of this investment and the relative advantage of the sum-of-years-digits method versus the double-declining method.

Please refer to the example terminal session which immediately follows this discussion. Since, in this case, the net cash flows by years are not known, the user is using the program to find them. The HELP command is entered at the first command level prompt in order to display all available commands. Next, the command SYDDEP is entered to select the sum-of-years-digits depreciation method. The SUMMARY command shows that the NPV is $6476 with an IRR of 18.71 percent. Next, we select the double-declining method using the DDDEP command. The SUMMARY command indicates a more attractive NPV of $6680 with an IRR of 19.33 percent. The DETAIL command is entered next so that we may obtain the detailed analysis. The payback period for this investment is 4 years for both types of depreciation method.

In order to illustrate the use of the program for the case when the net cash flows by periods are known, we will use the results of the analysis up to this point. The net cash flows of the investment from the last DETAIL command are:

Year	0	1	2	3	4	5
Cash flow	-23000	8000	7400	5940	6364	5796

The investment in year zero is equal to the initial cash investment of $20,000 plus the working capital of $3000, which equals $23,000. The working capital amount can be recovered together with the salvage value of the equipment of $2000 at the end of its economic life. Hence, the capital recovery amount is $3000+$2000=$5000 in year 5. In effect, the net cash flow in year 5 is increased from $5796 in the above table to $5796+$5000=$10,796. This is the amount that we have to use in our analysis in order to produce an equivalent result to the case when the net cash flows are unknown.

Let us now continue with the example terminal session. The RESTART command is entered in order to initiate a new problem. For the input data, we note that a positive response (Y) is given to the question concerning the knowledge of the net cash flows by periods. The DETAIL command shows that the results of this analysis are identical to the double-declining depreciation case as shown earlier.

CAPBUD EXAMPLE TERMINAL SESSION
================================

CAPITAL BUDGETING AND INVESTMENT DECISIONS

NEED INTRODUCTION (Y OR N)...!N

ENTER THE NAME OF THE INVESTMENT   !PUNCH PRESS

ENTER LIFE OF THE INVESTMENT IN WHOLE PERIOD   !5

ENTER PERCENT DISCOUNT RATE   !9

DO YOU KNOW THE NET CASH FLOWS BY PERIODS (Y OR N) ...!N

ENTER THE INITIAL INVESTMENT AMOUNT IN PERIOD 0   !20000

ENTER WORKING CAPITAL AMOUNT (EXPENSES)   !3000

ENTER CORPORATE TAX RATE IN % !50

```
ENTER REVENUE FOR PERIOD 1    !12000
ENTER REVENUE FOR PERIOD 2    !15000
ENTER REVENUE FOR PERIOD 3    !15000
ENTER REVENUE FOR PERIOD 4    !18000
ENTER REVENUE FOR PERIOD 5    !14000

ENTER TOTAL EXPENSES FOR PERIOD    1    !4000
ENTER TOTAL EXPENSES FOR PERIOD    2    !5000
ENTER TOTAL EXPENSES FOR PERIOD    3    !6000
ENTER TOTAL EXPENSES FOR PERIOD    4    !7000
ENTER TOTAL EXPENSES FOR PERIOD    5    !3000
```

ENTER THE SALVAGE VALUE OF THE INVESTMENT   !2000

PLEASE ENTER COMMAND OR TYPE HELP

COMMAND --> !HELP

AVAILABLE COMMANDS
==================
```
HELP      - PRINT THIS MESSAGE
CHANGE    - SELECT A NEW DISCOUNT RATE
SUMMARY   - SELECT SUMMARY OUTPUT OPTION
DETAIL    - SELECT DETAIL OUTPUT OPTION
SLDEP     - SELECT STRAIGHT-LINE DEPRECIATION METHOD
SYDDEP    - SELECT SUM-OF-YEARS-DIGITS DEPRE. METHOD
DDDEP     - SELECT DOUBLE-DECLINING DEPRE. METHOD
RESTART   - START A NEW PROBLEM
EXIT      - EXIT FROM THE PROGRAM
```

COMMAND --> !SYDDEP

COMMAND --> !SUMMARY

I N V E S T M E N T : PUNCH PRESS

DEPRECIATION METHOD = SUM-OF-YEARS-DIGITS DEPRECIATION

DISCOUNT RATE (%) = 9

I N T E R N A L   R A T E   O F   R E T U R N = 18.710%
P A Y B A C K   P E R I O D   =   4.000
PROJECT   P R O F I T A B I L I T Y   I N D E X =   1.282
N E T   P R E S E N T   V A L U E = $ 6476
COMMAND --> !DDDEP
COMMAND --> !SUMMARY

I N V E S T M E N T : PUNCH PRESS

DEPRECIATION METHOD = DOUBLE-DECLINING DEPRECIATION

DISCOUNT RATE (%) = 9

I N T E R N A L   R A T E   O F   R E T U R N = 19.330%
P A Y B A C K   P E R I O D   =   4.000
PROJECT   P R O F I T A B I L I T Y   I N D E X =   1.290
N E T   P R E S E N T   V A L U E = $ 6680
COMMAND --> !DETAIL

```
INVESTMENT : PUNCH PRESS

DEPRECIATION METHOD = DOUBLE-DECLINING DEPRECIATION

DISCOUNT RATE (%)    = 9

1. COST OF INVESTMENT ................. $ 20000
2. WORKING CAPITAL .................... $  3000
3. ECONOMIC LIFE (PERIODS) ...........    5
4. SALVAGE VALUE ..................... $  2000
5. DEPRECIATION BASE ................. $ 20000
6. CAPITAL RECOVERY .................. $  5000
7. TOTAL DEPRECIATION ................ $ 18000


N E T    C A S H    F L O W   DOUBLE-DECLINING DEPRECIATION
-----------------------------------------------------------------
PERIOD OPERATING  DEPRE-    TAXABLE      TAX    INCOME     CASH
       PROFIT     CIATION   INCOME               AFT.TAX   FLOW
------ --------- --------- --------- --------- --------- --------
   1     8000     8000         0         0         0      8000
   2    10000     4800      5200      2600      2600      7400
   3     9000     2880      6120      3060      3060      5940
   4    11000     1728      9272      4636      4636      6364
   5    11000      592     10408      5204      5204      5796

CAPITAL RECOVERY ..................................   5000

I N T E R N A L    R A T E    O F    R E T U R N  =  19.330%

P A Y B A C K    P E R I O D                       =   4.000

PROJECT   P R O F I T A B I L I T Y    I N D E X  =   1.290


NET PRESENT VALUE USING DISCOUNT RATE    9.000 %
-----------------------------------------------------
PERIOD    CASH   CUMULATIVE    PV OF    CUMULATIVE
          FLOW   CASH FLOW   CASH FLOW  PRESENT VALUE
------ --------- ---------- --------- -------------
   0 -  23000 -   23000 -   23000 -     23000
   1     8000     15000      7339 -     15661
   2     7400      7600      6228 -      9432
   3     5940      1660      4587 -      4845
   4     6364      4704      4508 -       337
   5     5796     10500      3767        3430

PRESENT VALUE OF CAPITAL RECOVERY   = $    3250

N E T    P R E S E N T    V A L U E = $    6680
```

```
COMMAND --> !RESTART

ENTER THE NAME OF THE INVESTMENT    !PUNCH PRESS - KNOWN NET CASH FLOWS
ENTER LIFE OF THE INVESTMENT IN WHOLE PERIOD    !5
ENTER PERCENT DISCOUNT RATE    !9
DO YOU KNOW THE NET CASH FLOWS BY PERIODS (Y OR N) ...!Y
ENTER THE INITIAL CASH INVESTMENT IN PERIOD 0    !23000
ENTER NET CASH FLOW FOR PERIOD    1    !8000
ENTER NET CASH FLOW FOR PERIOD    2    !7400
ENTER NET CASH FLOW FOR PERIOD    3    !5940
ENTER NET CASH FLOW FOR PERIOD    4    !6364
ENTER NET CASH FLOW FOR PERIOD    5    !10796

PLEASE ENTER COMMAND OR TYPE HELP

COMMAND --> !DETAIL


I N V E S T M E N T  :  PUNCH PRESS - KNOWN NET CASH FLOWS
DISCOUNT RATE (%)   = 9

I N T E R N A L     R A T E     O F     R E T U R N  =  19.330%
P A Y B A C K     P E R I O D                        =   4.000
PROJECT    PROFITABILITY    INDEX  =   1.290


NET PRESENT VALUE USING DISCOUNT RATE    9.000 %
----------------------------------------------------
PERIOD    CASH  CUMULATIVE    PV OF   CUMULATIVE
          FLOW  CASH FLOW   CASH FLOW PRESENT VALUE
------   ------ ---------- ---------- -------------
   0  -  23000 -   23000 -    23000 -    23000
   1      8000    15000        7339 -    15661
   2      7400     7600        6228 -     9432
   3      5940     1660        4587 -     4845
   4      6364     4704        4508 -      337
   5     10796    15500        7017       6680

N E T     P R E S E N T     V A L U E = $    6680

COMMAND --> !EXIT
```

```
10 REM      CAPBUD
20 REM
30 DIM H$(20),B$(3)
40 DIM E0(20),E1(20),E2(20)
50 DIM C(20),R(20),G(20),F(20),Q(20),U(20),V(20),O(20)
60 REM M0 IS THE MAXIMUM NUMBER OF PERIODS ALLOWED
70 M0=20
80 GOSUB 3420
90 K9=3
100 PRINT "CAPITAL BUDGETING AND INVESTMENT DECISIONS"
110 PRINT "--------------------------------------"
120 PRINT
130 PRINT "NEED INTRODUCTION (Y OR N)...";
140 INPUT Y$
150 IF Y$<>"Y" THEN 430
160 PRINT
170 PRINT "THIS PROGRAM CAN BE USED TO EVALUATE THE RETURN ON"
180 PRINT "AN INVESTMENT.  THREE TECHNIQUES ARE PROVIDED TO"
190 PRINT "PERFORM THE ANALYSIS :"
200 PRINT "    1. NPV = NET PRESENT VALUE"
210 PRINT "    2. IRR = INTERNAL RATE OF RETURN"
220 PRINT "    3. PBP = PAYBACK PERIOD"
230 PRINT "INPUT REQUIREMENTS"
240 PRINT "------------------"
250 PRINT "1. THE NAME OF THE INVESTMENT"
260 PRINT "2. THE LIFE OF THE INVESTMENT (IN WHOLE NUMBER)"
270 PRINT "3. THE PERCENT DISCOUNT RATE"
280 PRINT "4. THE INITIAL CASH INVESTMENT(CASH OUTFLOW)"
290 PRINT "5. IF THE NET CASH FLOWS BY PERIODS ARE UNKNOWN"
300 PRINT "   - THE WORKING CAPITAL AMOUNT"
310 PRINT "   - PERCENT CORPORATE TAX RATE"
320 PRINT "   - REVENUES BY PERIODS"
330 PRINT "   - TOTAL EXPENSES BY PERIODS"
340 PRINT "   - THE SALVAGE VALUE OF THE INVESTMENT"
350 PRINT "6. IF THE NET CASH FLOWS BY PERIODS ARE KNOWN"
360 PRINT "   - THE NET CASH FLOWS BY PERIODS"
370 PRINT
380 PRINT "A PROFITABILITY INDEX IS ALSO PROVIDED FOR EVALUATING"
390 PRINT "THE CAPITAL RATIONING DECISIONS.  THE PROGRAM CAN"
400 PRINT "ALSO BE USED TO ANALYZE THE BEST DEPRECIATION SCHEME"
410 PRINT "FOR THE INVESTMENT."
420 REM     ***  I N P U T   S E C T I O N  ***
430 PRINT
440 PRINT "ENTER THE NAME OF THE INVESTMENT    ";
450 INPUT P$
460 PRINT
470 PRINT "ENTER LIFE OF THE INVESTMENT IN WHOLE PERIOD
480 INPUT K
490 IF K <= M0 THEN 530
500 PRINT "MAXIMUM NUMBER OF PERIODS ALLOWED IS   ";M0
510 PRINT "PLEASE TRY AGAIN  !!!"
520 GOTO 460
530 PRINT
540 PRINT "ENTER PERCENT DISCOUNT RATE     ";
```

```
550 INPUT R1
560 K7=0
570 PRINT
580 PRINT "DO YOU KNOW THE NET CASH FLOWS BY PERIODS (Y OR N) ...";
590 INPUT W$
600 IF W$ = "N" THEN 740
610 REM   NET CASH FLOWS ARE KNOWN -- ENTER DATA
620 K7=1
630 PRINT
640 PRINT "ENTER THE INITIAL CASH INVESTMENT IN PERIOD 0  ";
650 INPUT W
660 C(0)=-W
670 PRINT
680 FOR J=1 TO K
690 PRINT "ENTER NET CASH FLOW FOR PERIOD    ";J;"     ";
700 INPUT C(J)
710 NEXT J
720 GOTO 990
730 REM
740 REM   NET CASH FLOWS UNKNOWN -- CALCULATED
750 REM   ---------------------
760 PRINT
770 PRINT "ENTER THE INITIAL INVESTMENT AMOUNT IN PERIOD 0   ";
780 INPUT G1
790 PRINT
800 PRINT "ENTER WORKING CAPITAL AMOUNT (EXPENSES)    ";
810 INPUT G2
820 PRINT
830 PRINT "ENTER CORPORATE TAX RATE IN % ";
840 INPUT R
850 PRINT
860 FOR J=1 TO K
870 PRINT "ENTER REVENUE FOR PERIOD ";J;"   ";
880 INPUT R(J)
890 NEXT J
900 PRINT
910 FOR J=1 TO K
920 PRINT "ENTER TOTAL EXPENSES FOR PERIOD   ";J;"     ";
930 INPUT G(J)
940 G(J)=-G(J)
950 NEXT J
960 PRINT
970 PRINT "ENTER THE SALVAGE VALUE OF THE INVESTMENT   ";
980 INPUT S0
990 K8=0
1000 G5=0
1010 REM     *** ACCEPT COMMANDS HERE ***
1020 PRINT
1030 PRINT "PLEASE ENTER COMMAND OR TYPE HELP"
1040 PRINT
1050 PRINT "COMMAND --> ";
1060 INPUT C$
1070 PRINT
1080 IF C$ <> "HELP" THEN 1110
```

```
1090 GOSUB 1340
1100 GOTO 1050
1110 IF C$ <> "CHANGE" THEN 1140
1120 GOSUB 3380
1130 GOTO 1040
1140 IF C$ <> "SLDEP" THEN 1170
1150 K9=1
1160 GOTO 1050
1170 IF C$ <> "SYDDEP" THEN 1200
1180 K9=2
1190 GOTO 1050
1200 IF C$ <> "DDDEP" THEN 1230
1210 K9=3
1220 GOTO 1050
1230 IF C$ <> "SUMMARY" THEN 1260
1240 K8=0
1250 GOTO 1280
1260 IF C$ <> "DETAIL" THEN 1300
1270 K8=1
1280 GOSUB 1480
1290 GOTO 1050
1300 IF C$= "RESTART" THEN 430
1310 IF C$ = "EXIT" THEN 3670
1320 PRINT " *** ERROR ***   PLEASE TRY AGAIN !"
1330 GOTO 1040
1340 REM    *** HELP COMMAND ***
1350 PRINT "AVAILABLE COMMANDS "
1360 PRINT "=================="
1370 PRINT "HELP      - PRINT THIS MESSAGE"
1380 PRINT "CHANGE    - SELECT A NEW DISCOUNT RATE"
1390 PRINT "SUMMARY   - SELECT SUMMARY OUTPUT OPTION"
1400 PRINT "DETAIL    - SELECT DETAIL OUTPUT OPTION"
1410 PRINT "SLDEP     - SELECT STRAIGHT-LINE DEPRECIATION METHOD"
1420 PRINT "SYDDEP    - SELECT SUM-OF-YEARS-DIGITS DEPRE. METHOD"
1430 PRINT "DDDEP     - SELECT DOUBLE-DECLINING DEPRE. METHOD"
1440 PRINT "RESTART   - START A NEW PROBLEM"
1450 PRINT "EXIT      - EXIT FROM THE PROGRAM"
1460 PRINT
1470 RETURN
1480 REM    *** SUMMARY/DETAIL COMMANDS ***
1490 PRINT
1500 PRINT
1510 PRINT "I N V E S T M E N T  :   ";P$
1520 IF K7=1 THEN 1550
1530 PRINT
1540 PRINT "DEPRECIATION METHOD = ";B$(K9)
1550 PRINT
1560 PRINT "DISCOUNT RATE (%)   = ";R1
1570 PRINT
1580 IF K7=1 THEN 2000
1590 IF K9=1 THEN GOSUB 2090
1600 IF K9=2 THEN GOSUB 2210
1610 IF K9=3 THEN GOSUB 2310
1620 REM    PRINT DEPRECIATION TABLE
```

```
1630 IF K8=0 THEN 1710
1640 PRINT "1. COST OF INVESTMENT ................. $ ";G1
1650 PRINT "2. WORKING CAPITAL .................... $ ";G2
1660 PRINT "3. ECONOMIC LIFE (PERIODS) ............    ";K
1670 PRINT "4. SALVAGE VALUE ...................... $ ";S0
1680 PRINT "5. DEPRECIATION BASE .................. $ ";G4
1690 PRINT "6. CAPITAL RECOVERY ................... $ ";G5
1700 PRINT "7. TOTAL DEPRECIATION ................. $ ";G0
1710 REM    CALCULATE NET CASH FLOW
1720 REM Q(.)=TAXABLE INCOME ; V(.)=INCOME AFTER TAXES
1730 REM F(.)=DEPRECIATION   ; O(.)=OPERATING PROFIT
1740 REM U(.)=TAX
1750 C(0)= -(G1+G2)
1760 FOR J=1 TO K
1770 O(J)=R(J)+G(J)
1780 Q(J)=O(J)-F(J)
1790 U(J)=R*Q(J)/100
1800 V(J)=Q(J)-U(J)
1810 C(J)=U(J)+F(J)
1820 NEXT J
1830 REM CAPITAL RECOVERY(IF ANY) OCCURS AT END OF ECONOMIC LIFE
1840 G5=S0+G2
1850 IF K8=0 THEN 2000
1860 REM    PRINT NET CASH FLOW TABLE
1870 REM    ------------------------
1880 PRINT
1890 PRINT
1900 PRINT USING H$(1),B$(K9)
1910 PRINT H$(2)
1920 PRINT H$(3)
1930 PRINT H$(4)
1940 PRINT H$(5)
1950 FOR J=1 TO K
1960 PRINT USING H$(6),J,O(J),F(J),Q(J),U(J),V(J),C(J)
1970 NEXT J
1980 PRINT
1990 PRINT USING H$(7),G5
2000 REM    FIND INTERNAL RATE OF RETURN
2010 GOSUB 2430
2020 REM    CALCULATE PAYBACK PERIOD
2030 GOSUB 3160
2040 REM    CALCULATE PROJECT PROFITABILITY INDEX
2050 GOSUB 3280
2060 REM    CALCULATE NET PRESENT VALUE
2070 GOSUB 2860
2080 RETURN
2090 REM    STRAIGHT-LINE DEPRECIATION
2100 REM    G4= DEPRECIATION BASE ,   G5= CAPITAL RECOVERY
2110 REM    G0 = TOTAL DEPRECIATION
2120 G4=G1-S0
2130 G5=G2+S0
2140 G0=0
2150 L1=G4/K
2160 FOR J=1 TO K
```

22

```
2170 F(J)=L1
2180 G0=G0+F(J)
2190 NEXT J
2200 RETURN
2210 REM SUM-OF-YEARS-DIGITS DEPRECIATION
2220 G4=G1-S0
2230 G5=G2+S0
2240 G0=0
2250 Y0=K*(K+1)/2
2260 FOR J=1 TO K
2270 F(J)=G4*(K+1-J)/Y0
2280 G0=G0+F(J)
2290 NEXT J
2300 RETURN
2310 REM DOUBLE-DECLINING DEPRECIATION
2320 G4=G1
2330 G5=G2+S0
2340 G0=0
2350 K0=K-1
2360 FOR J=1 TO K0
2370 F(J)=(G1-G0)*2/K
2380 G0=G0+F(J)
2390 NEXT J
2400 F(K)=G1-S0-G0
2410 G0=G0+F(K)
2420 RETURN
2430 REM    CALCULATING INTERNAL RATE OF RETURN
2440 REM    CALCULATION ACCURACY IS + OR - .01%
2450 PRINT
2460 D1=0
2470 D2=100
2480 REM    CHECK IF ACTUAL IRR FALLS OUTSIDE LOWER LIMIT
2490 A0=-C(0)
2500 A1=0
2510 FOR J=1 TO K
2520 A1=A1+C(J)*(1/(1+D1/100)**J)
2530 NEXT J
2540 A1=A1+G5*(1/(1+D1/100)**K)
2550 IF A0>A1 THEN 2820
2560 REM    CHECK IF ACTUAL IRR FALLS OUTSIDE UPPER LIMIT
2570 A1=0
2580 FOR J=1 TO K
2590 A1=A1+C(J)*(1/(1+D2/100)**J)
2600 NEXT J
2610 A1=A1+G5*(1/(1+D2/100)**K)
2620 IF A0<A1 THEN 2840
2630 L=D1
2640 H=D2
2650 S=1
2660 FOR Y=1 TO 3
2670 FOR X=L TO H STEP S
2680 A1=0
2690 FOR J=1 TO K
2700 A1=A1+C(J)*(1/(1+X/100)**J)
```

```
2710 NEXT J
2720 A1=A1+G5*(1/(1+X/100)**K)
2730 IF A0=A1 THEN 2800
2740 IF A0>A1 THEN 2760
2750 NEXT X
2760 L=X-S
2770 H=X
2780 S=S/10
2790 NEXT Y
2800 PRINT USING H$(8),X
2810 GOTO 2850
2820 PRINT USING H$(9),D1
2830 GO TO 2850
2840 PRINT USING H$(10),D2
2850 RETURN
2860 REM CALCULATE NET PRESENT VALUE
2870 E0(0)=C(0)
2880 E1(0)=C(0)
2890 E2(0)=C(0)
2900 FOR J=1 TO K
2910 E0(J)=E0(J-1)+C(J)
2920 E1(J)=C(J)*(1/(1+R1/100)**J)
2930 E2(J)=E2(J-1)+E1(J)
2940 NEXT J
2950 REM    PV OF CAPITAL RECOVERY
2960 V2=G5*(1/(1+R1/100)**K)
2970 V3=E2(K)+V2
2980 IF K8=0 THEN 3120
2990 PRINT
3000 PRINT
3010 PRINT USING H$(13),R1
3020 PRINT H$(14)
3030 PRINT H$(15)
3040 PRINT H$(16)
3050 PRINT H$(17)
3060 FOR J=0 TO K
3070 PRINT USING H$(18),J,C(J),E0(J),E1(J),E2(J)
3080 NEXT J
3090 PRINT
3100 IF K7=1 THEN 3130
3110 PRINT USING H$(19),V2
3120 PRINT
3130 PRINT USING H$(20),V3
3140 PRINT
3150 RETURN
3160 REM    CALCULATE PAYBACK PERIOD
3170 PRINT
3180 A0=-C(0)
3190 C1=0
3200 FOR J=1 TO K
3210 C1=C1+C(J)
3220 IF C1 > A0 THEN 3260
3230 NEXT J
3240 PRINT "**** PAYBACK PERIOD IS  > ECONOMIC LIFE *****"
```

```
3250 GOTO 3270
3260 PRINT USING H$(11),J
3270 RETURN
3280 REM CALCULATES PROJECT PROFITABILITY
3290 PRINT
3300 E2=0
3310 FOR J=1 TO K
3320 E2=E2+C(J)*(1/(1+R1/100)**J)
3330 NEXT J
3340 E2=E2+G5*(1/(1+R1/100)**K)
3350 E1=E2/(-C(0))
3360 PRINT USING H$(12),E1
3370 RETURN
3380 REM    *** CHANGE COMMAND ***
3390 PRINT "ENTER NEW DISCOUNT RATE % .............";
3400 INPUT R1
3410 RETURN
3420 REM    ***  I N I T I A L I Z A T I O N  ***
3430 H$(1)="N E T      C A S H      F L O W  ################################"
3440 H$(2)="--------------------------------------------------------"
3450 H$(3)="PERIOD OPERATING  DEPRE-    TAXABLE      TAX    INCOME    CASH"
3460 H$(4)="        PROFIT   CIATION   INCOME              AFT.TAX   FLOW"
3470 H$(5)="------  --------- -------  --------- --------  -------- --------"
3480 H$(6)="####  -######### -###### -######## -####### -####### -#######"
3490 H$(7)="CAPITAL RECOVERY ................................... -#######"
3500 H$(8)="I N T E R N A L    R A T E    O F    R E T U R N  = -###.####%"
3510 H$(9)="IRR IS BELOW THE LOWER LIMIT OF  -###.###%"
3520 H$(10)="IRR IS ABOVE THE UPPER LIMIT OF  - ###.###%"
3530 H$(11)="P A Y B A C K    P E R I O D                       = -###.###"
3540 H$(12)="PROJECT    P R O F I T A B I L I T Y    I N D E X  = -###.###"
3550 H$(13)="NET PRESENT VALUE USING DISCOUNT RATE -###.### %"
3560 H$(14)="--------------------------------------------------"
3570 H$(15)="PERIOD   CASH CUMULATIVE    PV OF   CUMULATIVE"
3580 H$(16)="         FLOW  CASH FLOW CASH FLOW PRESENT VALUE"
3590 H$(17)="------  -------- --------- -------- ----------"
3600 H$(18)="####  -######## -######### -######## -##########"
3610 H$(19)="PRESENT VALUE OF CAPITAL RECOVERY   = $ -#######"
3620 H$(20)="N E T      P R E S E N T      V A L U E = $ -#######"
3630 B$(1)="STRAIGHT-LINE DEPRECIATION"
3640 B$(2)="SUM-OF-YEARS-DIGITS DEPRECIATION"
3650 B$(3)="DOUBLE-DECLINING DEPRECIATION"
3660 RETURN
3670 END
```

II. Financing

After the acceptability of an investment is established, the method of financing can be explored. A number of financing alternatives exist. The investment project can be financed through equity, through lease financing, or through debt. Debt financing in our context means borrowing money from a financial institution to pay off (purchase) the initial cost of the investment and periodically paying back the loan (debt) with interest.

In this section, we will only consider the problem of financing an asset through leasing or through purchasing. The steps necessary to solve this problem are as follows:

1. Separate and remove any specific cost differences between the lease and the purchase options. A typical example is the maintenance cost, which is sometimes included as part of the lease payment.

2. Assume an equal life for both the lease and the loan so that both can be compared on the same basis. It is possible to assume that the depreciable life of the asset is less than the loan life for tax benefit purposes. However, it may not exceed the loan life.

3. The interest rate for computing the periodic debt payments of the loan can be different from the discount rate used in comparing the lease versus purchase alternatives.

4. The expected salvage value and the tax salvage value of the asset can be different. If the former is greater than the latter, the difference becomes a capital recovery amount and is taxed as ordinary income. Otherwise the difference is treated as a loss for tax purposes.

5. A higher risk-adjusted discount rate for the salvage value may be used because of the future uncertainty of any capital recovery.

6. Under lease financing, the asset is owned by the lessor who owns its title. The firm (lessee) makes periodic payments to the lessor throughout the lease life. The payments are made either in advance or in arrears. In this case, the firm obtains tax benefits for the entire lease payment. The lessor, on the other hand, benefits from the investment tax credit (ITC) and the depreciation expense. In some cases, the lessor may even pass along to the lessee the total or partial investment tax credit.

Let N be the life of the lease, L_t be the lease payment in period t, TR be the maginal tax rate of the firm, k be the after-tax discount rate, and ITC be the investment tax credit passed on to the lessee by the lessor. The present value of the after-tax cost of leasing (PVL) is

$$PVL = \sum_{t=1}^{N} L_t (1-TR)/(1+k)^t - (ITC)/(1+TR)$$

if the payments are made in arrears. If the payments are prepaid, the lower and upper limits of the summation sign are replaced by t=0 and N-1, respectively.

7. Under debt financing, it is assumed that the firm acquires a loan at a certain rate of interest to purchase the asset in full and thus holds the title to the asset. In return, it makes periodic payments to the lender to pay off the interest and the principal until the entire debt is paid. Under this method of financing, the firm can obtain tax benefits in terms of the investment tax credit, the depreciation expense, and the interest paid to the lender. In addition, it can recover the salvage value if there is any.

The present value of the after-tax cost of the purchase option (PVP) can be determined from

$$PVP = DO + \sum_{t=1}^{N} (P_t - TR \cdot I_t)/(1+k)^t - \sum_{t=1}^{N} (TR \cdot D_t)/(1+k)^t$$

$$- (ITC)/(1+k) - (SV)/(1+R_s)^M,$$

where   DO = Down payment,
        N = Loan life,
        P_t = Loan payment in period t,

        TR = Marginal tax rate,
        I_t = Interest paid in period t,

        k = After-tax discount rate,
        D_t = Amount of tax depreciation in period t,

        ITC = Investment tax credit taken,
        SV = Salvage value,
        R_s = After-tax discount rate for the salvage value,

and   M = Depreciable life of the asset.

The Program LEASE

The computational procedure used in this program is based on the approach taken above. The program also assumes that the lease payments and the loan payments are made in equal installments. The three most common methods of depreciation implemented in the program are straight-line depreciation (SLDEP), sum-of-years-digits depreciation (SYDDEP), and double-declining depreciation (DDDEP).

It should be noted that although the above formulas refer to all of the discount rates as after-tax discount rates, the program actually prompts for the pretax discount rates as input. The after-tax discount rate is equal to the pretax discount rate multiplied by (1-tax rate).

INPUT
-----
1. The name of the project
2. Life of the lease/loan (must be integer)
3. Marginal tax rate of the firm in percentage
4. Pretax discount rate in percentage, which may be the cost of capital or the desired rate of return

Input for Lease Option
----
5. Amount of each payment
6. Payments are prepaid or in arrears
7. ITC passed on by the lessor to the firm

Input for Purchase Option
----
8. Asset's purchase price
9. Initial down payment. The difference between the purchase price and the down payment is the amount to be financed.
10. Pretax borrowing rate in percentage
11. Amount of ITC taken
12. Tax salvage value
13. Expected salvage value
14. Pretax risk-adjusted discount rate in percentage for the salvage value
15. Depreciable life of the asset (must be integer)

OUTPUT

The depreciation method can be selected using the SLDEP, SYDDEP, and DDDEP commands, although the default method is the double-declining method (DDDEP). Once the depreciation method is selected, the command SUMMARY will produce the present values of the net cash flows for both the lease option and the purchase option. The DETAIL command, on the other hand, will produce the payment schedules for both options and their respective present values in addition to the summary information.

EXAMPLE

Zeta company wants to acquire a minicomputer that will cost $100,000 to purchase. The same computer configuration can be leased for $30,000 per year for five years. The lease payments can be made in arrears. If the purchase option is selected, the company can ascertain a loan equal to the cost of the equipment at an annual pretax borrowing rate of 14 percent. The loan should be paid in five years. Assume that Zeta company will use the straight-line method for depreciation with a $10,000 salvage value after five years for tax computation. The company believes that the expected salvage value will not be any higher and wants to use a higher risk-adjusted discount rate of 30 percent for the salvage value. The investment tax credit taken is $10,000 for both the lease option and the purchase option. The minimum desired pretax rate of return for such an investment is 15 percent, and the firm's marginal tax rate is 50 percent.

The solution to the above example is shown in the following example terminal session. After all of the data have been entered, the HELP command is used which lists all available commands. Next, the SLDEP command for selecting the straight-line depreciation method is used. The DETAIL command produces detailed analysis of the two financing options. The results indicate that the purchase option is more favorable, since the difference in the net present values is $3446. The EXIT command is used at the end to terminate the program.

```
LEASE EXAMPLE TERMINAL SESSION
==============================

LEASE/PURCHASE ANALYSIS
-----------------------

NEED INTRODUCTION (Y OR N)....!N

ENTER THE NAME OF THE PROJECT
!MINICOMPUTER ACQUISITION
ENTER DURATION OF LEASE/LOAN (PERIODS) ... !5
ENTER MARGINAL TAX RATE (%) ... !50
ENTER PRETAX DISCOUNT RATE (%) ... !15

INPUT FOR LEASE OPTION
----------------------
AMOUNT OF EACH PAYMENT ($) ... !30000
ARE PAYMENTS PREPAID (Y OR N) ... !N
ITC PASSED ON BY THE LESSOR ($) ... !10000

INPUT FOR PURCHASE OPTION
-------------------------
PURCHASE PRICE ($) ... !100000
INITIAL DOWN PAYMENT ($) ... !0
PRETAX BORROWING RATE FOR LOAN (%) ... !14
ITC TAKEN ($) ... !10000
SALVAGE VALUE FOR TAX CALCULATION ($) ... !10000
EXPECTED SALVAGE VALUE ($) ... !10000
PRETAX DISCOUNT RATE APPLIED TO SALVAGE VALUE (%) ... !30
DEPRECIABLE LIFE OF THE ASSET (PERIODS) ... !5

PLEASE ENTER COMMAND OR TYPE HELP

COMMAND --> !HELP

AVAILABLE COMMANDS
==================
HELP     - PRINT THIS MESSAGE
CHANGE   - SELECT A NEW PRETAX DISCOUNT RATE
DETAIL   - SELECT DETAIL OUTPUT OPTION
SUMMARY  - SELECT SUMMARY OUTPUT OPTION
SLDEP    - SELECT STRAIGHT-LINE DEPRECIATION METHOD
SYDDEP   - SELECT SUM-OF-YEARS-DIGITS DEPRE. METHOD
DDDEP    - SELECT DOUBLE-DECLINING DEPRE. METHOD
RESTART  - START A NEW PROBLEM
EXIT     - EXIT FROM THE PROGRAM


COMMAND --> !SLDEP

COMMAND --> !DETAIL
```

P R O J E C T : MINICOMPUTER ACQUISITION

```
DURATION OF LEASE/PURCHASE, PERIODS.....5
MARGINAL TAX RATE % ....................50
PRETAX DISCOUNT RATE % .................15
```

L E A S E   O P T I O N

```
AMOUNT OF EACH PAYMENT ($)    30000
ITC TAKEN ($)    10000

PV OF TOTAL CASH OUTFLOW ............$60688
PV OF ITC TAKEN .....................$9302
PV OF NET CASH OUTFLOW ..............$51385
```

CASH OUTFLOW FOR LEASE ALTERNATIVE

END OF PERIOD	LEASE PAYMENT	TAX SHIELD	AFT-TAX CASHFLO	PV OF CASHFLO
0	0	0	0	0
1	30,000	15,000	15,000	13,953
2	30,000	15,000	15,000	12,980
3	30,000	15,000	15,000	12,074
4	30,000	15,000	15,000	11,232
5	30,000	15,000	15,000	10,448

P U R C H A S E   O P T I O N

```
DEPRECIATION METHOD : STRAIGHT-LINE
PURCHASE PRICE ($)    100000
INITIAL PAYMENT ($)    0
AMOUNT OF LOAN ($)    100000
PRETAX BORROWING RATE FOR LOAN (%)    14
ITC TAKEN ($)    10000
TAX SALVAGE VALUE ($)    10000
EXPECTED SALVAGE VALUE ($)    10000
PRETAX DISCOUNT RATE FOR SALVAGE VALUE (%) 30
DEPRECIABLE LIFE (PERIODS)    5

PV OF TOTAL CASH OUTFLOW ............$62213
PV OF INITIAL DOWN PAYMENT ..........$0
PV OF ITC TAKEN .....................$9302
PV OF CAPITAL RECOVERY ..............$4971
PV OF NET CASH OUTFLOW ..............$47939
```

```
            CASH OUTFLOW OF PURCHASE ALTERNATIVE

END OF  PAYMENT                          TAX   AFT-TAX   PV OF
PERIOD  AMOUNT  INTRST.  DEPREC.       SHIELD  CASHFLO  CASHFLO
------  -------  -------  -------      -------  -------  -------
     1  29,128   14,000   18,000       16,000   13,128   12,212
     2  29,128   11,882   18,000       14,941   14,187   12,277
     3  29,128    9,468   18,000       13,734   15,395   12,392
     4  29,128    6,715   18,000       12,358   16,771   12,558
     5  29,128    3,577   18,000       10,789   18,340   12,775


                  LOAN PAYMENT SCHEDULE

         BEGINNG  PAYMENT  INTRST.  PRINCPL  ENDING
PERIOD   PRINCPL  AMOUNT   AMOUNT   REDUCED  PRINCPL
------   -------  -------  -------  -------  -------
     1   100,000   29,128   14,000   15,128   84,872
     2    84,872   29,128   11,882   17,246   67,625
     3    67,625   29,128    9,468   19,661   47,965
     4    47,965   29,128    6,715   22,413   25,551
     5    25,551   29,128    3,577   25,551        0

     *** ADVANTAGE OF PURCHASE OPTION *** $3446

COMMAND --> !EXIT
```

```
10 REM     LEASE/PURCHASE ANALYSIS PROGRAM
20 REM
30 DIM L(25),T(25),L1(25),P1(25)
40 DIM L3(25),I1(25),G3(25),L4(25),P2(25),T1(25)
50 DIM S1(25),S2(25),S3(25),F$(10),B$(3)
60 REM: M0=MAXIMUM NUMBER OF PERIODS FOR LEASE/LOAN
70 M0=25
80 GOSUB 3590
90 K9=3
100 PRINT "LEASE/PURCHASE ANALYSIS"
110 PRINT "----------------------"
120 PRINT
130 PRINT "NEED INTRODUCTION (Y OR N)....";
140 INPUT Y$
150 IF Y$= "N" THEN 490
160 PRINT
170 PRINT "THIS PROGRAM PERFORMS LEASE VERSUS PURCHASE ANALYSIS WHEN"
180 PRINT "AN ASSET CAN BE ACQUIRED USING EITHER OPTION. IT DETERMINES"
190 PRINT "THE PRESENT VALUE OF NET CASH OUTFLOWS FOR BOTH OPTIONS."
200 PRINT "THE ANALYSIS IS MADE OVER THE LIFE OF THE ASSET WHICH IS "
210 PRINT "ASSUMED TO BE EQUAL TO THE LOAN PERIOD IF THE PURCHASE"
220 PRINT "OPTION IS SELECTED AND ALSO EQUAL TO THE LEASE PERIOD."
230 PRINT
240 PRINT "INPUT REQUIREMENTS"
250 PRINT "------------------"
260 PRINT "1. THE NAME OF THE PROJECT"
270 PRINT "2. DURATION OF LEASE/LOAN IN PERIODS (MUST BE INTEGER)"
280 PRINT "3. MARGINAL TAX RATE (%) OF THE FIRM"
290 PRINT "4. DISCOUNT RATE (%) WHICH MAY BE THE COST OF CAPITAL OR"
300 PRINT "   THE MINIMUM DESIRED RATE OF RETURN"
310 PRINT
320 PRINT "LEASE OPTION"
330 PRINT "------------"
340 PRINT "5. AMOUNT OF EACH PAYMENT ($)"
350 PRINT "6. PAYMENTS ARE PREPAID OR IN ARREARS"
360 PRINT "7. AMOUNT OF ITC PASSED ON BY THE LESSOR TO THE LESSEE"
370 PRINT
380 PRINT "PURCHASE OPTION"
390 PRINT "---------------"
400 PRINT "8. ASSET'S PURCHASE PRICE ($)"
410 PRINT "9. INITIAL DOWN PAYMENT ($).  THE DIFFERENCE BETWEEN"
420 PRINT "   THE PURCHASE PRICE AND DOWN PAYMENT IS FINANCED"
430 PRINT "10. PRETAX BORROWING RATE (%)"
440 PRINT "11. AMOUNT OF ITC TAKEN ($)"
450 PRINT "12. TAX SALVAGE VALUE ($)"
460 PRINT "13. EXPECTED SALVAGE VALUE ($)"
470 PRINT "14. PRETAX RISK-ADJUSTED DISCOUNT RATE (%) FOR SALVAGE VALUE"
480 PRINT "15. DEPRECIABLE LIFE OF THE ASSET (PERIODS)"
490 REM     I N P U T     S E C T I O N
500 PRINT
510 PRINT "ENTER THE NAME OF THE PROJECT"
520 INPUT P$
530 PRINT "ENTER DURATION OF LEASE/LOAN (PERIODS) ... ";
540 INPUT N
```

```
550 IF N <= M0 THEN 580
560 PRINT "*** ERROR *** MAX ALLOWABLE NO. OF PERIODS = ";M0
570 GOTO 530
580 PRINT "ENTER MARGINAL TAX RATE (%) ... ";
590 INPUT R1
600 PRINT "ENTER PRETAX DISCOUNT RATE (%) ... ";
610 INPUT K1
620 PRINT
630 PRINT "INPUT FOR LEASE OPTION"
640 PRINT "----------------------"
650 PRINT "AMOUNT OF EACH PAYMENT ($) ... ";
660 INPUT L0
670 PRINT "ARE PAYMENTS PREPAID (Y OR N) ... ";
680 INPUT D$
690 PRINT "ITC PASSED ON BY THE LESSOR ($) ... ";
700 INPUT I1
710 PRINT
720 PRINT "INPUT FOR PURCHASE OPTION"
730 PRINT "-------------------------"
740 PRINT "PURCHASE PRICE ($) ... ";
750 INPUT G1
760 PRINT "INITIAL DOWN PAYMENT ($) ... ";
770 INPUT F
780 IF F <= G1 THEN 810
790 PRINT "*** ERROR ***   DOWN PAYMENT CANNOT EXCEED PURCHASE PRICE !"
800 GOTO 760
810 PRINT "PRETAX BORROWING RATE FOR LOAN (%) ... ";
820 INPUT K0
830 PRINT "ITC TAKEN ($) ... ";
840 INPUT I2
850 PRINT "SALVAGE VALUE FOR TAX CALCULATION ($) ... ";
860 INPUT S0
870 PRINT "EXPECTED SALVAGE VALUE ($) ... ";
880 INPUT S1
890 PRINT "PRETAX DISCOUNT RATE APPLIED TO SALVAGE VALUE (%) ... ";
900 INPUT K5
910 PRINT "DEPRECIABLE LIFE OF THE ASSET (PERIODS) ... ";
920 INPUT M
930 IF M > 0  THEN 990
940 PRINT "*** ERROR ***   DEPRECIABLE LIFE MUST BE GREATER THAN 0 !"
950 GOTO 910
960 IF M < = N THEN 990
970 PRINT "*** ERROR ***   DEPRECIABLE LIFE MAY NOT EXCEED LOAN LIFE !"
980 GOTO 910
990 F0 = G1 - F
1000 K8=0
1010 PRINT
1020 PRINT "PLEASE ENTER COMMAND OR TYPE HELP"
1030 PRINT
1040 PRINT "COMMAND --> ";
1050 INPUT C$
1060 IF C$<> "HELP" THEN 1090
1070 GOSUB 1340
1080 GOTO 1030
```

```
1090 IF C$ <> "CHANGE" THEN 1120
1100 GOSUB 3540
1110 GOTO 1030
1120 IF C$ <> "SLDEP" THEN 1150
1130 K9=1
1140 GOTO 1030
1150 IF C$ <> "SYDDEP" THEN 1180
1160 K9=2
1170 GOTO 1030
1180 IF C$ <> "DDDEP" THEN 1210
1190 K9=3
1200 GOTO 1030
1210 IF C$ <> "DETAIL" THEN 1240
1220 K8=1
1230 GOTO 1260
1240 IF C$ <> "SUMMARY" THEN 1280
1250 K8=0
1260 GOSUB 1690
1270 GOTO 1030
1280 IF C$ = "RESTART" THEN 500
1290 PRINT
1300 IF C$="EXIT" THEN 3780
1310 PRINT "*** ERROR ***    PLEASE TRY AGAIN !!!"
1320 GOTO 1030
1330 PRINT
1340 REM *** HELP COMMAND ***
1350 PRINT
1360 PRINT "AVAILABLE COMMANDS"
1370 PRINT "=================="
1380 PRINT "HELP      - PRINT THIS MESSAGE"
1390 PRINT "CHANGE    - SELECT A NEW PRETAX DISCOUNT RATE"
1400 PRINT "DETAIL    - SELECT DETAIL OUTPUT OPTION"
1410 PRINT "SUMMARY   - SELECT SUMMARY OUTPUT OPTION"
1420 PRINT "SLDEP     - SELECT STRAIGHT-LINE DEPRECIATION METHOD"
1430 PRINT "SYDDEP    - SELECT SUM-OF-YEARS-DIGITS DEPRE. METHOD"
1440 PRINT "DDDEP     - SELECT DOUBLE-DECLINING DEPRE. METHOD"
1450 PRINT "RESTART   - START A NEW PROBLEM"
1460 PRINT "EXIT      - EXIT FROM THE PROGRAM"
1470 PRINT
1480 RETURN
1490 REM     *** DATA OUTPUT ROUTINES ***
1500 GOSUB 3740
1510 PRINT "P R O J E C T : ";P$
1520 PRINT
1530 PRINT TAB(5);"DURATION OF LEASE/PURCHASE, PERIODS.....";N
1540 PRINT TAB(5);"MARGINAL TAX RATE % ...................";R1
1550 PRINT TAB(5);"PRETAX DISCOUNT RATE % ................";K1
1560 RETURN
1570 PRINT
1580 PRINT TAB(5);"DEPRECIATION METHOD : ";B$(K9)
1590 PRINT TAB(5);"PURCHASE PRICE ($)      ";G1
1600 PRINT TAB(5);"INITIAL PAYMENT ($)     ";F
1610 PRINT TAB(5);"AMOUNT OF LOAN ($)     ";F0
1620 PRINT TAB(5);"PRETAX BORROWING RATE FOR LOAN (%)  ";K0
```

```
1630 PRINT TAB(5);"ITC TAKEN ($)    ";I2
1640 PRINT TAB(5);"TAX SALVAGE VALUE ($)     ";S0
1650 PRINT TAB(5);"EXPECTED SALVAGE VALUE ($)     ";S1
1660 PRINT TAB(5);"PRETAX DISCOUNT RATE FOR SALVAGE VALUE (%) ";K5
1670 PRINT TAB(5);"DEPRECIABLE LIFE (PERIODS)    ";M
1680 RETURN
1690 REM    *** SUMMARY/DETAIL COMMANDS ***
1700 REM    *** LEASE OPTION CALCULATIONS ***
1710 GOSUB 1500
1720 GOSUB 3740
1730 PRINT "L E A S E    O P T I O N"
1740 PRINT "-----------------------"
1750 PRINT
1760 PRINT TAB(5);"AMOUNT OF EACH PAYMENT ($)    ";L0
1770 PRINT TAB(5);"ITC TAKEN ($)    ";I1
1780 R = R1/100
1790 K2 = K1 *(1-R)/100
1800 REM: K2=AFTER-TAX BORROWING RATE
1810 REM L(0) = LEASE PAYMENT PER PERIOD
1820 REM T(0) = TAX PER PERIOD
1830 REM L1(0)= CASH FLOW AFTER TAX ; P1(0)=PV OF CASH FLOW
1840 REM V1=TOTAL CASH FLOW  ;  V2=NET CASH FLOW AFTER ITC
1850 IF D$ <> "N" THEN 2000
1860 REM    *** LEASE PAYMENTS ARE NOT PREPAID ***
1870 V1 =0
1880 L(0)=0
1890 T(0)=0
1900 FOR J= 1 TO N
1910 L(J) = L0
1920 NEXT J
1930 FOR J=1 TO N
1940 T(J) = L(J)*R
1950 L1(J) = L(J)-T(J)
1960 P1(J) = L1(J)/(1+K2)**J
1970 V1 = V1+P1(J)
1980 NEXT J
1990 GOTO 2170
2000 REM    *** LEASE PAYMENTS ARE PREPAID ***
2010 N0= N-1
2020 FOR J = 0 TO N0
2030 L(J) = L0
2040 NEXT J
2050 T(0)=0
2060 L(N) = 0
2070 L1(0)=L(0)-T(0)
2080 P1(0) = L1(0)
2090 V1 = P1(0)
2100 FOR J = 1 TO N
2110 J1 = J-1
2120 T(J) = L(J1)*R
2130 L1(J) = L(J)-T(J)
2140 P1(J) = L1(J)/((1+K2)**J)
2150 V1 = V1 + P1(J)
2160 NEXT J
```

```
2170 I3 = I1/(1+K2)
2180 V2 = V1-I3
2190 PRINT
2200 PRINT TAB(5);"PV OF TOTAL CASH OUTFLOW ............$";INT(V1)
2210 PRINT TAB(5);"PV OF ITC TAKEN ....................$";INT(I3)
2220 PRINT TAB(5);"PV OF NET CASH OUTFLOW .............$";INT(V2)
2230 IF K8=0 THEN 2250
2240 GOSUB 3310
2250 REM     *** PURCHASE OPTION CALCULATIONS ***
2260 F0 = G1-F
2270 K7=K0/100
2280 K=N
2290 A0 = ((K7*(1+K7)**K)/((1+K7)**K-1))*F0
2300 FOR J = 1 TO K
2310 L3(J) = A0
2320 NEXT J
2330 REM CALCULATING INTEREST SCHEDULE
2340 S1(1) = F0
2350 FOR J = 1 TO K
2360 I1(J) = S1(J)*K7
2370 S2(J) = L3(J) - I1(J)
2380 S3(J) = S1(J) - S2(J)
2390 J1 = J + 1
2400 S1(J1) = S3(J)
2410 NEXT J
2420 REM     *** DEPRECIATION CALCULATION ***
2430 IF K9=1 THEN 2460
2440 IF K9=2 THEN 2550
2450 IF K9=3 THEN 2640
2460 REM     G4=DEPREC. BASE FOR STRAIGHT-LINE
2470 G4 = G1-S0
2480 G0 = 0
2490 D0 = G4/M
2500 FOR J = 1 TO M
2510 G3(J) = D0
2520 G0 = G0+G3(J)
2530 NEXT J
2540 GOTO 2720
2550 REM CALCULATING SUM-OF-YEARS-DIGITS DEPRECIATION
2560 G0 = 0
2570 G4 = G1-S0
2580 Y0 = M*(M+1)/2
2590 FOR J = 1 TO M
2600 G3(J) = G4*(M+1-J)/Y0
2610 G0 = G0+G3(J)
2620 NEXT J
2630 GOTO 2720
2640 REM:CALCULATING DOUBLE-DECLINING DEPRECIATION
2650 G0=0
2660 K0=M-1
2670 FOR J=1 TO K0
2680 G3(J)=(G1-G0)*2/M
2690 G0=G0+G3(J)
2700 NEXT J
```

```
2710 G3(M)=G1-S0-G0
2720 IF M = N THEN 2770
2730 M4=M+1
2740 FOR J = M4 TO N
2750 G3(J)=0
2760 NEXT J
2770 REM: CALCULATE TAX SHIELD
2780 V3=0
2790 REM : T1(.) = TAX SHIELD FOR PURCHASE
2800 REM : L4(.) = CASH OUTFLOW AFTER TAX
2810 REM : P2(.) = PV OF CASH OUT FLOW
2820 FOR J= 1 TO N
2830 T1(J)=(G3(J)+I1(J))*R
2840 L4(J)=L3(J)-T1(J)
2850 P2(J)=L4(J)/((1+K2)**J)
2860 V3=V3+P2(J)
2870 NEXT J
2880 REM    PV OF ITC TAKEN
2890 I4=I2/(1+K2)
2900 REM:CAPITAL RECOVERY
2910 K6=K5*(1-R)/100
2920 REM:AND TAX-BASIS SALVAGE VALUE
2930 IF S1 < S0 THEN 2960
2940 C1= (S1-(S1-S0)*R)/((1+K6)**M)
2950 GOTO 2970
2960 C1= (S1+(S0-S1)*R)/((1+K6)**M)
2970 V4=V3 + F - I4 - C1
2980 GOSUB 3740
2990 PRINT "P U R C H A S E    O P T I O N"
3000 PRINT "------------------------------"
3010 GOSUB 1570
3020 PRINT
3030 PRINT TAB(5);"PV OF TOTAL CASH OUTFLOW ............$";INT(V3)
3040 PRINT TAB(5);"PV OF INITIAL DOWN PAYMENT ..........$";INT(F)
3050 PRINT TAB(5);"PV OF ITC TAKEN .....................$";INT(I4)
3060 PRINT TAB(5);"PV OF CAPITAL RECOVERY ..............$";INT(C1)
3070 PRINT TAB(5);"PV OF NET CASH OUTFLOW ..............$";INT(V4)
3080 PRINT
3090 IF K8=0 THEN 3120
3100 GOSUB 3200
3110 GOSUB 3420
3120 IF V2 < V4 THEN 3160
3130 V5=V2 - V4
3140 PRINT TAB(5);"*** ADVANTAGE OF PURCHASE OPTION *** $";INT(V5)
3150 GOTO 3190
3160 V5=V4 - V2
3170 PRINT TAB(5);"*** ADVANTAGE OF LEASE OPTION ***    $";INT(V5)
3180 PRINT
3190 RETURN
3200 REM    *** CASH OUTFLOW FOR PURCHASE OPTION ***
3210 GOSUB 3740
3220 PRINT TAB(9);"CASH OUTFLOW OF PURCHASE ALTERNATIVE"
3230 PRINT
3240 PRINT F$(1)
```

```
3250 PRINT F$(2)
3260 PRINT F$(3)
3270 FOR J=1 TO N
3280 PRINT USING F$(4),J,L3(J),I1(J),G3(J),T1(J),L4(J),P2(J)
3290 NEXT J
3300 RETURN
3310 REM    *** CASH OUTFLOW FOR LEASE OPTION ***
3320 GOSUB 3740
3330 PRINT TAB(2);"CASH OUTFLOW FOR LEASE ALTERNATIVE"
3340 PRINT
3350 PRINT F$(5)
3360 PRINT F$(6)
3370 PRINT F$(7)
3380 FOR J = 0 TO N
3390 PRINT USING F$(4),J,L(J),T(J),L1(J),P1(J)
3400 NEXT J
3410 RETURN
3420 REM    *** LOAN PAYMENT FOR PURCHASE OPTION ***
3430 GOSUB 3740
3440 PRINT TAB(13);"LOAN PAYMENT SCHEDULE"
3450 PRINT
3460 PRINT F$(8)
3470 PRINT F$(9)
3480 PRINT F$(10)
3490 FOR J = 1 TO K
3500 PRINT USING F$(4),J,S1(J),L3(J),I1(J),S2(J),S3(J)
3510 NEXT J
3520 PRINT
3530 RETURN
3540 REM    *** CHANGE COMMAND ***
3550 PRINT
3560 PRINT "ENTER NEW PRETAX DISCOUNT RATE (%) ... ";
3570 INPUT K1
3580 RETURN
3590 REM *** INITIALIZATION ***
3600 F$(1)="END OF PAYMENT                             TAX AFT-TAX   PV OF"
3610 F$(2)="PERIOD   AMOUNT  INTRST. DEPREC.  SHIELD CASHFLO CASHFLO"
3620 F$(3)="------  ------- ------- ------- ------- ------- -------"
3630 F$(4)="######  ###,### ###,### ###,### ###,### ###,### ###,###"
3640 F$(5)="END OF    LEASE      TAX AFT-TAX    PV OF"
3650 F$(6)="PERIOD PAYMENT    SHIELD CASHFLO CASHFLO"
3660 F$(7)="------ ------- ------- ------- -------"
3670 F$(8)="        BEGINNG PAYMENT INTRST. PRINCPL  ENDING"
3680 F$(9)="PERIOD PRINCPL  AMOUNT  AMOUNT REDUCED PRINCPL"
3690 F$(10)="------ ------- ------- ------- ------- -------"
3700 B$(1)="STRAIGHT-LINE"
3710 B$(2)="SUM-OF-YEARS-DIGITS"
3720 B$(3)="DOUBLE-DECLINING"
3730 RETURN
3740 FOR J5=1 TO 3
3750 PRINT
3760 NEXT J5
3770 RETURN
3780 END
```

III. Dimensional Specifications

Program CAPBUD

The number of time periods, n, in the current program is limited to 20. In order to change the program to accommodate other sizes, the following statements in CAPBUD should be modified:

Line No.
----
40 DIM E0(n),E1(n),E2(n)
50 DIM C(n),R(n),G(n),F(n),Q(n),U(n),V(n),O(n)
70 M0=n

Program LEASE

Similarly, the maximum number of time periods that LEASE can handle is currently set at 25. To increase or decrease this limit, the variable n in the following statements must be replaced with the appropriate value:

Line No.
----
30 DIM L(n),T(n),L1(n),P1(n)
40 DIM L3(n),I1(n),G3(n),L4(n),P2(n),T1(n)
50 DIM S1(n),S2(n),S3(n),F$(10),B$(3)
70 M0=n

IV. References
----
1. Clark, J.J., M.T. Clark, and P.T. Elgers, FINANCIAL MANAGEMENT, Holbrook Press, Boston, 1976.

2. Horngren, C.T., COST ACCOUNTING: A MANAGERIAL EMPHASIS, 5th ed., Prentice-Hall, Englewood Cliffs, N.J., 1982.

3. Pritchard, R.E., and T.J. Hindeland, THE LEASE/BUY DECISION, AMACOM(Amer. Mgmt. Assoc.), New York, 1980.

4. Vancil, R.F., LEASING OF INDUSTRIAL EQUIPMENT, McGraw-Hill, New York, 1963.

5. Van Horne, J.C., FINANCIAL MANAGEMENT AND POLICY, 6th ed., Prentice-Hall, Englewood Cliffs, N.J., 1983.

6. Stevens, G.T., Jr., ECONOMIC AND FINANCIAL ANALYSIS OF CAPITAL INVESTMENTS, John Wiley, New York, 1979.

CHAPTER 3

FORECASTING

I. Eight curves

- The Program 8CURVS
- Input
- Output
- Example
- 8CURVS Example Terminal Session
- 8CURVS Program Listing

II. Forecasting

- The Program FORCST
- Input
- Output
- Example 1
- Example 2
- FORCST Example Terminal Session 1
- FORCST Example Terminal Session 2
- FORCST Program Listing

III. Dimensional Specifications

IV. References

Forecasting is one of the most important functions in business and governmental organizations. The need for good forecasts is obvious because they are used for planning, budgeting, and setting goals. Typical use in a business organization is the sales forecasts by product lines.

Depending on the applications and the availability of the data, forecasts may be generated using simple or sophisticated models or human judgment. Simple models such as exponential smoothing models are used in forecasting demands for inventory items. Other more sophisticated models may contain hundreds of interrelated equations, as in some econometric models.

In this chapter, we will concentrate on very simple and yet useful forecasting models. These models are well known and are described in many forecasting textbooks (see references). Two computer programs are provided with sample terminal sessions to demonstrate the use of these programs.

The first program is called "8CURVS." As the name implies 8CURVS is a curve-fitting program that can fit eight different curves to a set of observations between two variables, X and Y. All of the curves are based on some transformations of the X and the Y variables. The program has a general use in that X and Y need not be time series data.

The second program, FORCST, is a small time series analysis package. It contains a number of forecasting models and other techniques that will help the user to identify the best model for the time series being analyzed. For example, it can produce an autocorrelation function which is used to detect any trend or seasonal pattern in the data. The models in this program are the most common types of exponential smoothing, moving averages and regression. For the exponential smoothing models, an optimization capability is also provided which will allow the user to determine the best combination of the smoothing constants for the time series.

The listings of these two programs are somewhat longer than those of the other programs in this book. This is due to the many techniques that we have combined into these programs. The reason for doing this is that we feel that the users will benefit more from the program by having a variety of techniques at their fingertips while performing the analysis rather than having a large number of small programs.

I. Eight Curves

When a causal relationship exists between two variables X and Y where X is the independent variable and Y is the dependent variable, we may wish to investigate the relationship using some equation form. The simplest form would be a linear relationship. We have identified some other useful relationships that result from transformations of the X and the Y variables such that the resulting transformations will be linear. Table 1 shows the eight different curves and their corresponding linear transformations.

Table 1. Eight Curves and Their Linear Transformations

Curve	Equation Form	Linear Transformation
1	$y = a + bx$	$y = a + bx$
2	$y = a e^{bx}$	$\log y = \log a + bx$
3	$y = a x^b$	$\log y = \log a + b \log x$
4	$y = a + b/x$	$y = a + b/x$
5	$y = 1/(a + bx)$	$1/y = a + bx$
6	$y = x/(ax + b)$	$1/y = a + b/x$
7	$y = a + b \log x$	$y = a + b \log x$
8	$y = e^{(a+b/x)}$	$\log y = a + b/x$

When the observation values are all positive, all of the transformations can be performed and the coefficients a and b of every curve can be calculated using the method of least squares. However, if some observations are negative, we cannot perform the logarithmic transformation. When some observation values are zeros, the inverse transformation cannot be performed. It is obvious that if X is considered to be the time scale, the curves can be used to represent time series models. For example, curve number 8 can be used in the analysis of an S-curve for long-term forecasting [3]. If the data for X are replaced by 1/X, the same curve can be applied to an exponential growth model [3]. Other uses of this program in operations management are plentiful. One example may be to use curve 3 in developing the learning curve useful in assembly-line balancing analyses.

The Program 8CURVS

The program 8CURVS solves for the coefficients a and b for the eight curves in Table 1. A minimum number of three observations is required. The allowable number of observations is currently set at fifty. See Section III, "Dimensional Specifications," on how to change it.

INPUT
-----
1. The number of the observations
2. Observation values entered as X,Y

OUTPUT

A number of output-producing commands are available. The SOLVE command produces a table of coefficients a and b for the curves together with an analysis of variance table. The RESIDUAL command produces a table of fitted values and residual errors, percentage errors, mean absolute percentage error, and mean squared error. The FORECAST command allows the user to calculate the fitted Y value of a specified curve by entering X. The program also contains a PLOT command which will produce an X-Y plot of the observations and the fitted curve. This will aid in interpreting the results in conjunction with the statistics r-squared and the adjusted r-squared. Note that the SOLVE command must be issued before the FORECAST, PLOT, or RESIDUAL commands will function.

EXAMPLE

The demand for a product has been steadily increasing for the past ten years. Forecast the total demand for the next year. The following data are available:

Year	Sales (1000 units)	Year	Sales (1000 units)
1	48.2	6	82.1
2	57.9	7	83.3
3	70.7	8	85.0
4	75.4	9	85.5
5	80.2	10	86.0

Based on the available information, we suspect that an S-curve type of a relationship may be appropriate. From the example terminal session, the SOLVE-command output indicates four curves that give very good fit based on the adjusted r-squared statistic. They are curves number 3, 6, 7, and 8. Any one of these curves should produce satisfactory results for the next-year forecast. Assuming that we select curve number 6, then the next-year forecast is 86,551 units.

8CURVS EXAMPLE TERMINAL SESSION
================================

E I G H T   C U R V E S

NEED INTRODUCTION (Y OR N) ... !N

ENTER NUMBER OF OBSERVATIONS (>=3) -->!10

ENTER OBSERVATIONS IN PAIR, I.E. X,Y

OBSN   1   !1,48.2
OBSN   2   !2,57.9
OBSN   3   !3,70.7
OBSN   4   !4,75.4
OBSN   5   !5,80.2
OBSN   6   !6,82.1
OBSN   7   !7,83.3
OBSN   8   !8,85.0
OBSN   9   !9,85.5
OBSN   10   !10,86.0

PLEASE ENTER COMMAND OR TYPE HELP FOR COMMAND DESCRIPTIONS

COMMAND -->!HELP

AVAILABLE COMMANDS
==================
  CHANGE   - CHANGE THE VALUES OF AN OBSERVATION
  DELETE   - DELETE AN OBSERVATION
  CURVES   - SHOW EQUATION FORMS OF THE EIGHT CURVES
  EXIT   - EXIT FROM THE PROGRAM
  FORECAST - GIVEN X, CALCULATE THE FITTED Y-VALUE
  HELP   - PRINT THIS MESSAGE
  INSERT   - INSERT AN OBSERVATION AFTER ANOTHER
  LIST   - LIST ALL OBSNS AND SUMMARY STATISTICS
  PLOT   - PLOT THE FITTED CURVE AND THE OBSNS
  RESIDUAL - PRINT RESIDUAL TABLE AND FITTED Y-VALUE
  SOLVE   - SOLVE FOR THE COEFFICIENTS OF A AND B
              AND VARIOUS STATISTICS

COMMAND -->!LIST

OBSERVATION	X	Y
1	1.0000	48.2000
2	2.0000	57.9000
3	3.0000	70.7000
4	4.0000	75.4000
5	5.0000	80.2000
6	6.0000	82.1000
7	7.0000	83.3000
8	8.0000	85.0000
9	9.0000	85.5000

```
         10     10.0000        86.0000

    MEAN        5.5000        75.4300
 STD DEV        3.0277        12.9316
 MINIMUM        1.0000        48.2000
 MAXIMUM       10.0000        86.0000
```

COMMAND -->!SOLVE

```
CURVE  EQUATION-FORM              A                 B       R-SQUARED  ADJ. R**2
  1    Y = A+B*X              54.413333         3.821212    .80040153   .77545172
  2    Y = A*EXP(B*X)         54.679031          .055654    .75105349   .71993517
  3    Y = A*(X**B)           50.251438          .258558    .95021791   .94399515
  4    Y = A+B/X              88.560563 -      44.829993    .92389195   .91437845
  5    Y = 1/(A+B*X)           .018285 -         .000831    .69800056   .66025063
  6    Y = X/(A*X+B)           .010580           .010708    .97278569   .96938390
  7    Y = A+B*LOG(X)         49.219171         17.353094   .96757414   .96352091
  8    Y = EXP(A+B/X)          4.508009 -        .684315    .95229892   .94633628
```

A N A L Y S I S   O F   V A R I A N C E

```
CURVE  SS-REGRESSION   SS-RESIDUAL    SS-TOTAL       F-VALUE
  1      1204.6371       300.4039     1505.0410      32.0805
  2          .2555          .0847         .3402      24.1354
  3          .3233          .0169         .3402     152.7004
  4      1390.4953       114.5457     1505.0410      97.1137
  5          .0001          .0000         .0001      18.4901
  6          .0001          .0000         .0001     285.9630
  7      1456.2388        48.8022     1505.0410     238.7167
  8          .3240          .0162         .3402     159.7111
D.F.           1              8             9
```

COMMAND -->!RESIDUAL

ENTER CURVE NUMBER -->!6

```
CURVE  EQUATION-FORM              A                 B       R-SQUARED  ADJ. R**2
  6    Y = X/(A*X+B)           .010580           .010708    .97278569   .96938390

OBSN         X            Y       FITTED-Y    RESID.ERROR    %ERROR
  1       1.0000       48.2000    46.9739         1.2261      2.54
  2       2.0000       57.9000    62.7573  -      4.8573      8.39
  3       3.0000       70.7000    70.6727         .0273        .04
  4       4.0000       75.4000    75.4296  -      .0296        .04
  5       5.0000       80.2000    78.6041        1.5959       1.99
  6       6.0000       82.1000    80.8731        1.2269       1.49
  7       7.0000       83.3000    82.5757         .7243        .87
  8       8.0000       85.0000    83.9004        1.0996       1.29
```

```
 9       9.0000    85.5000    84.9606       .5394    .63
10      10.0000    86.0000    85.8282       .1718    .20

MEAN ABSOLUTE % ERROR   1.748896703396
MEAN SQUARE ERROR       3.12048311935


COMMAND -->!PLOT

ENTER CURVE NUMBER -->!6


CURVE  EQUATION-FORM           A           B      R-SQUARED  ADJ. R**2
  6    Y = X/(A*X+B)        .010580    .010708     .97278569  .96938390
```

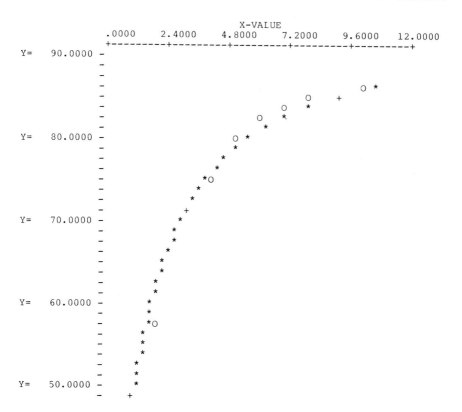

```
            -    *
            -     *
            -      *
            -       *
            -    *
            -      *
Y=   40.0000 -   *
             +---------+---------+---------+---------+---------+
              .0000    2.4000   4.8000    7.2000   9.6000   12.0000
                                 X-VALUE
```

COMMAND --->!FORECAST

ENTER CURVE NUMBER -->!6

CURVE EQUATION-FORM   A   B   R-SQUARED ADJ. R**2
  6   Y = X/(A*X+B)   .010580   .010708 .97278569 .96938390

ENTER X-VALUE (OR -99999 TO RETURN) -->!11

X = 11   Y = 86.55130801392

ENTER X-VALUE (OR -99999 TO RETURN) -->!-99999

COMMAND -->!EXIT

```
10 REM     E I G H T    C U R V E S
20 REM
30 REM     THIS PROGRAM PERFORMS CURVE FITTING OF TWO VARIABLES
40 REM     UP TO EIGHT TYPES OF CURVES.  X = INDEPENDENT VARIABLE.
50 REM     Y = DEPENDENT VARIABLE.
60 REM
70 DIM U(50),V(50),X(50),Y(50),P(50)
80 DIM A(8),B(8),C(8),F(8),R(8),T(8),W(8),Z(12)
90 DIM F$(11),Q$(8),P$(50)
100 N0=50
110 J0=8
120 PRINT
130 PRINT "E I G H T    C U R V E S"
140 PRINT "------------------------"
150 GOSUB 4360
160 PRINT
170 PRINT "NEED INTRODUCTION (Y OR N) ... ";
180 INPUT A$
190 IF A$ <> "Y" THEN 280
200 PRINT
210 PRINT "THIS PROGRAM CALCULATES THE LEAST SQUARES COEFFICIENTS"
220 PRINT "A AND B FOR THE FOLLOWING EIGHT CURVES :"
230 GOSUB 4810
240 PRINT "INPUT REQUIREMENTS"
250 PRINT "------------------"
260 PRINT " 1. NUMBER OF OBSERVATIONS (>=3)"
270 PRINT " 2. OBSERVATION VALUES ENTERED AS   X,Y"
280 GOTO 700
290 PRINT
300 PRINT "PLEASE ENTER COMMAND OR TYPE HELP FOR COMMAND DESCRIPTIONS"
310 PRINT
320 PRINT "COMMAND -->";
330 INPUT C$
340 IF C$ <> "HELP" THEN 370
350 GOSUB 4630
360 GOTO 310
370 IF C$ <> "CURVES" THEN 400
380 GOSUB 4810
390 GOTO 310
400 IF C$ <> "CHANGE" THEN 430
410 GOSUB 1090
420 GOTO 310
430 IF C$ <> "DELETE" THEN 460
440 GOSUB 1090
450 GOTO 310
460 IF C$ <> "EXIT" THEN 480
470 STOP
480 IF C$ <> "FORECAST" THEN 510
490 GOSUB 4890
500 GOTO 310
510 IF C$ <> "INSERT" THEN 540
520 GOSUB 890
530 GOTO 310
540 IF C$ <> "LIST" THEN 570
```

```
550 GOSUB 1330
560 GOTO 310
570 IF C$ <> "PLOT" THEN 600
580 GOSUB 5030
590 GOTO 310
600 IF C$ <> "RESIDUAL" THEN 630
610 GOSUB 2550
620 GOTO 310
630 IF C$ <> "SOLVE" THEN 660
640 GOSUB 1730
650 GOTO 310
660 PRINT
670 PRINT "*** INVALID COMMAND ***   PLEASE TRY AGAIN !!!"
680 GOTO 290
690 REM
700 REM   INPUT DATA HERE FOR THE FIRST TIME
710 PRINT
720 PRINT "ENTER NUMBER OF OBSERVATIONS (>=3) -->";
730 INPUT N
740 IF N <= 2 THEN 760
750 IF N <= N0 THEN 790
760 PRINT
770 PRINT "*** ERROR *** PLEASE TRY AGAIN !!!"
780 GOTO 710
790 PRINT
800 PRINT "ENTER OBSERVATIONS IN PAIR, I.E.   X,Y"
810 PRINT "-----------------------------------"
820 FOR I=1 TO N
830 PRINT "OBSN",I," ";
840 INPUT X(I),Y(I)
850 NEXT I
860 K9=0
870 GOTO 290
880 REM
890 REM    INSERT    C O M M A N D
900 REM    --------------------
910 PRINT
920 PRINT "INSERT AFTER OBSN NUMBER -->";
930 INPUT N1
940 IF N1 < 0 THEN 1080
950 IF N1 > N THEN 1080
960 IF N >= N0 THEN 1080
970 PRINT
980 PRINT "ENTER OBSN ";N1+1," AS X,Y ->";
990 INPUT A(1),B(1)
1000 IF N1 = N THEN 1050
1010 FOR I=N TO N1+1 STEP -1
1020 X(I+1)=X(I)
1030 Y(I+1)=Y(I)
1040 NEXT I
1050 X(N1+1)=A(1)
1060 Y(N1+1)=B(1)
1070 N=N+1
1080 RETURN
```

```
1090 REM   CHANGE OR DELETE  C O M M A N D S
1100 REM   --------------------------------
1110 PRINT "OBSERVATION NUMBER -->";
1120 INPUT K
1130 IF K <= 0 THEN 1160
1140 IF K <= N THEN 1170
1150 PRINT "LAST OBSERVATION NUMBER IS ",N
1160 RETURN
1170 IF C$ = "DELETE" THEN 1240
1180 PRINT "ENTER  X,Y -->";
1190 INPUT X1,Y1
1200 X(K)=X1
1210 Y(K)=Y1
1220 K9=0
1230 RETURN
1240 IF N = 1 THEN 470
1250 FOR I=K TO N-1
1260 X(I)=X(I+1)
1270 Y(I)=Y(I+1)
1280 NEXT I
1290 N=N-1
1300 PRINT "OBSERVATION",K,"DELETED !!!"
1310 K9=0
1320 RETURN
1330 REM  LIST    C O M M A N D   - LIST X,Y AND STATISTICS
1340 REM  --------------------
1350 X1=0
1360 X2=0
1370 Y1=0
1380 Y2=0
1390 X8=E7
1400 X9=E8
1410 Y8=E7
1420 Y9=E8
1430 FOR I=1 TO N
1440 X1=X1+X(I)
1450 Y1=Y1+Y(I)
1460 X2=X2+X(I)*X(I)
1470 Y2=Y2+Y(I)*Y(I)
1480 IF X(I) >= X8 THEN 1500
1490 X8=X(I)
1500 IF X(I) <= X9 THEN 1520
1510 X9=X(I)
1520 IF Y(I) >= Y8 THEN 1540
1530 Y8=Y(I)
1540 IF Y(I) <= Y9 THEN 1560
1550 Y9=Y(I)
1560 NEXT I
1570 PRINT
1580 PRINT "OBSERVATION        X            Y"
1590 FOR I=1 TO N
1600 PRINT USING F$(1),I,X(I),Y(I)
1610 NEXT I
1620 IF N <= 1 THEN 1720
```

```
1630 X2=SQR((X2-X1*X1/N)/(N-1))
1640 Y2=SQR((Y2-Y1*Y1/N)/(N-1))
1650 X1=X1/N
1660 Y1=Y1/N
1670 PRINT
1680 PRINT USING F$(1),"    MEAN",X1,Y1
1690 PRINT USING F$(1),"STD DEV",X2,Y2
1700 PRINT USING F$(1),"MINIMUM",X8,Y8
1710 PRINT USING F$(1),"MAXIMUM",X9,Y9
1720 RETURN
1730 REM   SOLVE   C O M M A N D
1740 REM   ---------------------
1750 IF K9=1 THEN 2250
1760 IF N >= 3 THEN 1800
1770 PRINT
1780 PRINT "NOT ENOUGH OBSERVATIONS !!! PLEASE ADD MORE DATA !!!"
1790 RETURN
1800 GOSUB 3440
1810 REM
1820 REM   CALCULATE COEFFICIENTS FOR THE EIGHT CURVES
1830 R(1)=Z(7)-Z(1)*Z(1)/N
1840 B(1)=(W(1)-Z(1)*Z(2)/N)/R(1)
1850 A(1)=(Z(2)-B(1)*Z(1))/N
1860 T(1)=Z(8)-Z(2)*Z(2)/N
1870 IF C(2) < 0.5 THEN 1930
1880 R(2)=R(1)
1890 B(2)=(W(2)-Z(1)*Z(6)/N)/R(2)
1900 A(2)=(Z(6)-B(2)*Z(1))/N
1910 A(2)=EXP(A(2))
1920 T(2)=Z(12)-Z(6)*Z(6)/N
1930 IF C(3) < 0.5 THEN 1990
1940 R(3)=Z(11)-Z(5)*Z(5)/N
1950 B(3)=(W(3)-Z(5)*Z(6)/N)/R(3)
1960 A(3)=(Z(6)-B(3)*Z(5))/N
1970 A(3)=EXP(A(3))
1980 T(3)=T(2)
1990 IF C(4) < 0.5 THEN 2040
2000 R(4)=Z(9)-Z(3)*Z(3)/N
2010 B(4)=(W(4)-Z(3)*Z(2)/N)/R(4)
2020 A(4)=(Z(2)-B(4)*Z(3))/N
2030 T(4)=T(1)
2040 IF C(5) < 0.5 THEN 2090
2050 R(5)=R(1)
2060 B(5)=(W(5)-Z(1)*Z(4)/N)/R(5)
2070 A(5)=(Z(4)-B(5)*Z(1))/N
2080 T(5)=Z(10)-Z(4)*Z(4)/N
2090 IF C(6) < 0.5 THEN 2140
2100 R(6)=R(4)
2110 B(6)=(W(6)-Z(3)*Z(4)/N)/R(6)
2120 A(6)=(Z(4)-B(6)*Z(3))/N
2130 T(6)=T(5)
2140 IF C(7) < 0.5 THEN 2190
2150 R(7)=Z(11)-Z(5)*Z(5)/N
2160 B(7)=(W(7)-Z(5)*Z(2)/N)/R(7)
```

```
2170 A(7)=(Z(2)-B(7)*Z(5))/N
2180 T(7)=T(1)
2190 IF C(8) < 0.5 THEN 2240
2200 R(8)=R(4)
2210 B(8)=(W(8)-Z(6)*Z(3)/N)/R(8)
2220 A(8)=(Z(6)-B(8)*Z(3))/N
2230 T(8)=T(2)
2240 K9=1
2250 PRINT
2260 PRINT
2270 PRINT "CURVE EQUATION-FORM                A                B ";
2280 PRINT " R-SQUARED ADJ. R**2"
2290 FOR J=1 TO J0
2300 IF C(J) < 0.5 THEN 2350
2310 PRINT USING F$(2),J,Q$(J);
2320 R1=B(J)*B(J)*R(J)/T(J)
2330 R2=1-(1-R1)*(N-1)/(N-2)
2340 PRINT USING F$(3),A(J),B(J),R1,R2
2350 NEXT J
2360 PRINT
2370 PRINT
2380 PRINT "A N A L Y S I S    O F    V A R I A N C E"
2390 PRINT
2400 PRINT "CURVE SS-REGRESSION    SS-RESIDUAL";
2410 PRINT "         SS-TOTAL          F-VALUE"
2420 FOR J=1 TO J0
2430 IF C(J) < 0.5 THEN 2500
2440 S1=B(J)*B(J)*R(J)
2450 S2=T(J)-S1
2460 S3=S1/S2*(N-2)
2470 PRINT USING F$(4),J;
2480 PRINT USING F$(5),S1,S2;
2490 PRINT USING F$(5),T(J),S3
2500 NEXT J
2510 PRINT USING F$(6)," D.F.",1,N-2,N-1
2520 PRINT
2530 PRINT
2540 RETURN
2550 REM    RESIDUAL   C O M M A N D
2560 REM    ----------------------
2570 GOSUB 2830
2580 IF I0 > 0.5 THEN 2600
2590 RETURN
2600 S1=0
2610 S2=0
2620 PRINT "OBSN           X            Y       FITTED-Y";
2630 PRINT "    RESID.ERROR    %ERROR"
2640 FOR I=1 TO N
2650 X0=X(I)
2660 GOSUB 3140
2670 E0=Y(I)-Y0
2680 S2=S2+E0*E0
2690 P1=999
2700 IF Y(I) = 0 THEN 2720
```

```
2710 P1=ABS(E0/Y(I)*100)
2720 S1=S1+P1
2730 PRINT USING F$(7),I,X(I),Y(I),Y0,E0;
2740 PRINT USING F$(8),P1
2750 NEXT I
2760 S1=S1/N
2770 S2=S2/N
2780 PRINT
2790 PRINT "MEAN ABSOLUTE % ERROR   ";S1
2800 PRINT "MEAN SQUARE ERROR       ";S2
2810 PRINT
2820 RETURN
2830 REM   SCREENING PROCESS
2840 REM
2850 I0=0
2860 IF K9 > 0.5 THEN 2900
2870 PRINT
2880 PRINT "USE THE SOLVE COMMAND, FIRST !!!"
2890 RETURN
2900 PRINT
2910 PRINT "ENTER CURVE NUMBER -->";
2920 INPUT K0
2930 IF K0 <= 0 THEN 2990
2940 IF K0 > J0 THEN 2990
2950 IF C(K0) > 0.5 THEN 3020
2960 PRINT
2970 PRINT "DATA NOT ELIGIBLE FOR SOLUTION OF CURVE",K0
2980 RETURN
2990 PRINT
3000 PRINT "*** INVALID CURVE NUMBER ***"
3010 RETURN
3020 I0=1
3030 PRINT
3040 PRINT
3050 PRINT
3060 PRINT "CURVE EQUATION-FORM               A           B ";
3070 PRINT "  R-SQUARED ADJ. R**2"
3080 PRINT USING F$(2),K0,Q$(K0);
3090 R1=B(K0)*B(K0)*R(K0)/T(K0)
3100 R2=1-(1-R1)*(N-1)/(N-2)
3110 PRINT USING F$(3),A(K0),B(K0),R1,R2
3120 PRINT
3130 RETURN
3140 REM   CALCULATE FITTED-Y VALUE ==> Y0
3150 Y0=0
3160 IF K0=8 THEN 3410
3170 IF K0=7 THEN 3390
3180 IF K0=6 THEN 3360
3190 IF K0=5 THEN 3330
3200 IF K0=4 THEN 3300
3210 IF K0=3 THEN 3270
3220 IF K0=2 THEN 3250
3230 Y0=A(1)+B(1)*X0
3240 RETURN
```

```
3250 Y0=A(2)*EXP(B(2)*X0)
3260 RETURN
3270 IF X0 < 0 THEN 3290
3280 Y0=A(3)*X0**B(3)
3290 RETURN
3300 IF X0=0 THEN 3320
3310 Y0=A(4)+B(4)/X0
3320 RETURN
3330 IF A(5)+B(5)*X0 = 0 THEN 3350
3340 Y0=1/(A(5)+B(5)*X0)
3350 RETURN
3360 IF A(6)*X0+B(6) = 0 THEN 3380
3370 Y0=X0/(A(6)*X0+B(6))
3380 RETURN
3390 IF X0 <= 0 THEN 3410
3400 Y0=A(7)+B(7)*LOG(X0)
3410 IF X0=0 THEN 3430
3420 Y0=EXP(A(8)+B(8)/X0)
3430 RETURN
3440 REM   DETERMINE ELIGIBILITY FOR CALCULATION OF EACH CURVE
3450 REM   ----------------------------------------------------
3460 FOR J=1 TO J0
3470 C(J)=1
3480 W(J)=0
3490 NEXT J
3500 X8=0
3510 X9=0
3520 Y8=0
3530 Y9=0
3540 FOR I=1 TO N
3550 IF ABS(X(I)) > E9 THEN 3570
3560 X8=1
3570 IF X(I) > E9 THEN 3590
3580 X9=1
3590 IF ABS(Y(I)) > E9 THEN 3610
3600 Y8=1
3610 IF Y(I) > E9 THEN 3630
3620 Y9=1
3630 NEXT I
3640 IF Y9 < 0.5 THEN 3680
3650 C(2)=0
3660 C(3)=0
3670 C(8)=0
3680 IF X9 < 0.5 THEN 3710
3690 C(3)=0
3700 C(7)=0
3710 IF Y8 < 0.5 THEN 3770
3720 C(2)=0
3730 C(3)=0
3740 C(5)=0
3750 C(6)=0
3760 C(8)=0
3770 IF X8 < 0.5 THEN 3830
3780 C(3)=0
```

```
3790 C(4)=0
3800 C(6)=0
3810 C(7)=0
3820 C(8)=0
3830 REM
3840 REM   CALCULATE LOG(X) AND LOG(Y)
3850 IF Y9 > 0.5 THEN 3890
3860 FOR I=1 TO N
3870 V(I)=LOG(Y(I))
3880 NEXT I
3890 IF X9 > 0.5 THEN 3930
3900 FOR I=1 TO N
3910 U(I)=LOG(X(I))
3920 NEXT I
3930 FOR I=1 TO 12
3940 Z(I)=0
3950 NEXT I
3960 FOR I=1 TO N
3970 Z(1)=Z(1)+X(I)
3980 Z(2)=Z(2)+Y(I)
3990 IF X8 > 0.5 THEN 4030
4000 Z0=1/X(I)
4010 Z(3)=Z(3)+Z0
4020 Z(9)=Z(9)+Z0*Z0
4030 IF Y8 > 0.5 THEN 4070
4040 Z0=1/Y(I)
4050 Z(4)=Z(4)+Z0
4060 Z(10)=Z(10)+Z0*Z0
4070 IF X9 > 0.5 THEN 4100
4080 Z(5)=Z(5)+U(I)
4090 Z(11)=Z(11)+U(I)*U(I)
4100 IF Y9 > 0.5 THEN 4130
4110 Z(6)=Z(6)+V(I)
4120 Z(12)=Z(12)+V(I)*V(I)
4130 Z(7)=Z(7)+X(I)*X(I)
4140 Z(8)=Z(8)+Y(I)*Y(I)
4150 NEXT I
4160 REM
4170 REM   CALCULATE CROSS-PRODUCT TERMS
4180 FOR I=1 TO N
4190 W(1)=W(1)+X(I)*Y(I)
4200 IF C(2) < 0.5 THEN 4220
4210 W(2)=W(2)+V(I)*X(I)
4220 IF C(3) < 0.5 THEN 4240
4230 W(3)=W(3)+U(I)*V(I)
4240 IF C(4) < 0.5 THEN 4260
4250 W(4)=W(4)+Y(I)/X(I)
4260 IF C(5) < 0.5 THEN 4280
4270 W(5)=W(5)+X(I)/Y(I)
4280 IF C(6) < 0.5 THEN 4300
4290 W(6)=W(6)+1/(X(I)*Y(I))
4300 IF C(7) < 0.5 THEN 4320
4310 W(7)=W(7)+Y(I)*U(I)
4320 IF C(8) < 0.5 THEN 4340
```

```
4330 W(8)=W(8)+V(I)/X(I)
4340 NEXT I
4350 RETURN
4360 REM   I N I T I A L I Z A T I O N
4370 REM   --------------------------
4380 N=0
4390 K9=0
4400 E9=1.E-8
4410 E8=-1.E-20
4420 E7=1.E20
4430 Q$(1)="Y = A+B*X"
4440 Q$(2)="Y = A*EXP(B*X)"
4450 Q$(3)="Y = A*(X**B)"
4460 Q$(4)="Y = A+B/X"
4470 Q$(5)="Y = 1/(A+B*X)"
4480 Q$(6)="Y = X/(A*X+B)"
4490 Q$(7)="Y = A+B*LOG(X)"
4500 Q$(8)="Y = EXP(A+B/X)"
4510 F$(1)="####### -#####.#### -#####.####"
4520 F$(2)="###   ##############"
4530 F$(3)=" -######.###### -######.###### .######## .########"
4540 F$(4)="#####"
4550 F$(5)=" ########.#### ########.####"
4560 F$(6)="##### ############ ############ ############"
4570 F$(7)="####   -####.#### -####.#### -####.####  -####.####"
4580 F$(8)="    ###.##"
4590 F$(9)="         -####.####"
4600 F$(10)=" -###.####"
4610 F$(11)="Y=-####.####"
4620 RETURN
4630 REM  HELP    C O M M A N D
4640 REM  -------------------
4650 PRINT
4660 PRINT "AVAILABLE COMMANDS"
4670 PRINT "=================="
4680 PRINT " CHANGE   - CHANGE THE VALUES OF AN OBSERVATION"
4690 PRINT " DELETE   - DELETE AN OBSERVATION"
4700 PRINT " CURVES   - SHOW EQUATION FORMS OF THE EIGHT CURVES"
4710 PRINT " EXIT     - EXIT FROM THE PROGRAM"
4720 PRINT " FORECAST - GIVEN X, CALCULATE THE FITTED Y-VALUE"
4730 PRINT " HELP     - PRINT THIS MESSAGE"
4740 PRINT " INSERT   - INSERT AN OBSERVATION AFTER ANOTHER"
4750 PRINT " LIST     - LIST ALL OBSNS AND SUMMARY STATISTICS"
4760 PRINT " PLOT     - PLOT THE FITTED CURVE AND THE OBSNS"
4770 PRINT " RESIDUAL - PRINT RESIDUAL TABLE AND FITTED Y-VALUE"
4780 PRINT " SOLVE    - SOLVE FOR THE COEFFICIENTS OF A AND B"
4790 PRINT "            AND VARIOUS STATISTICS"
4800 RETURN
4810 REM  CURVES    C O M M A N D
4820 REM  ---------------------
4830 PRINT
4840 FOR J=1 TO J0
4850 PRINT "CURVE",J,Q$(J)
4860 NEXT J
```

```
4870 PRINT
4880 RETURN
4890 REM    FORECAST   C O M M A N D
4900 REM    ------------------------
4910 GOSUB 2830
4920 IF I0 > 0.5 THEN 4950
4930 RETURN
4940 PRINT
4950 PRINT "ENTER X-VALUE (OR -99999 TO RETURN) -->";
4960 INPUT X0
4970 IF X0=-99999 THEN 5020
4980 GOSUB 3140
4990 PRINT
5000 PRINT "X = ";X0;"    Y = ";Y0
5010 GOTO 4940
5020 RETURN
5030 REM    P L O T    C O M M A N D
5040 REM    ------------------------
5050 GOSUB 2830
5060 IF I0 > 0.5 THEN 5080
5070 RETURN
5080 REM
5090 REM    SORT Y(I) HIGH TO LOW
5100 FOR I=1 TO N
5110 P(I)=I
5120 NEXT I
5130 FOR I=1 TO N-1
5140 Z0=Y(P(I))
5150 FOR J=I+1 TO N
5160 IF Z0 >= Y(P(J)) THEN 5210
5170 Z0=Y(P(J))
5180 K1=P(I)
5190 P(I)=P(J)
5200 P(J)=K1
5210 NEXT J
5220 NEXT I
5230 REM
5240 REM    FIND MIN/MAX OF X(I), Y(I)
5250 X8=E7
5260 X9=E8
5270 Y8=E7
5280 Y9=E8
5290 FOR I=1 TO N
5300 IF X(I) >= X8 THEN 5320
5310 X8=X(I)
5320 IF X(I) <= X9 THEN 5340
5330 X9=X(I)
5340 IF Y(I) >= Y8 THEN 5360
5350 Y8=Y(I)
5360 IF Y(I) <= Y9 THEN 5380
5370 Y9=Y(I)
5380 NEXT I
5390 REM
5400 REM    DETERMINE SCALING FOR Y-AXIS
```

```
5410 A0=Y8
5420 A1=Y9
5430 GOSUB 6810
5440 IF L5 > 0 THEN 5460
5450 RETURN
5460 Y1=L0
5470 Y2=U0
5480 Y3=H0
5490 Y4=Y3/8
5500 REM
5510 REM    DETERMINE SCALING FOR X-AXIS
5520 A0=X8
5530 A1=X9
5540 GOSUB 6810
5550 IF L5 > 0 THEN 5570
5560 RETURN
5570 X1=L0
5580 X2=U0
5590 X3=H0
5600 X4=X3/10
5610 PRINT
5620 PRINT TAB(37);"X-VALUE"
5630 GOSUB 5840
5640 GOSUB 5950
5650 I=1
5660 M=8
5670 Y5=Y2-Y4/2
5680 Y6=Y5+Y4
5690 L0=Y2
5700 GOSUB 6030
5710 FOR L=1 TO 5
5720 L0=L0-Y3
5730 FOR M=1 TO 8
5740 Y6=Y5
5750 Y5=Y6-Y4
5760 GOSUB 6030
5770 NEXT M
5780 NEXT L
5790 GOSUB 5950
5800 GOSUB 5840
5810 PRINT TAB(37);"X-VALUE"
5820 PRINT
5830 RETURN
5840 REM
5850 REM    PRINT AXIS LABEL
5860 H0=X1
5870 PRINT USING F$(9),H0;
5880 FOR J=1 TO 5
5890 H0=H0+X3
5900 PRINT USING F$(10),H0;
5910 NEXT J
5920 PRINT
5930 RETURN
5940 REM
```

```
5950 REM    PRINT AXIS
5960 PRINT TAB(15);"+";
5970 FOR J=1 TO 5
5980 PRINT "---------+";
5990 NEXT J
6000 PRINT
6010 RETURN
6020 REM
6030 REM    CALCULATE PLOT LINE
6040 FOR J=0 TO 50
6050 P$(J)=" "
6060 NEXT J
6070 J9=0
6080 IF I > N THEN 6220
6090 J2=I
6100 FOR J1=J2 TO N
6110 IF Y(P(J1)) < Y5 THEN 6190
6120 N9=INT((X(P(J1))-X1)/X4+0.5)
6130 P$(N9)="O"
6140 IF N9 <= J9 THEN 6160
6150 J9=N9
6160 IF J1 <> N THEN 6210
6170 I=N+1
6180 GOTO 6210
6190 I=J1
6200 GOTO 6240
6210 NEXT J1
6220 REM
6230 REM    CALCULATE FITTED VALUE OF X
6240 Y7=(Y5+Y6)/2
6250 IF K0 = 8 THEN 6540
6260 IF K0 = 7 THEN 6510
6270 IF K0 = 6 THEN 6480
6280 IF K0 = 5 THEN 6450
6290 IF K0 = 4 THEN 6420
6300 IF K0 = 3 THEN 6350
6310 IF K0 = 2 THEN 6350
6320 IF B(K0) = 0 THEN 6690
6330 X0=(Y7-A(K0))/B(K0)
6340 GOTO 6590
6350 IF B(K0) = 0 THEN 6690
6360 IF A(K0) = 0 THEN 6690
6370 IF Y7/A(K0) <= 0 THEN 6690
6380 X0=LOG(Y7/A(K0))/B(K0)
6390 IF K0 = 2 THEN 6590
6400 X0=EXP(X0)
6410 GOTO 6590
6420 IF Y7-A(K0) = 0 THEN 6690
6430 X0=B(K0)/(Y7-A(K0))
6440 GOTO 6590
6450 IF Y7*B(K0) = 0 THEN 6690
6460 X0=(1-A(K0)*Y7)/(Y7*B(K0))
6470 GOTO 6590
6480 IF 1-A(K0)*Y7 = 0 THEN 6690
```

```
6490 X0=B(K0)*Y7/(1-A(K0)*Y7)
6500 GOTO 6590
6510 IF B(K0) = 0 THEN 6690
6520 X0=EXP((Y7-A(K0))/B(K0))
6530 GOTO 6590
6540 IF Y7 <= 0 THEN 6690
6550 X0=LOG(Y7) - A(K0)
6560 IF X0 = 0 THEN 6690
6570 X0=B(K0)/X0
6580 REM
6590 REM    CALCULATE FITTED PLOT POSITION
6600 IF X0 < X1 THEN 6690
6610 IF X0 > X2 THEN 6690
6620 N9=INT((X0-X1)/X4+0.5)
6630 IF P$(N9) = "O" THEN 6660
6640 P$(N9)="*"
6650 GOTO 6670
6660 P$(N9)="+"
6670 IF N9 <= J9 THEN 6690
6680 J9=N9
6690 REM
6700 REM    PLOT THE LINE
6710 IF M <> 8 THEN 6740
6720 PRINT USING F$(11),L0;
6730 GOTO 6750
6740 PRINT "              ";
6750 PRINT " -";
6760 FOR J=0 TO J9
6770 PRINT P$(J);
6780 NEXT J
6790 PRINT
6800 RETURN
6810 REM    S C A L I N G    R O U T I N E
6820 REM    ------------------------------
6830 L5=0
6840 A8=A0
6850 A9=A1
6860 A2=(A1-A0)/4
6870 IF A2 <= 0 THEN 7200
6880 L5=L5+1
6890 A3=LOG(A2)/LOG(10)
6900 REM
6910 REM    ROUND DOWN A3 TO A4
6920 A4=INT(A3)
6930 IF A4 <= A3 THEN 6950
6940 A4=A4-1
6950 A5=10**A4
6960 A6=A2/A5
6970 REM
6980 REM    ROUND UP A6 TO A7
6990 A7=INT(A6+1)
7000 H0=A5*A7
7010 REM    ROUND DOWN L0
7020 L0=INT(A0/H0)
```

```
7030 IF L0 <= A0/H0 THEN 7050
7040 L0=L0-1
7050 L0=L0*H0
7060 U0=L0+5*H0
7070 IF U0 >= A1 THEN 7100
7080 A1=A1+A2/2
7090 GOTO 6860
7100 IF L0 <= A0 THEN 7130
7110 A0=A0-A2/2
7120 GOTO 6860
7130 IF U0-H0 <= A9 THEN 7160
7140 U0=U0-H0
7150 GOTO 7130
7160 IF L0+H0 >= A8 THEN 7190
7170 L0=L0+H0
7180 GOTO 7160
7190 H0=(U0-L0)/5
7200 RETURN
7210 END
```

II. Forecasting

One of the first things that a forecaster may want to do when analyzing a time series is to determine whether or not there is an underlying trend and/or seasonality in the data. This can be done by examining the autocorrelation function of the time series and perhaps the autocorrelation function of the first difference of the series.

Let x_t, $t=1,\ldots,T$ be the time series. Then the autocorrelation of lag k is

$$(1) \quad r_k = \frac{\sum_{t=1}^{T-k} (x_t - \bar{x})(x_{t+k} - \bar{x})}{\sum_{t=1}^{T} (x_t - \bar{x})^2}$$

where $\bar{x}$ is the mean of the observations. The autocorrelation function of a time series up to lag k consists of $r_0, r_1, \ldots, r_k$. Let $y_t = x_t - x_{t-1}$, $t=2,\ldots,T$. Then y_t is the first difference of x_t. The autocorrelation function of the first difference of the series will be the autocorrelation of the original series with the trend removed.

We will now describe the formulas used in the forecasting models contained in the program FORCST.

1. Single Moving Average (SMAVE)

The model assumes that the underlying process is a constant -i.e., a horizontal model [7]. Let N be the number of terms (or periods) in the moving average M_t. Then

$$(2) \quad M_t = (x_{t-N+1} + \ldots + x_{t-1} + x_t)/N$$

or

$$(3) \quad M_t = M_{t-1} + (x_t - x_{t-N})/N.$$

The k-period-ahead forecast, given that we are in period t, is

$$(4) \quad F_t(k) = M_t, \quad k=1,2,\ldots$$

2. Linear Moving Average (LMAVE)

The model assumes a linear trend process -i.e., there is a constant component and a slope component. Formulas (2) and (3) are also used in LMAVE. In addition, a double moving average is required.

(5) $\quad M_t^{(2)} = (M_{t-N+1} + \ldots + M_{t-1} + M_t)/N$

or

(6) $\quad M_t^{(2)} = M_{t-1}^{(2)} + (M_t - M_{t-N})/N$.

The slope b_t and the intercept a_t and the k-period-ahead forecast are calculated as follows:

(7) $\quad b_t = 2(M_t - M_t^{(2)})/(N-1)$,

(8) $\quad a_t = 2 M_t - M_t^{(2)}$,

and

(9) $\quad F_t(k) = a_t + b_t k$, $k=1,2,\ldots$

Note that LMAVE requires a minimum of 2N observations before a forecast can be generated.

3. Linear Regression with Fixed Periods (REGRESS)

In this model, a simple linear regression is performed on all observations taking N observations at a time. The result is a moving projection of a linear trend. The slope b_t and the intercept a_t are

(10) $\quad b_t = \dfrac{-0.5(N-1) \sum_{j=0}^{N-1} x_{t-j} + \sum_{j=0}^{N-1} (j x_{t-j})}{N(N-1)^2/4 - N(N-1)(2N-1)/6}$

(11) $\quad a_t = \bar{x}_t + b_t (N-1)/2$

where $\bar{x}_t$ is the mean of the last N observations before time t. Equation (9) is used in forecasting future periods. See [7].

4. Single Exponential Smoothing (X)

This is perhaps the most widely used forecasting model because of its simplicity. The underlying process is assumed to be a constant. Equation (12) shows how the Smoothed Average (S_t) is updated in every period:

$$(12) \qquad S_t = \alpha x_t + (1-\alpha) S_{t-1}$$

where α (ALPHA) is a smoothing constant for the model and is a value between 0 and 1. ALPHA is the weight given to the most recent observation. It is usually set between 0.1 and 0.3. A high value of ALPHA causes the forecasts to fluctuate more so than a lower value of ALPHA. The k-period-ahead forecast is given by

$$(13) \qquad F_t(k) = S_t, \quad k=1,2,\ldots$$

The program FORCST initializes the Smoothed Average with the first observation, otherwise the user must supply its initial value.

5. Double Exponential Smoothing (XX)

Since the underlying process assumed in the Double Exponential Smoothing model is a linear trend, the slope (b) and the intercept (a) will have to be updated every period. The necessary equations are:

$$(14) \qquad a_t = x_t + (1-\alpha)^2 e_t,$$

$$(15) \qquad b_t = b_{t-1} - \alpha^2 e_t$$

where

$$(16) \qquad e_t = F_{t-1}(1) - x_t$$

is the one-period-ahead forecast error at time t [7]. Again, equation (9) is used for forecasting future periods. The program FORCST prompts for the initial estimates of the intercept (a) and the slope (b) if a manual initialization option is selected. Otherwise the program initially sets the intercept to the first observation and the slope is set to zero.

6. Triple Exponential Smoothing (XXX)

In this case, the underlying process is assumed to be a quadratic model:

$$(17) \qquad F_t(k) = a_t + b_t k + 0.5 c_t k^2, \quad k=1,2,\ldots$$

The parameters a, b, and c are updated using equations (18), (19), and (20):

$$(18) \qquad a_t = x_t + (1-\alpha)^3 e_t$$

$$(19) \qquad b_t = b_{t-1} + c_{t-1} - 1.5\alpha^2 (2-\alpha) e_t$$

$$(20) \qquad c_t = c_{t-1} - \alpha^3 e_t$$

where e_t is the usual one-period-ahead forecast error. If the manual initialization method is selected, the program FORCST expects the availability of the initial estimates of a, b, and c. Otherwise the program initializes a with the first observation and sets b and c to zeros. Equation (17) is used in the forecast.

7. Adaptive Response Rate Exponential Smoothing (ADAPT)

This method was developed by Trigg and Leach [8]. It also assumes a constant process as in Single Exponential Smoothing. However, the smoothing parameter α is dynamically determined so that the forecasts can adapt more quickly when there is a sudden shift in the constant process.

Two smoothed errors are utilized in the method. They are

$$(21) \qquad E_t = \beta e_t + (1-\beta) E_{t-1}$$

and

$$(22) \qquad A_t = \beta |e_t| + (1-\beta) A_{t-1}$$

where $e_t = x_t - F_{t-1}(1)$, which is slightly different from the definition for the one-period-ahead forecast error of (16). E_t is called the Smoothed Error, and A_t is called the Smoothed

Absolute Error; β is a smoothing parameter which lies between zero and one. The ratio of E_t and A_t is called the Tracking Signal. Its absolute value becomes the smoothing parameter α, which changes from period to period and is used in updating the constant process -i.e., the smoothed average

(23) $$S_t = \alpha_t x_t + (1-\alpha_t) S_{t-1}.$$

The program initializes S_1 with x_1 and sets e_1, E_1, A_1 to zeros. This will cause the first two consecutive periods to have the same forecast value in this program. Other initialization techniques are possible but if the number of observations is large enough the resulting forecasts should all converge to the same results regardless of the initialization procedures. If manual initialization is selected, the program will prompt for the initial values of S_0, E_0, and A_0. The forecasting equation is (13).

8. Holt's 2-Parameters Linear Exponential Smoothing (HOLT)

This is also a linear trend model. Equations (24) and (25) show the method of updating the intercept and the slope:

(24) $$a_t = \alpha x_t + (1-\alpha)(a_{t-1} + b_{t-1})$$

(25) $$b_t = \beta(a_t - a_{t-1}) + (1-\beta)b_{t-1}$$

where α and β are the smoothing parameters. The method of initialization employed in the program is the same as that in Makridakis and Wheelwright [4]:

$$a_1 = x_1$$

and

$$b_1 = (x_2 - x_1)/2 + (x_4 - x_3)/2.$$

This implies that at least four observation values are required if manual initialization is not utilized. Equation (9) is used in generating the forecasts.

9. Winters's 3-Parameters Linear Exponential Smoothing (WINTERS)

This is the only seasonal model available in the program FORCST. The underlying process is assumed to be a multiplicative linear trend model [7, 9]. The necessary updating equations are

(26) $$a_t = \alpha(x_t/r_t) + (1-\alpha)(a_{t-1} + b_{t-1}),$$

(27) $$b_t = \beta(a_t - a_{t-1}) + (1-\beta)b_{t-1}$$

and

(28) $$r_{t+p} = \gamma(x_t/a_t) + (1-\gamma)r_t$$

where α, β, γ are the smoothing constants; p is the periodicity of the time series; and r_t is the seasonal factor for period t, such that

$$\sum_{i=1}^{p} r_i = p \ .$$

The forecasting equation for this model is

(29) $$F_t(k) = (a_t + b_t k)r_{t+k}, \quad k=1,2,\ldots$$

The initialization procedure adopted in the program requires at least 2p periods of observations. The method of semi-averages is used to provide initial estimates of the intercept (a) and the slope (b). When manual initialization is set, the program expects the initial values of a and b and all p values of seasonal factors.

The Program FORCST

Much has already been written about the procedures and methods used in FORCST. Operationally, the program is command-driven. It consists of 29 commands. The default values of α, β, and γ are set at 0.2, 0.1, and 0.1, respectively. It should be noted that although we have used the same symbols for ALPHA(α), BETA(β), and GAMMA(γ) in many models, their meanings may not necessarily be the same in all models. There are commands ALPHA, BETA, and GAMMA to change the default values of these parameters. The commands INITON and INITOFF are used to turn the manual initialization option on and off, respectively. Commands are also provided for editing the data.

There is also an optimization capability in FORCST that only applies to exponential smoothing models. The commands OPTON and OPTOFF allow the user to set the optimization option on or off. When this option is turned on, the program will prompt for the minimum, the maximum, and the step size of the appropriate parameters to search. The optimal solution will be based on the minimum mean squared error found based on the specified grids for searching. One word of caution when using this option: Since the program evaluates all possible combinations of the smoothing parameters, it is wise to set the grid size relatively coarse at the beginning and refine it later, as it may take a considerable amount of computer time for the evaluations to be complete. Based on our experience on the PRIME-400 computer, one should perhaps limit the number of combinations for each search to about 100.

The default number of periods to forecast is initially set at five. This can easily be changed by using the command FPERIOD. In generating the forecasts for all constant models -i.e., those that use equations (4) and (13), we have used the so-called Bootstrapping technique [4, p. 18]. This means that for forecasts beyond the historical data, we will use the last observation value as though it were equal to all future observations in computing the forecasts. Therefore the resulting forecasts will not appear as a constant, as equation (4) or equation (13) implies, for these models.

INPUT
-----
1. The name of the time series
2. The number of observations
3. All observation values

OUTPUT

There are three main types of outputs available:

1. The autocorrelation function of the time series and the autocorrelation function of the first difference of the time series.
2. A table of the observations and their forecasts and the error statistics. The table can be turned on or off using the commands TABLEON or TABLEOFF.
3. A plot of the observations and the forecasts which can be turned on or off using the commands PLOTON or PLOTOFF.

EXAMPLE 1

Monthly sales figures for High Speed Drills Model A3 are available for 15 periods as follows:

Month	Sales	Month	Sales	Month	Sales
1	350	6	325	11	390
2	326	7	355	12	389
3	340	8	345	13	405
4	319	9	385	14	412
5	298	10	374	15	403 .

Determine the forecasts for the next five periods using the program FORCST.

The example terminal session 1 below shows the dialog of the commands when we perform the analysis. Note that observation number 1 is entered incorrectly as 250. After all data are entered, we issue the command HELP which prints out all available commands. Then we proceed to edit the data by deleting observation 1 and we insert the correct value of 350 after observation 0. The LIST command is issued next to make sure that the data are entered correctly. When the user specifies the complete range to list the data, the program also reports the minimum, the maximum, the mean, and the standard deviation of the observations. Otherwise only observations are listed.

We are now ready to perform some analysis. The AUTO command is given next to see the autocorrelation function for 10 lags. The autocorrelation function indicates that some trend exists in the data. To remove the trend and detect any seasonality, the command AUTODIF is entered and no seasonal pattern appears to exist. This may be due to the relatively small number of observations available.

We proceed to issue the commands TABLEOFF and PLOTOFF to suppress lengthy outputs. Then the commands SMAVE, LMAVE, REGRESS, X, XX, XXX, ADAPT, HOLT, and WINTERS are given so that we may examine the statistics generated by these techniques using the default parameters. Note that for the SMAVE and the LMAVE commands, we have specified the number of periods in the moving averages to be 3. This is an attempt to eliminate the randomness in the data and to obtain better estimates of the underlying process. The number of periods in the REGRESS command is also set at 3. The WINTERS command does not function because of insufficient data.

By examining the statistics provided, we can see that LMAVE performs better than SMAVE and XX performs better than X based on the mean squared error. However, XXX also appears to perform very well. Obviously, other combinations of parameters will produce different results. The number of error observations is also an important consideration when making comparisons using these statistics. The number should be roughly the same for more meaningful comparisons.

Assuming that we wish to examine the forecasts of LMAVE and also determine the best value of ALPHA in the XXX model, we entered commands TABLEON and PLOTON to turn the full output options on. The command OPTON is entered to activate the optimization feature. Next, LMAVE is entered at the command level. The number of periods in the moving average is set at 3, and the full output table and the plot are generated.

The XXX command is issued next, and a range of .05 to .5 for ALPHA is set together with the step size of .05. This represents 10 evaluations of the Triple Exponential Smoothing Model. The best value of ALPHA that minimizes the mean squared error is 0.3.

EXAMPLE 2

Previous analysis was made on sales data of small diesel engines using Winters's model. The analyst has kept the following statistics of the last computer run. The intercept and the slope are 114.3 and 0.535, respectively. The twelve seasonal factors are .477, .493, .918, 1.012, 1.628, 1.724, 1.718, 1.350, 1.021, .498, .604, and .557, respectively. New sales figures were compiled during the last 12 months after the first analysis. They are 72, 58, 137, 137, 176, 231, 244, 190, 155, 78, 60, and 70, respectively. Management wishes to have the next-year projection of the sales.

Please refer to the FORCST example terminal session 2 below for details of the sample run.

After the monthly sales data are entered, we turn the optimization option on. We also turn the manual initialization mode on. In addition, we wish to see the plot, and we also set the number of periods to forecast to 12. Next the WINTERS command is entered and the program prompts for the initial estimates of the intercept and the slope and the previous values of the seasonal factors. Then the ranges of ALPHA, BETA, and GAMMA are entered for optimization. This represents 125 combinations of evaluations, and the best combination found for ALPHA, BETA, and GAMMA is 0.1, 0.4, and 0.5, respectively.

```
FORCST EXAMPLE TERMINAL SESSION 1
===================================

FORECASTING SEASONAL AND NONSEASONAL MODELS
-------------------------------------------

NEED INTRODUCTION (Y OR N) ... !N

ENTER TIME SERIES DESCRIPTION
!SALES OF HIGH SPEED DRILLS MODEL A3

ENTER NUMBER OF OBSERVATIONS (>=3) -->!15

ENTER OBSERVATIONS
------------------
OBSN           1                          !250
OBSN           2                          !326
OBSN           3                          !340
OBSN           4                          !319
OBSN           5                          !298
OBSN           6                          !325
OBSN           7                          !355
OBSN           8                          !345
OBSN           9                          !385
OBSN          10                          !374
OBSN          11                          !390
OBSN          12                          !389
OBSN          13                          !405
OBSN          14                          !412
OBSN          15                          !403
PLEASE ENTER COMMAND OR TYPE HELP FOR COMMAND DESCRIPTIONS

COMMAND -->!HELP


AVAILABLE COMMANDS
==================
AUTO     - AUTOCORRELATION OF THE TIME SERIES
AUTODIF  - AUTOCOR. OF FIRST DIFFERENCE OF THE SERIES
SMAVE    - SINGLE MOVING AVERAGE
LMAVE    - LINEAR MOVING AVERAGE
REGRESS  - LINEAR REGRESSION WITH FIXED PERIODS
X        - SINGLE EXPONENTIAL SMOOTHING
XX       - DOUBLE EXPONENTIAL SMOOTHING
XXX      - TRIPLE EXPONENTIAL SMOOTHING
ADAPT    - ADAPTIVE RESPONSE RATE EXPO. SMOOTH.
HOLT     - HOLT 2-PARAMETERS LINEAR EXPO. SMOOTH.
WINTERS  - WINTERS 3-PARAMETERS LINEAR EXPO. SMOOTH.
HELP     - PRINT THIS MESSAGE
LIST     - LIST OBSERVATIONS WITHIN A RANGE
DELETE   - DELETE AN OBSERVATION
INSERT   - INSERT AN OBSERVATION AFTER AN OBSN
PLOTON   - SET OPTION PLOT ON
PLOTOFF  - SET OPTION PLOT OFF
```

```
INITON   - MANUAL INITIALIZATION OF MODEL CONSTANTS
INITOFF  - USE DEFAULT METHOD OF INITIALIZATION
FPERIOD  - SPECIFY NO. OF PERIODS TO FORECAST
TABLEON  - SET OPTION TO PRODUCE RESIDUAL TABLE ON
TABLEOFF - SET OPTION TO PRODUCE RESIDUAL TABLE OFF
OPTON    - SET OPTIMIZATION OPTION ON
OPTOFF   - SET OPTIMIZATION OPTION OFF
ALPHA    - SPECIFY SMOOTHING PARAMETER - ALPHA
BETA     - SPECIFY SMOOTHING PARAMETER - BETA
GAMMA    - SPECIFY SMOOTHING PARAMETER - GAMMA
RESTART  - START A NEW PROBLEM
EXIT     - EXIT FROM THE PROGRAM

COMMAND -->!DELETE

ENTER OBSN NUMBER -->!1
OBSN 1              250  DELETED !!!

COMMAND -->!INSERT

INSERT AFTER OBSN NUMBER -->!0

OBSN            1                  ENTER VALUE -->!350

COMMAND -->!LIST

TOTAL NO. OF OBSNS = 15
ENTER RANGE OF OBSNS TO LIST (FROM,TO) -->!1,15

OBSN            1                  350
OBSN            2                  326
OBSN            3                  340
OBSN            4                  319
OBSN            5                  298
OBSN            6                  325
OBSN            7                  355
OBSN            8                  345
OBSN            9                  385
OBSN           10                  374
OBSN           11                  390
OBSN           12                  389
OBSN           13                  405
OBSN           14                  412
OBSN           15                  403
MINIMUM = 298
MAXIMUM = 412
MEAN    = 361.0666666667
STD DEV = 35.63599036822

COMMAND -->!AUTO

HOW MANY LAGS -->!10

TIME SERIES : SALES OF HIGH SPEED DRILLS MODEL A3
```

```
                 -   1.00 -      .50       .00       .50      1.00
     LAG   AUTOCORR.  +---------+---------+---------+---------+
      0   1.0000000                        I*********************
      1    .7854799                        I*****************
      2    .6048202                        I*************
      3    .4038817                        I*********
      4    .1737815                        I***
      5  - .0579125                       *I
      6  - .1873656                    ****I
      7  - .3644743                 *******I
      8  - .4142601                ********I
      9  - .4679081               *********I
     10  - .3944244                ********I
     LAG   AUTOCORR.  +---------+---------+---------+---------+
                 -   1.00 -      .50       .00       .50      1.00

COMMAND -->!AUTODIF

HOW MANY LAGS -->!10

TIME SERIES : FIRST DIFFERENCE OF SALES OF HIGH SPEED DRILLS MODEL A3
                 -   1.00 -      .50       .00       .50      1.00
     LAG   AUTOCORR.  +---------+---------+---------+---------+
      0   1.0000000                        I*********************
      1  - .2857557                  ******I
      2    .0963272                        I**
      3    .1170147                        I**
      4  - .1174847                      **I
      5  - .1992584                    ****I
      6    .0934737                        I**
      7  - .1437559                     ***I
      8    .0163924                        I
      9  - .1924659                    ****I
     10    .0892422                        I**
     LAG   AUTOCORR.  +---------+---------+---------+---------+
                 -   1.00 -      .50       .00       .50      1.00

COMMAND -->!TABLEOFF


COMMAND -->!PLOTOFF


COMMAND -->!SMAVE

ENTER NUMBER OF PERIODS IN THE MOVING AVERAGE -->!3

SMAVE    - SINGLE MOVING AVERAGE
TIME SERIES : SALES OF HIGH SPEED DRILLS MODEL A3

NUMBER OF ERROR OBSNS     12
MEAN % ERROR OR BIAS      2.821108435629
MEAN ABSOLUTE % ERROR     5.545120133257
```

```
MEAN SQUARED ERROR (MSE)   555.277777778
MEAN ABSOLUTE ERROR        19.8888888889

COMMAND -->!LMAVE

ENTER NUMBER OF PERIODS IN THE MOVING AVERAGES -->!3

LMAVE     - LINEAR MOVING AVERAGE
TIME SERIES : SALES OF HIGH SPEED DRILLS MODEL A3

NUMBER OF ERROR OBSNS      10
MEAN % ERROR OR BIAS       1.626444916409
MEAN ABSOLUTE % ERROR      4.665837017181
MEAN SQUARED ERROR (MSE)   507.2148148151
MEAN ABSOLUTE ERROR        17.04444444445

COMMAND -->!REGRESS

ENTER NUMBER OF PERIODS IN THE LINEAR REGRESSION -->!3

REGRESS - LINEAR REGRESSION WITH FIXED PERIODS
NUMBER OF PERIODS IN THE REGRESSION = 3
TIME SERIES : SALES OF HIGH SPEED DRILLS MODEL A3

NUMBER OF ERROR OBSNS      12
MEAN % ERROR OR BIAS       -.07840239827636
MEAN ABSOLUTE % ERROR      5.56366592577
MEAN SQUARED ERROR (MSE)   586.5833333333
MEAN ABSOLUTE ERROR        19.41666666667

COMMAND -->!X


X       - SINGLE EXPONENTIAL SMOOTHING
TIME SERIES : SALES OF HIGH SPEED DRILLS MODEL A3
ALPHA = .2
INITIAL SMOOTHED AVERAGE = 350
LAST PERIOD ESTIMATE OF SMOOTHED AVERAGE = 385.4746511342

NUMBER OF ERROR OBSNS      14
MEAN % ERROR OR BIAS       2.818504990496
MEAN ABSOLUTE % ERROR      7.446503226083
MEAN SQUARED ERROR (MSE)   902.4367140387
MEAN ABSOLUTE ERROR        27.15386111937

COMMAND -->!XX


XX      - DOUBLE EXPONENTIAL SMOOTHING
TIME SERIES : SALES OF HIGH SPEED DRILLS MODEL A3
ALPHA = .2
INITIAL ESTIMATE OF INTERCEPT = 350
INITIAL ESTIMATE OF SLOPE     = 0
LAST PERIOD ESTIMATE OF INTERCEPT = 405.9160226339
```

```
LAST PERIOD ESTIMATE OF SLOPE      = 5.110342874926

NUMBER OF ERROR OBSNS     14
MEAN % ERROR OR BIAS      2.133043433487
MEAN ABSOLUTE % ERROR     5.814508217203
MEAN SQUARED ERROR (MSE)  596.7551771625
MEAN ABSOLUTE ERROR       20.75822447171

COMMAND -->!XXX


XXX     - TRIPLE EXPONENTIAL SMOOTHING
TIME SERIES : SALES OF HIGH SPEED DRILLS MODEL A3
ALPHA = .2
INITIAL ESTIMATE OF A = 350
INITIAL ESTIMATE OF B = 0
INITIAL ESTIMATE OF C = 0
LAST PERIOD ESTIMATE OF A = 413.8733662002
LAST PERIOD ESTIMATE OF B = 9.337681644526
LAST PERIOD ESTIMATE OF C = .4973339728939

NUMBER OF ERROR OBSNS     14
MEAN % ERROR OR BIAS      1.081690439997
MEAN ABSOLUTE % ERROR     5.051962905778
MEAN SQUARED ERROR (MSE)  510.953131277
MEAN ABSOLUTE ERROR       17.73883511896

COMMAND -->!ADAPT


ADAPT   - ADAPTIVE RESPONSE RATE EXPO. SMOOTH.
TIME SERIES : SALES OF HIGH SPEED DRILLS MODEL A3
BETA = .1
INITIAL SMOOTHED AVERAGE = 350
INITIAL SMOOTHED ERROR = 0
INITIAL SMOOTHED ABSOLUTE ERROR = 0
LAST PERIOD ESTIMATE OF SMOOTHED AVERAGE = 404.5395838084
LAST PERIOD SMOOTHED ERROR = 15.34200772476
LAST PERIOD SMOOTHED ABSOLUTE ERROR = 20.49519612668

NUMBER OF ERROR OBSNS     14
MEAN % ERROR OR BIAS      3.930112090535
MEAN ABSOLUTE % ERROR     7.728891627545
MEAN SQUARED ERROR (MSE)  1126.705131524
MEAN ABSOLUTE ERROR       27.75545996542

COMMAND -->!HOLT


HOLT    - HOLT 2-PARAMETERS LINEAR EXPO. SMOOTH.
TIME SERIES : SALES OF HIGH SPEED DRILLS MODEL A3
ALPHA = .2
BETA  = .1
INITIAL ESTIMATE OF INTERCEPT = 350
```

```
INITIAL ESTIMATE OF SLOPE = -22.5
LAST PERIOD ESTIMATE OF INTERCEPT = 349.2310929767
LAST PERIOD ESTIMATE OF SLOPE = -1.692688373628

NUMBER OF ERROR OBSNS     14
MEAN % ERROR OR BIAS      19.95833157947
MEAN ABSOLUTE % ERROR     20.02406339367
MEAN SQUARED ERROR (MSE)  6906.64713413
MEAN ABSOLUTE ERROR       74.52611295138

COMMAND -->!WINTERS

ENTER TIME SERIES PERIODICITY (12=MONTHLY DATA) -->!12

*** ERROR *** NOT ENOUGH OBSERVATIONS !!!

COMMAND -->!TABLEON


COMMAND -->!PLOTON


COMMAND -->!OPTON


COMMAND -->!LMAVE

ENTER NUMBER OF PERIODS IN THE MOVING AVERAGES -->!3

LMAVE    - LINEAR MOVING AVERAGE
TIME SERIES : SALES OF HIGH SPEED DRILLS MODEL A3

PERIOD    ACTUAL    FORECAST       ERROR    % ABS ERROR
  6       325.00    299.67         25.33      7.79
  7       355.00    301.11         53.89     15.18
  8       345.00    338.67          6.33      1.84
  9       385.00    370.56         14.44      3.75
 10       374.00    398.78 -       24.78      6.63
 11       390.00    389.78           .22       .06
 12       389.00    407.22 -       18.22      4.68
 13       405.00    396.11          8.89      2.19
 14       412.00    409.33          2.67       .65
 15       403.00    418.67 -       15.67      3.89
 16                 417.78
 17                 423.33
 18                 428.89
 19                 434.44
 20                 440.00

NUMBER OF ERROR OBSNS     10
MEAN % ERROR OR BIAS      1.626444916409
MEAN ABSOLUTE % ERROR     4.665837017181
MEAN SQUARED ERROR (MSE)  507.2148148151
MEAN ABSOLUTE ERROR       17.04444444445
```

```
                      280.00    330.00    380.00    430.00    480.00
PERIOD  ACTUAL  FORECAST  +---------+---------+---------+---------+
  1     350.00      -                        O
  2     326.00      -              O
  3     340.00      -                O
  4     319.00      -            O
  5     298.00      -       O
  6     325.00    299.67 -   *     O
  7     355.00    301.11 -   *          O
  8     345.00    338.67 -            *O
  9     385.00    370.56 -                     *   O
 10     374.00    398.78 -                        O    *
 11     390.00    389.78 -                           +
 12     389.00    407.22 -                           O   *
 13     405.00    396.11 -                            *  O
 14     412.00    409.33 -                               +
 15     403.00    418.67 -                               O  *
 16              417.78 -                                  *
 17              423.33 -                                    *
 18              428.89 -                                      *
 19              434.44 -                                        *
 20              440.00 -                                          *
PERIOD  ACTUAL  FORECAST  +---------+---------+---------+---------+
                      280.00    330.00    380.00    430.00    480.00

COMMAND -->!XXX


ENTER LOW ALPHA, HIGH ALPHA, ALPHA STEP -->!.05,.5,.05

*********************
* OPTIMAL SOLUTION *
*********************
SEARCH ALPHA FROM .05 TO .5 STEP .05

XXX      - TRIPLE EXPONENTIAL SMOOTHING
TIME SERIES : SALES OF HIGH SPEED DRILLS MODEL A3
ALPHA = .3
INITIAL ESTIMATE OF A = 350
INITIAL ESTIMATE OF B = 0
INITIAL ESTIMATE OF C = 0
LAST PERIOD ESTIMATE OF A = 410.8865422788
LAST PERIOD ESTIMATE OF B = 6.992126187917
LAST PERIOD ESTIMATE OF C = .1055176811445

   PERIOD    ACTUAL    FORECAST      ERROR    % ABS ERROR
      2      326.00     350.00  -    24.00       7.36
      3      340.00     328.40       11.60       3.41
      4      319.00     332.36  -    13.36       4.19
      5      298.00     316.34  -    18.34       6.15
      6      325.00     291.90       33.10      10.19
      7      355.00     308.10       46.90      13.21
      8      345.00     344.47         .53        .15
```

```
 9        385.00      351.47       33.53      8.71
10        374.00      389.29 -     15.29      4.09
11        390.00      393.20 -      3.20       .82
12        389.00      405.76 -     16.76      4.31
13        405.00      406.72 -      1.72       .43
14        412.00      418.08 -      6.08      1.48
15        403.00      425.99 -     22.99      5.71
16                    417.93
17                    425.08
18                    432.34
19                    439.70
20                    447.17

NUMBER OF ERROR OBSNS       14
MEAN % ERROR OR BIAS        .08147689072495
MEAN ABSOLUTE % ERROR       5.014031532767
MEAN SQUARED ERROR (MSE)    481.292076145
MEAN ABSOLUTE ERROR         17.67127709539

                         280.00     330.00     380.00     430.00     480.00
PERIOD ACTUAL  FORECAST  +---------+---------+---------+---------+
  1     350.00      -                              O
  2     326.00    350.00 -             O     *
  3     340.00    328.40 -              *    O
  4     319.00    332.36 -           O  *
  5     298.00    316.34 -      O    *
  6     325.00    291.90 -  *        O
  7     355.00    308.10 -       *             O
  8     345.00    344.47 -                  +
  9     385.00    351.47 -                   *      O
 10     374.00    389.29 -                       O   *
 11     390.00    393.20 -                         O*
 12     389.00    405.76 -                         O  *
 13     405.00    406.72 -                            +
 14     412.00    418.08 -                            O  *
 15     403.00    425.99 -                            O   *
 16              417.93 -                                *
 17              425.08 -                                *
 18              432.34 -                                 *
 19              439.70 -                                   *
 20              447.17 -                                     *
PERIOD ACTUAL  FORECAST  +---------+---------+---------+---------+
                         280.00     330.00     380.00     430.00     480.00

COMMAND -->!EXIT
```

FORCST EXAMPLE TERMINAL SESSION 2
==================================

FORECASTING SEASONAL AND NONSEASONAL MODELS

NEED INTRODUCTION (Y OR N) ... !N

ENTER TIME SERIES DESCRIPTION
!SALES OF SMALL DIESEL ENGINES

ENTER NUMBER OF OBSERVATIONS (>=3) -->!12

ENTER OBSERVATIONS

OBSN   1   !72
OBSN   2   !58
OBSN   3   !137
OBSN   4   !137
OBSN   5   !176
OBSN   6   !231
OBSN   7   !244
OBSN   8   !190
OBSN   9   !155
OBSN   10   !78
OBSN   11   !60
OBSN   12   !70
PLEASE ENTER COMMAND OR TYPE HELP FOR COMMAND DESCRIPTIONS

COMMAND -->!OPTON

COMMAND -->!INITON

COMMAND -->!PLOTON

COMMAND -->!FPERIOD

ENTER NO. OF PERIODS TO FORECAST -->!12

COMMAND -->!WINTERS

ENTER TIME SERIES PERIODICITY (12=MONTHLY DATA) -->!12
ENTER INITIAL ESTIMATES OF INTERCEPT, SLOPE -->!114.3,.535
ENTER INITIAL SET OF 12 SEASONAL FACTORS
PERIOD 1 SEASONAL FACTOR = !.477
PERIOD 2 SEASONAL FACTOR = !.493
PERIOD 3 SEASONAL FACTOR = !.918
PERIOD 4 SEASONAL FACTOR = !1.012
PERIOD 5 SEASONAL FACTOR = !1.628
PERIOD 6 SEASONAL FACTOR = !1.724
PERIOD 7 SEASONAL FACTOR = !1.718

```
PERIOD  8 SEASONAL FACTOR = !1.350
PERIOD  9 SEASONAL FACTOR = !1.021
PERIOD 10 SEASONAL FACTOR = !.498
PERIOD 11 SEASONAL FACTOR = !.604
PERIOD 12 SEASONAL FACTOR = !.557

ENTER LOW ALPHA, HIGH ALPHA, ALPHA STEP -->!.1,.5,.1

ENTER LOW BETA, HIGH BETA, BETA STEP -->!.1,.5,.1

ENTER LOW GAMMA, HIGH GAMMA, GAMMA STEP -->!.1,.5,.1

*********************
* OPTIMAL SOLUTION *
*********************
SEARCH ALPHA FROM .1 TO .5 STEP .1
SEARCH BETA  FROM .1 TO .5 STEP .1
SEARCH GAMMA FROM .1 TO .5 STEP .1

WINTERS - WINTERS 3-PARAMETERS LINEAR EXPO. SMOOTH.
TIME SERIES : SALES OF SMALL DIESEL ENGINES
ALPHA = .1
BETA  = .4
GAMMA = .5
INITIAL ESTIMATE OF INTERCEPT = 114.3
INITIAL ESTIMATE OF SLOPE     = .535
LAST PERIOD ESTIMATE OF INTERCEPT = 146.5028061986
LAST PERIOD ESTIMATE OF SLOPE     = 1.010284821449
```

SEASONAL FACTOR	INITIAL EST.	FINAL EST.
1	.477000	.540075
2	.493000	.486928
3	.918000	1.005872
4	1.012000	1.039993
5	1.628000	1.495097
6	1.724000	1.729402
7	1.718000	1.753839
8	1.350000	1.358157
9	1.021000	1.052487
10	.498000	.513947
11	.604000	.507263
12	.557000	.516940

PERIOD	ACTUAL	FORECAST	ERROR	% ABS ERROR
1	72.00	54.78	17.22	23.92
2	58.00	59.05 -	1.05	1.81
3	137.00	111.52	25.48	18.60
4	137.00	127.80	9.20	6.71
5	176.00	211.94 -	35.94	20.42
6	231.00	227.20	3.80	1.65
7	244.00	230.93	13.07	5.36
8	190.00	185.68	4.32	2.27
9	155.00	143.63	11.37	7.34
10	78.00	72.10	5.90	7.57

11	60.00	90.37	−	30.37	50.62
12	70.00	82.21	−	12.21	17.45
13		79.67
14		72.32
15		150.41
16		156.56
17		226.59
18		263.85
19		269.35
20		209.95
21		163.76
22		80.49
23		79.95
24		82.00

```
NUMBER OF ERROR OBSNS      12
MEAN % ERROR OR BIAS       -1.406588742608
MEAN ABSOLUTE % ERROR      13.64186023752
MEAN SQUARED ERROR (MSE)   313.5478620074
MEAN ABSOLUTE ERROR        14.16114611154
```

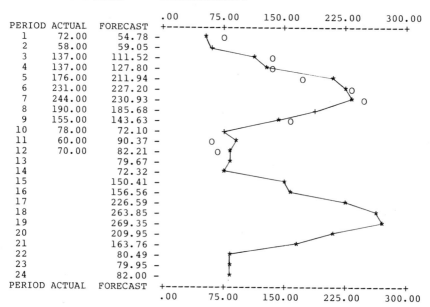

```
                                .00       75.00     150.00    225.00    300.00
   PERIOD ACTUAL  FORECAST      +---------+---------+---------+---------+
      1    72.00    54.78 -
      2    58.00    59.05 -
      3   137.00   111.52 -
      4   137.00   127.80 -
      5   176.00   211.94 -
      6   231.00   227.20 -
      7   244.00   230.93 -
      8   190.00   185.68 -
      9   155.00   143.63 -
     10    78.00    72.10 -
     11    60.00    90.37 -
     12    70.00    82.21 -
     13            79.67 -
     14            72.32 -
     15           150.41 -
     16           156.56 -
     17           226.59 -
     18           263.85 -
     19           269.35 -
     20           209.95 -
     21           163.76 -
     22            80.49 -
     23            79.95 -
     24            82.00 -
   PERIOD ACTUAL  FORECAST      +---------+---------+---------+---------+
                                .00       75.00     150.00    225.00    300.00
```

COMMAND --->!EXIT

```
10  REM     FORECASTING SEASONAL AND NONSEASONAL MODELS
20  REM
30  REM     THIS PROGRAM FORECASTS TIME SERIES WHICH ARE NONSEASONAL
40  REM     OR SEASONAL IN NATURE.  THE AVAILABLE TECHNIQUES INCLUDE
50  REM     MOVING AVERAGES AND EXPONENTIAL SMOOTHING.  OPTIMIZATION
60  REM     WITHIN A SMOOTHING MODEL IS POSSIBLE BASED ON THE MINIMUM
70  REM     MEAN SQUARED ERROR (MSE).
80  REM
90  DIM X(50),Y(50),Z(50),F(50),W(3,3),R(12),S(12)
100 DIM F$(29),P$(40),W$(3),G$(8)
110 N0=50
120 S0=12
130 PRINT
140 PRINT "FORECASTING SEASONAL AND NONSEASONAL MODELS"
150 PRINT "-----------------------------------------"
160 GOSUB 1460
170 PRINT
180 PRINT "NEED INTRODUCTION (Y OR N) ... ";
190 INPUT A$
200 IF A$ <> "Y" THEN 1260
210 PRINT
220 PRINT "THIS PROGRAM PERFORMS TIME SERIES ANALYSIS."
230 PRINT "THE AVAILABLE TECHNIQUES ARE :"
240 PRINT
250 FOR I=1 TO 11
260 PRINT I;" ";F$(I)
270 NEXT I
280 PRINT
290 PRINT "INPUT REQUIREMENTS"
300 PRINT "------------------"
310 PRINT "1. DESCRIPTION OF THE TIME SERIES"
320 PRINT "2. NUMBER OF OBSERVATIONS AND THEIR VALUES"
330 GOTO 1260
340 PRINT "PLEASE ENTER COMMAND OR TYPE HELP FOR COMMAND DESCRIPTIONS"
350 PRINT
360 PRINT "COMMAND -->";
370 INPUT C$
380 PRINT
390 IF C$ <> "HELP" THEN 420
400 GOSUB 1970
410 GOTO 350
420 IF C$ <> "AUTO" THEN 450
430 GOSUB 2970
440 GOTO 350
450 IF C$ <> "AUTODIF" THEN 480
460 GOSUB 3770
470 GOTO 350
480 IF C$ <> "SMAVE" THEN 510
490 GOSUB 3930
500 GOTO 350
510 IF C$ <> "LMAVE" THEN 540
520 GOSUB 4260
530 GOTO 350
540 IF C$ <> "REGRESS" THEN 570
```

```
550 GOSUB 4730
560 GOTO 350
570 IF C$ <> "X" THEN 600
580 GOSUB 5130
590 GOTO 350
600 IF C$ <> "XX" THEN 630
610 GOSUB 5650
620 GOTO 350
630 IF C$ <> "XXX" THEN 660
640 GOSUB 6230
650 GOTO 350
660 IF C$ <> "ADAPT" THEN 690
670 GOSUB 6900
680 GOTO 350
690 IF C$ <> "HOLT" THEN 720
700 GOSUB 7710
710 GOTO 350
720 IF C$ <> "WINTERS" THEN 750
730 GOSUB 8390
740 GOTO 350
750 IF C$ <> "LIST" THEN 780
760 GOSUB 2060
770 GOTO 350
780 IF C$ <> "DELETE" THEN 810
790 GOSUB 2420
800 GOTO 350
810 IF C$ <> "INSERT" THEN 840
820 GOSUB 2560
830 GOTO 350
840 IF C$ <> "PLOTON" THEN 870
850 P0=1
860 GOTO 350
870 IF C$ <> "PLOTOFF" THEN 900
880 P0=0
890 GOTO 350
900 IF C$ <> "INITON" THEN 930
910 I0=1
920 GOTO 350
930 IF C$ <> "INITOFF" THEN 960
940 I0=0
950 GOTO 350
960 IF C$ <> "FPERIOD" THEN 990
970 GOSUB 2730
980 GOTO 350
990 IF C$ <> "TABLEON" THEN 1020
1000 T0=1
1010 GOTO 350
1020 IF C$ <> "TABLEOFF" THEN 1050
1030 T0=0
1040 GOTO 350
1050 IF C$ <> "OPTON" THEN 1080
1060 R0=1
1070 GOTO 350
1080 IF C$ <> "OPTOFF" THEN 1110
```

```
1090 R0=0
1100 GOTO 350
1110 IF C$ <> "ALPHA" THEN 1140
1120 GOSUB 2800
1130 GOTO 350
1140 IF C$ <> "BETA" THEN 1170
1150 GOSUB 2800
1160 GOTO 350
1170 IF C$ <> "GAMMA" THEN 1200
1180 GOSUB 2800
1190 GOTO 350
1200 IF C$ = "RESTART" THEN 1260
1210 IF C$ = "EXIT" THEN 11570
1220 PRINT
1230 PRINT "*** INVALID COMMAND *** PLEASE TRY AGAIN !!!"
1240 GOTO 340
1250 REM
1260 REM    INPUT DATA HERE FOR THE FIRST TIME
1270 PRINT
1280 PRINT "ENTER TIME SERIES DESCRIPTION"
1290 INPUT S$
1300 PRINT
1310 PRINT "ENTER NUMBER OF OBSERVATIONS (>=3) -->";
1320 INPUT N
1330 IF N <= 2 THEN 1350
1340 IF N <= N0 THEN 1380
1350 PRINT
1360 PRINT "*** ERROR *** PLEASE TRY AGAIN !!!"
1370 GOTO 1270
1380 PRINT
1390 PRINT "ENTER OBSERVATIONS"
1400 PRINT "------------------"
1410 FOR I=1 TO N
1420 PRINT "OBSN",I,"  ";
1430 INPUT X(I)
1440 NEXT I
1450 GOTO 340
1460 REM    I N I T I A L I Z A T I O N
1470 F$(1)="AUTO     - AUTOCORRELATION OF THE TIME SERIES"
1480 F$(2)="AUTODIF  - AUTOCOR. OF FIRST DIFFERENCE OF THE SERIES"
1490 F$(3)="SMAVE    - SINGLE MOVING AVERAGE"
1500 F$(4)="LMAVE    - LINEAR MOVING AVERAGE"
1510 F$(5)="REGRESS  - LINEAR REGRESSION WITH FIXED PERIODS"
1520 F$(6)="X        - SINGLE EXPONENTIAL SMOOTHING"
1530 F$(7)="XX       - DOUBLE EXPONENTIAL SMOOTHING"
1540 F$(8)="XXX      - TRIPLE EXPONENTIAL SMOOTHING"
1550 F$(9)="ADAPT    - ADAPTIVE RESPONSE RATE EXPO. SMOOTH."
1560 F$(10)="HOLT    - HOLT 2-PARAMETERS LINEAR EXPO. SMOOTH."
1570 F$(11)="WINTERS - WINTERS 3-PARAMETERS LINEAR EXPO. SMOOTH."
1580 F$(12)="HELP    - PRINT THIS MESSAGE"
1590 F$(13)="LIST    - LIST OBSERVATIONS WITHIN A RANGE"
1600 F$(14)="DELETE  - DELETE AN OBSERVATION"
1610 F$(15)="INSERT  - INSERT AN OBSERVATION AFTER AN OBSN"
1620 F$(16)="PLOTON  - SET OPTION PLOT ON"
```

```
1630 F$(17)="PLOTOFF  - SET OPTION PLOT OFF"
1640 F$(18)="INITON   - MANUAL INITIALIZATION OF MODEL CONSTANTS"
1650 F$(19)="INITOFF  - USE DEFAULT METHOD OF INITIALIZATION"
1660 F$(20)="FPERIOD  - SPECIFY NO. OF PERIODS TO FORECAST"
1670 F$(21)="TABLEON  - SET OPTION TO PRODUCE RESIDUAL TABLE ON"
1680 F$(22)="TABLEOFF- SET OPTION TO PRODUCE RESIDUAL TABLE OFF"
1690 F$(23)="OPTON    - SET OPTIMIZATION OPTION ON"
1700 F$(24)="OPTOFF   - SET OPTIMIZATION OPTION OFF"
1710 F$(25)="ALPHA    - SPECIFY SMOOTHING PARAMETER - ALPHA"
1720 F$(26)="BETA     - SPECIFY SMOOTHING PARAMETER - BETA"
1730 F$(27)="GAMMA    - SPECIFY SMOOTHING PARAMETER - GAMMA"
1740 F$(28)="RESTART  - START A NEW PROBLEM"
1750 F$(29)="EXIT     - EXIT FROM THE PROGRAM"
1760 G$(1)="TIME SERIES : "
1770 G$(2)="  -#####.##"
1780 G$(3)="### -#.#######    "
1790 G$(4)="####                -#####.##"
1800 G$(5)="####   -######.## -#####.## -#####.## -###.##"
1810 G$(6)="###            -#####.##"
1820 G$(7)="### -#####.## -#####.##"
1830 G$(8)="         ###           -##.######    -##.######"
1840 W$(1)="ALPHA"
1850 W$(2)="BETA"
1860 W$(3)="GAMMA"
1870 P0=0
1880 R0=0
1890 T0=1
1900 N9=5
1910 I0=0
1920 K9=0
1930 A9=0.2
1940 B9=0.1
1950 G9=0.1
1960 RETURN
1970 REM    HELP   C O M M A N D
1980 REM    --------------------
1990 PRINT
2000 PRINT "AVAILABLE COMMANDS"
2010 PRINT "=================="
2020 FOR I=1 TO 29
2030 PRINT F$(I)
2040 NEXT I
2050 RETURN
2060 REM    LIST   C O M M A N D
2070 REM    --------------------
2080 PRINT "TOTAL NO. OF OBSNS = ";N
2090 PRINT "ENTER RANGE OF OBSNS TO LIST (FROM,TO) -->";
2100 INPUT N1,N2
2110 IF N1 > N2 THEN 2410
2120 IF N1 < 1 THEN 2410
2130 IF N2 > N THEN 2410
2140 PRINT
2150 IF N=1 THEN 2380
2160 IF N1 <> 1 THEN 2380
```

```
2170 IF N2 <> N THEN 2380
2180 A0=1.E20
2190 A1=-1.E20
2200 A2=0
2210 A3=0
2220 FOR I=N1 TO N2
2230 A2=A2+X(I)
2240 A3=A3+X(I)*X(I)
2250 PRINT "OBSN",I,X(I)
2260 IF X(I) >= A0 THEN 2280
2270 A0=X(I)
2280 IF X(I) <= A1 THEN 2300
2290 A1=X(I)
2300 NEXT I
2310 A3=(A3-A2*A2/N)/(N-1)
2320 A2=A2/N
2330 PRINT "MINIMUM = ";A0
2340 PRINT "MAXIMUM = ";A1
2350 PRINT "MEAN    = ";A2
2360 PRINT "STD DEV = ";SQR(A3)
2370 GOTO 2410
2380 FOR I=N1 TO N2
2390 PRINT "OBSN",I,X(I)
2400 NEXT I
2410 RETURN
2420 REM     DELETE   C O M M A N D
2430 REM     ---------------------
2440 PRINT "ENTER OBSN NUMBER -->";
2450 INPUT N1
2460 IF N1 < 1 THEN 2550
2470 IF N1 > N THEN 2550
2480 IF N = 1 THEN 2550
2490 PRINT "OBSN ";N1,X(N1);"   DELETED !!!"
2500 IF N1 = N THEN 2540
2510 FOR I=N1 TO N-1
2520 X(I)=X(I+1)
2530 NEXT I
2540 N=N-1
2550 RETURN
2560 REM     INSERT   C O M M A N D
2570 REM     ---------------------
2580 PRINT "INSERT AFTER OBSN NUMBER -->";
2590 INPUT N1
2600 IF N1 < 0 THEN 2720
2610 IF N1 > N THEN 2720
2620 IF N >= N0 THEN 2720
2630 PRINT
2640 PRINT "OBSN",N1+1,"ENTER VALUE -->";
2650 INPUT N2
2660 IF N1 = N THEN 2700
2670 FOR I=N TO N1+1 STEP -1
2680 X(I+1)=X(I)
2690 NEXT I
2700 X(N1+1)=N2
```

```
2710 N=N+1
2720 RETURN
2730 REM     FPERIOD   C O M M A N D
2740 REM     -----------------------
2750 PRINT "ENTER NO. OF PERIODS TO FORECAST -->";
2760 INPUT N9
2770 IF N9 >= 0 THEN 2790
2780 N9=0
2790 RETURN
2800 REM     ALPHA  OR  BETA  OR  GAMMA    C O M M A N D S
2810 REM     --------------------------------------------
2820 PRINT "ENTER VALUE OF ";C$;" BETWEEN 0 AND 1 -->";
2830 INPUT N1
2840 IF N1 <= 0 THEN 2940
2850 IF N1 >= 1 THEN 2940
2860 IF C$ = "BETA" THEN 2900
2870 IF C$ = "GAMMA" THEN 2920
2880 A9=N1
2890 GOTO 2960
2900 B9=N1
2910 GOTO 2960
2920 G9=N1
2930 GOTO 2960
2940 PRINT
2950 PRINT "*** ERROR ***    IGNORED !!!"
2960 RETURN
2970 REM     AUTO   C O M M A N D
2980 REM     --------------------
2990 PRINT "HOW MANY LAGS -->";
3000 INPUT K0
3010 IF K0 <= 0 THEN 3120
3020 IF K0 <= N-1 THEN 3040
3030 K0=N-1
3040 FOR I=1 TO N
3050 Y(I)=X(I)
3060 NEXT I
3070 N2=N
3080 GOSUB 3560
3090 PRINT
3100 PRINT G$(1);S$
3110 GOSUB 3130
3120 RETURN
3130 REM     PLOT AUTOCORRELATION FUNCTION
3140 PRINT
3150 GOSUB 3420
3160 GOSUB 3490
3170 FOR I=1 TO K0+1
3180 FOR J=0 TO 40
3190 P$(J)=" "
3200 NEXT J
3210 N1=INT((Z(I)+1)/0.05+0.5)
3220 IF N1 >= 20 THEN 3260
3230 N2=20
3240 J9=N2
```

```
3250 GOTO 3290
3260 J9=N1
3270 N1=20
3280 N2=J9
3290 FOR J=N1 TO N2
3300 P$(J)="*"
3310 NEXT J
3320 P$(20)="I"
3330 PRINT USING G$(3),I-1,Z(I);
3340 FOR J=0 TO J9
3350 PRINT P$(J);
3360 NEXT J
3370 PRINT
3380 NEXT I
3390 GOSUB 3490
3400 GOSUB 3420
3410 RETURN
3420 REM    PRINT LABEL
3430 PRINT "            ";
3440 FOR J=-1 TO 1 STEP 0.5
3450 PRINT USING G$(2),J;
3460 NEXT J
3470 PRINT
3480 RETURN
3490 REM    PRINT AXIS OF AUTOCORRELATION FUNCTION
3500 PRINT "LAG    AUTOCORR.  +";
3510 FOR J=1 TO 4
3520 PRINT "---------+";
3530 NEXT J
3540 PRINT
3550 RETURN
3560 REM    CALCULATE AUTOCORRELATION FUNCTION UP TO LAG K0
3570 Y0=0
3580 Y1=0
3590 FOR I=1 TO N2
3600 Y0=Y0+Y(I)
3610 Y1=Y1+Y(I)*Y(I)
3620 NEXT I
3630 Y1=Y1-Y0*Y0/N2
3640 Y0=Y0/N2
3650 Z(1)=1
3660 FOR K=1 TO K0
3670 Z(K+1)=0
3680 NEXT K
3690 IF Y1=0 THEN 3760
3700 FOR K=1 TO K0
3710 FOR T=1 TO N2-K
3720 Z(K+1)=Z(K+1)+(Y(T)-Y0)*(Y(T+K)-Y0)
3730 NEXT T
3740 Z(K+1)=Z(K+1)/Y1
3750 NEXT K
3760 RETURN
3770 REM    AUTODIF   C O M M A N D
3780 REM    ----------------------
```

```
3790 PRINT "HOW MANY LAGS -->";
3800 INPUT K0
3810 IF K0 <= 0 THEN 3920
3820 IF K0 <= N-2 THEN 3840
3830 K0=N-2
3840 FOR I=2 TO N
3850 Y(I-1)=X(I)-X(I-1)
3860 NEXT I
3870 N2=N-1
3880 GOSUB 3560
3890 PRINT
3900 PRINT G$(1);"FIRST DIFFERENCE OF ";S$
3910 GOSUB 3130
3920 RETURN
3930 REM      SMAVE     C O M M A N D
3940 REM      ---------------------
3950 PRINT "ENTER NUMBER OF PERIODS IN THE MOVING AVERAGE -->";
3960 INPUT M0
3970 IF M0 < 1 THEN 4250
3980 IF M0 > N THEN 4250
3990 M1=M0+1
4000 F(M1)=X(1)
4010 FOR I=2 TO M0
4020 F(M1)=F(M1)+X(I)
4030 NEXT I
4040 F(M1)=F(M1)/M0
4050 IF M1 >= N THEN 4090
4060 FOR I=M1+1 TO N
4070 F(I)=F(I-1)+(X(I-1)-X(I-1-M0))/M0
4080 NEXT I
4090 REM      FORECAST N9 PERIODS
4100 M2=N9
4110 IF M2+N <= N0 THEN 4130
4120 M2=N0-N
4130 IF M2 <= 0 THEN 4180
4140 FOR I=N+1 TO M2+N
4150 F(I)=F(I-1)+(X(I-1)-X(I-1-M0))/M0
4160 X(I)=X(N)
4170 NEXT I
4180 PRINT
4190 PRINT F$(3)
4200 PRINT G$(1);S$
4210 N8=M1
4220 GOSUB 9990
4230 IF P0 < 0.5 THEN 4250
4240 GOSUB 10410
4250 RETURN
4260 REM      LMAVE     C O M M A N D
4270 REM      ---------------------
4280 PRINT "ENTER NUMBER OF PERIODS IN THE MOVING AVERAGES -->";
4290 INPUT M0
4300 IF M0 < 2 THEN 4720
4310 IF N-2*M0 < 0 THEN 4720
4320 Y(M0)=X(1)
```

```
4330 FOR I=2 TO M0
4340 Y(M0)=Y(M0)+X(I)
4350 NEXT I
4360 Y(M0)=Y(M0)/M0
4370 FOR I=M0+1 TO N
4380 Y(I)=Y(I-1)+(X(I)-X(I-M0))/M0
4390 NEXT I
4400 N1=2*M0-1
4410 Z(N1)=Y(M0)
4420 FOR I=M0+1 TO N1
4430 Z(N1)=Z(N1)+Y(I)
4440 NEXT I
4450 Z(N1)=Z(N1)/M0
4460 IF N1 >= N THEN 4550
4470 FOR I=N1+1 TO N
4480 Z(I)=Z(I-1)+(Y(I)-Y(I-M0))/M0
4490 NEXT I
4500 FOR I=N1+1 TO N
4510 A0=2*Y(I-1)-Z(I-1)
4520 A1=2*(Y(I-1)-Z(I-1))/(M0-1)
4530 F(I)=A0+A1
4540 NEXT I
4550 REM     FORECAST LINEAR(DOUBLE) MOVING AVE - N9 PERIODS
4560 M2=N9
4570 IF M2+N <= N0 THEN 4590
4580 M2=N0-N
4590 IF M2 <= 0 THEN 4650
4600 A0=2*Y(N)-Z(N)
4610 A1=2*(Y(N)-Z(N))/(M0-1)
4620 FOR I=1 TO M2
4630 F(N+I)=A0+I*A1
4640 NEXT I
4650 PRINT
4660 PRINT F$(4)
4670 PRINT G$(1);S$
4680 N8=N1+1
4690 GOSUB 9990
4700 IF P0 < 0.5 THEN 4720
4710 GOSUB 10410
4720 RETURN
4730 REM     REGRESS   C O M M A N D
4740 REM     ----------------------
4750 PRINT "ENTER NUMBER OF PERIODS IN THE LINEAR REGRESSION -->";
4760 INPUT M0
4770 IF M0 < 2 THEN 5120
4780 IF M0 > N THEN 5120
4790 Y2=(M0-1)/2
4800 Y3=M0*Y2*Y2-M0*Y2*(2*M0-1)/3
4810 FOR I=M0 TO N
4820 Y0=0
4830 Y1=0
4840 N1=I-M0+1
4850 N2=M0
4860 FOR J=N1 TO I
```

```
4870 N2=N2-1
4880 Y0=Y0+X(J)
4890 Y1=Y1+N2*X(J)
4900 NEXT J
4910 A1=(-Y2*Y0+Y1)/Y3
4920 A0=Y0/M0+A1*Y2
4930 IF I=N THEN 4950
4940 F(I+1)=A0+A1
4950 NEXT I
4960 REM    FORECAST N9 PERIODS
4970 M2=N9
4980 IF M2+N <= N0 THEN 5000
4990 M2=N0-N
5000 IF M2 <= 0 THEN 5040
5010 FOR I=1 TO M2
5020 F(N+I)=A0+I*A1
5030 NEXT I
5040 PRINT
5050 PRINT F$(5)
5060 PRINT "NUMBER OF PERIODS IN THE REGRESSION = ";M0
5070 PRINT G$(1);S$
5080 N8=M0+1
5090 GOSUB 9990
5100 IF P0 < 0.5 THEN 4720
5110 GOSUB 10410
5120 RETURN
5130 REM    X   C O M M A N D
5140 REM    -----------------
5150 IF I0 < 0.5 THEN 5210
5160 PRINT "ENTER INITIAL SMOOTHED AVERAGE -->";
5170 INPUT F(1)
5180 N1=2
5190 N8=1
5200 GOTO 5240
5210 F(2)=X(1)
5220 N1=3
5230 N8=2
5240 IF R0 < 0.5 THEN 5380
5250 K9=1
5260 GOSUB 9720
5270 E0=1.E20
5280 FOR W1=W(1,1) TO W(1,2) STEP W(1,3)
5290 GOSUB 5610
5300 GOSUB 9990
5310 IF Q2 >= E0 THEN 5340
5320 V1=W1
5330 E0=Q2
5340 NEXT W1
5350 W1=V1
5360 GOSUB 9870
5370 GOTO 5390
5380 W1=A9
5390 K9=0
5400 GOSUB 5610
```

```
5410 REM     FORECAST N9 PERIODS
5420 M2=N9
5430 IF M2+N <= N0 THEN 5450
5440 M2=N0-N
5450 IF M2 <= 0 THEN 5500
5460 FOR I=N+1 TO N+M2
5470 F(I)=W1*X(I-1)+(1-W1)*F(I-1)
5480 X(I)=X(N)
5490 NEXT I
5500 PRINT
5510 PRINT F$(6)
5520 PRINT G$(1);S$
5530 PRINT "ALPHA = ";W1
5540 PRINT "INITIAL SMOOTHED AVERAGE = ";F(N1-1)
5550 A0=W1*X(N)+(1-W1)*F(N)
5560 PRINT "LAST PERIOD ESTIMATE OF SMOOTHED AVERAGE = ";A0
5570 GOSUB 9990
5580 IF P0 < 0.5 THEN 5600
5590 GOSUB 10410
5600 RETURN
5610 FOR I=N1 TO N
5620 F(I)=W1*X(I-1)+(1-W1)*F(I-1)
5630 NEXT I
5640 RETURN
5650 REM     XX   C O M M A N D
5660 REM     ------------------
5670 IF I0 < 0.5 THEN 5730
5680 PRINT "ENTER INITIAL ESTIMATES OF INTERCEPT,SLOPE -->";
5690 INPUT Y0,Y1
5700 N1=1
5710 N8=1
5720 GOTO 5770
5730 Y0=X(1)
5740 Y1=0
5750 N1=2
5760 N8=2
5770 IF R0 < 0.5 THEN 5910
5780 K9=1
5790 GOSUB 9720
5800 E0=1.E20
5810 FOR W1=W(1,1) TO W(1,2) STEP W(1,3)
5820 GOSUB 6140
5830 GOSUB 9990
5840 IF Q2 >= E0 THEN 5870
5850 V1=W1
5860 E0=Q2
5870 NEXT W1
5880 W1=V1
5890 GOSUB 9870
5900 GOTO 5920
5910 W1=A9
5920 K9=0
5930 GOSUB 6140
5940 REM     FORECAST N9 PERIODS
```

```
5950 M2=N9
5960 IF M2+N <= N0 THEN 5980
5970 M2=N0-N
5980 IF M2 <= 0 THEN 6020
5990 FOR I=1 TO M2
6000 F(N+I)=A0+I*A1
6010 NEXT I
6020 PRINT
6030 PRINT F$(7)
6040 PRINT G$(1);S$
6050 PRINT "ALPHA = ";W1
6060 PRINT "INITIAL ESTIMATE OF INTERCEPT = ";Y0
6070 PRINT "INITIAL ESTIMATE OF SLOPE     = ";Y1
6080 PRINT "LAST PERIOD ESTIMATE OF INTERCEPT = ";A0
6090 PRINT "LAST PERIOD ESTIMATE OF SLOPE     = ";A1
6100 GOSUB 9990
6110 IF P0 < 0.5 THEN 6130
6120 GOSUB 10410
6130 RETURN
6140 A0=Y0
6150 A1=Y1
6160 A2=(1-W1)*(1-W1)
6170 FOR I=N1 TO N
6180 F(I)=A0+A1
6190 A0=X(I)+A2*(F(I)-X(I))
6200 A1=A1-W1*W1*(F(I)-X(I))
6210 NEXT I
6220 RETURN
6230 REM    XXX   C O M M A N D
6240 REM    -------------------
6250 IF I0 < 0.5 THEN 6320
6260 PRINT "MODEL : F(T+M)=A(T)+B(T)*M+0.5*C(T)*M**2"
6270 PRINT "ENTER INITIAL ESTIMATES OF COEFFICIENTS A,B,C -->";
6280 INPUT Y0,Y1,Y2
6290 N1=1
6300 N8=1
6310 GOTO 6370
6320 Y0=X(1)
6330 Y1=0
6340 Y2=0
6350 N1=2
6360 N8=2
6370 IF R0 < 0.5 THEN 6510
6380 K9=1
6390 GOSUB 9720
6400 E0=1.E20
6410 FOR W1=W(1,1) TO W(1,2) STEP W(1,3)
6420 GOSUB 6760
6430 GOSUB 9990
6440 IF Q2 >= E0 THEN 6470
6450 V1=W1
6460 E0=Q2
6470 NEXT W1
6480 W1=V1
```

```
6490 GOSUB 9870
6500 GOTO 6520
6510 W1=A9
6520 K9=0
6530 GOSUB 6760
6540 REM     FORECAST N9 PERIODS
6550 M2=N9
6560 IF M2+N <= N0 THEN 6580
6570 M2=N0-N
6580 IF M2 <= 0 THEN 6620
6590 FOR I=1 TO M2
6600 F(N+I)=A0+I*A1+0.5*A2*I*I
6610 NEXT I
6620 PRINT
6630 PRINT F$(8)
6640 PRINT G$(1);S$
6650 PRINT "ALPHA = ";W1
6660 PRINT "INITIAL ESTIMATE OF A = ";Y0
6670 PRINT "INITIAL ESTIMATE OF B = ";Y1
6680 PRINT "INITIAL ESTIMATE OF C = ";Y2
6690 PRINT "LAST PERIOD ESTIMATE OF A = ";A0
6700 PRINT "LAST PERIOD ESTIMATE OF B = ";A1
6710 PRINT "LAST PERIOD ESTIMATE OF C = ";A2
6720 GOSUB 9990
6730 IF P0 < 0.5 THEN 6750
6740 GOSUB 10410
6750 RETURN
6760 A0=Y0
6770 A1=Y1
6780 A2=Y2
6790 A3=1-W1
6800 A4=1.5*W1*W1*(1+A3)
6810 A5=W1**3
6820 A3=A3**3
6830 FOR I=N1 TO N
6840 F(I)=A0+A1+0.5*A2
6850 A0=X(I)+A3*(F(I)-X(I))
6860 A1=A1+A2-A4*(F(I)-X(I))
6870 A2=A2-A5*(F(I)-X(I))
6880 NEXT I
6890 RETURN
6900 REM     ADAPT   C O M M A N D
6910 REM     --------------------
6920 IF I0 < 0.5 THEN 7030
6930 PRINT "ENTER INITIAL SMOOTHED AVERAGE -->";
6940 INPUT Y0
6950 PRINT "ENTER INITIAL SMOOTHED ERROR -->";
6960 INPUT Y1
6970 PRINT "ENTER INITIAL SMOOTHED ABSOLUTE ERROR -->";
6980 INPUT Y2
6990 IF Y2 < 0 THEN 6970
7000 N1=1
7010 N8=1
7020 GOTO 7080
```

```
7030 Y0=X(1)
7040 Y1=0
7050 Y2=0
7060 N1=2
7070 N8=2
7080 IF R0 < 0.5 THEN 7220
7090 K9=2
7100 GOSUB 9720
7110 E0=1.E20
7120 FOR W2=W(2,1) TO W(2,2) STEP W(2,3)
7130 GOSUB 7540
7140 GOSUB 9990
7150 IF Q2 >= E0 THEN 7180
7160 V2=W2
7170 E0=Q2
7180 NEXT W2
7190 W2=V2
7200 GOSUB 9870
7210 GOTO 7230
7220 W2=B9
7230 K9=0
7240 GOSUB 7540
7250 REM     FORECAST N9 PERIODS
7260 M2=N9
7270 IF M2+N <= N0 THEN 7290
7280 M2=N0-N
7290 IF M2 <= 0 THEN 7400
7300 FOR I=N+1 TO M2+N
7310 F(I)=A0
7320 A2=W2*(X(I)-F(I))+(1-W2)*A2
7330 A3=W2*ABS(X(I)-F(I))+(1-W2)*A3
7340 A1=0
7350 IF A3 <= 0 THEN 7370
7360 A1=ABS(A2/A3)
7370 A0=A1*X(I-1)+(1-A1)*A0
7380 X(I)=X(N)
7390 NEXT I
7400 PRINT
7410 PRINT F$(9)
7420 PRINT G$(1);S$
7430 PRINT "BETA = ";W2
7440 PRINT "INITIAL SMOOTHED AVERAGE = ";Y0
7450 PRINT "INITIAL SMOOTHED ERROR = ";Y1
7460 PRINT "INITIAL SMOOTHED ABSOLUTE ERROR = ";Y2
7470 PRINT "LAST PERIOD ESTIMATE OF SMOOTHED AVERAGE = ";A4
7480 PRINT "LAST PERIOD SMOOTHED ERROR = ";A5
7490 PRINT "LAST PERIOD SMOOTHED ABSOLUTE ERROR = ";A6
7500 GOSUB 9990
7510 IF P0 < 0.5 THEN 7530
7520 GOSUB 10410
7530 RETURN
7540 A0=Y0
7550 A2=Y1
7560 A3=Y2
```

```
7570 FOR I=N1 TO N
7580 F(I)=A0
7590 A2=W2*(X(I)-F(I))+(1-W2)*A2
7600 A3=W2*ABS(X(I)-F(I))+(1-W2)*A3
7610 A1=0
7620 IF A3 <= 0 THEN 7640
7630 A1=ABS(A2/A3)
7640 A0=A1*X(I-1)+(1-A1)*A0
7650 NEXT I
7660 A0=A1*X(N)+(1-A1)*A0
7670 A4=A0
7680 A5=A2
7690 A6=A3
7700 RETURN
7710 REM     HOLT   C O M M A N D
7720 REM     --------------------
7730 IF I0 < 0.5 THEN 7790
7740 PRINT "ENTER INITIAL ESTIMATES OF INTERCEPT,SLOPE -->";
7750 INPUT Y0,Y1
7760 N1=1
7770 N8=1
7780 GOTO 7870
7790 IF N < 4 THEN 7850
7800 Y0=X(1)
7810 Y1=(X(2)-X(1))/2+(X(4)-X(3))/2
7820 N1=2
7830 N8=2
7840 GOTO 7870
7850 PRINT "NUMBER OF OBSERVATIONS MUST >= 4  *** IGNORED !!!"
7860 RETURN
7870 IF R0 < 0.5 THEN 8050
7880 K9=2
7890 GOSUB 9720
7900 E0=1.E20
7910 FOR W1=W(1,1) TO W(1,2) STEP W(1,3)
7920 FOR W2=W(2,1) TO W(2,2) STEP W(2,3)
7930 GOSUB 8300
7940 GOSUB 9990
7950 IF Q2 >= E0 THEN 7990
7960 V1=W1
7970 V2=W2
7980 E0=Q2
7990 NEXT W2
8000 NEXT W1
8010 W1=V1
8020 W2=V2
8030 GOSUB 9870
8040 GOTO 8070
8050 W1=A9
8060 W2=B9
8070 K9=0
8080 GOSUB 8300
8090 REM     FORECAST N9 PERIODS
8100 M2=N9
```

```
8110 IF M2+N <= N0 THEN 8130
8120 M2=N0-N
8130 IF M2 <= 0 THEN 8170
8140 FOR I=1 TO M2
8150 F(N+I)=A0+I*A1
8160 NEXT I
8170 PRINT
8180 PRINT F$(10)
8190 PRINT G$(1);S$
8200 PRINT "ALPHA = ";W1
8210 PRINT "BETA  = ";W2
8220 PRINT "INITIAL ESTIMATE OF INTERCEPT = ";Y0
8230 PRINT "INITIAL ESTIMATE OF SLOPE = ";Y1
8240 PRINT "LAST PERIOD ESTIMATE OF INTERCEPT = ";A0
8250 PRINT "LAST PERIOD ESTIMATE OF SLOPE = ";A1
8260 GOSUB 9990
8270 IF P0 < 0.5 THEN 8290
8280 GOSUB 10410
8290 RETURN
8300 A0=Y0
8310 A1=Y1
8320 FOR I=N1 TO N
8330 F(I)=A0+A1
8340 A3=A0
8350 A0=W1*X(I)+(1-W1)*(A0+A1)
8360 A1=W2*(A0-A3)+(1-W2)*A1
8370 NEXT I
8380 RETURN
8390 REM    WINTERS   C O M M A N D
8400 REM    ---------------------
8410 PRINT "ENTER TIME SERIES PERIODICITY (12=MONTHLY DATA) -->";
8420 INPUT S9
8430 IF S9 <= 0 THEN 8410
8440 IF S9 > S0 THEN 8410
8450 N1=1
8460 N8=1
8470 IF I0 < 0.5 THEN 8560
8480 PRINT "ENTER INITIAL ESTIMATES OF INTERCEPT, SLOPE -->";
8490 INPUT Y0,Y1
8500 PRINT "ENTER INITIAL SET OF ";S9;" SEASONAL FACTORS"
8510 FOR J=1 TO S9
8520 PRINT "PERIOD ";J;" SEASONAL FACTOR = ";
8530 INPUT R(J)
8540 NEXT J
8550 GOTO 8760
8560 IF 2*S9 <= N THEN 8600
8570 PRINT
8580 PRINT "*** ERROR *** NOT ENOUGH OBSERVATIONS !!!"
8590 RETURN
8600 A3=0
8610 A4=0
8620 FOR J=1 TO S9
8630 A3=A3+X(J)
8640 A4=A4+X(J+S9)
```

```
8650 NEXT J
8660 A3=A3/S9
8670 A4=A4/S9
8680 Y1=(A4-A3)/S9
8690 Y0=A3-(S9/2+0.5)*Y1
8700 FOR I=1 TO S9
8710 J=I+S9
8720 Y(I)=Y0+Y1*I
8730 Y(J)=Y0+Y1*J
8740 R(I)=(X(I)/Y(I)+X(J)/Y(J))/2
8750 NEXT I
8760 REM    NORMALIZE SEASONAL FACTORS
8770 A3=0
8780 FOR I=1 TO S9
8790 A3=A3+R(I)
8800 NEXT I
8810 A3=A3/S9
8820 FOR I=1 TO S9
8830 R(I)=R(I)/A3
8840 NEXT I
8850 IF R0 < 0.5 THEN 9070
8860 K9=3
8870 GOSUB 9720
8880 E0=1.E20
8890 FOR W1=W(1,1) TO W(1,2) STEP W(1,3)
8900 FOR W2=W(2,1) TO W(2,2) STEP W(2,3)
8910 FOR W3=W(3,1) TO W(3,2) STEP W(3,3)
8920 GOSUB 9440
8930 GOSUB 9990
8940 IF Q2 >= E0 THEN 8990
8950 V1=W1
8960 V2=W2
8970 V3=W3
8980 E0=Q2
8990 NEXT W3
9000 NEXT W2
9010 NEXT W1
9020 W1=V1
9030 W2=V2
9040 W3=V3
9050 GOSUB 9870
9060 GOTO 9100
9070 W1=A9
9080 W2=B9
9090 W3=G9
9100 K9=0
9110 GOSUB 9440
9120 REM    FORECAST N9 PERIODS
9130 M2=N9
9140 IF M2+N <= N0 THEN 9160
9150 M2=N0-N
9160 IF M2 <= 0 THEN 9250
9170 T=INT(N/S9)
9180 T=N-T*S9
```

```
9190 FOR I=1 TO M2
9200 T=T+1
9210 IF T <= S9 THEN 9230
9220 T=1
9230 F(N+I)=(A0+A1*I)*S(T)
9240 NEXT I
9250 PRINT
9260 PRINT F$(11)
9270 PRINT G$(1);S$
9280 PRINT "ALPHA = ";W1
9290 PRINT "BETA  = ";W2
9300 PRINT "GAMMA = ";W3
9310 PRINT "INITIAL ESTIMATE OF INTERCEPT = ";Y0
9320 PRINT "INITIAL ESTIMATE OF SLOPE     = ";Y1
9330 PRINT "LAST PERIOD ESTIMATE OF INTERCEPT = ";A0
9340 PRINT "LAST PERIOD ESTIMATE OF SLOPE     = ";A1
9350 PRINT
9360 PRINT "SEASONAL FACTOR   INITIAL EST.      FINAL EST."
9370 FOR I=1 TO S9
9380 PRINT USING G$(8),I,R(I),S(I)
9390 NEXT I
9400 GOSUB 9990
9410 IF P0 < 0.5 THEN 9430
9420 GOSUB 10410
9430 RETURN
9440 A0=Y0
9450 A1=Y1
9460 FOR I=1 TO S9
9470 S(I)=R(I)
9480 NEXT I
9490 T=0
9500 FOR I=N1 TO N
9510 T=T+1
9520 IF T <= S9 THEN 9540
9530 T=1
9540 F(I)=(A0+A1)*S(T)
9550 A2=A0
9560 A0=W1*X(I)/S(T)+(1-W1)*(A0+A1)
9570 A1=W2*(A0-A2)+(1-W2)*A1
9580 S(T)=W3*(X(I)/A0)+(1-W3)*S(T)
9590 GOSUB 9620
9600 NEXT I
9610 RETURN
9620 REM    NORMALIZE SEASONAL FACTORS
9630 A3=0
9640 FOR J=1 TO S9
9650 A3=A3+S(J)
9660 NEXT J
9670 A3=A3/S9
9680 FOR J=1 TO S9
9690 S(J)=S(J)/A3
9700 NEXT J
9710 RETURN
9720 REM    PROMPT FOR ALPHA, OR BETA, OR GAMMA RANGES
```

```
9730 FOR I=1 TO K9
9740 PRINT
9750 IF I <> 1 THEN 9770
9760 IF C$ = "ADAPT" THEN 9850
9770 PRINT "ENTER LOW ";W$(I);", HIGH ";W$(I);
9780 PRINT ", ";W$(I);" STEP -->";
9790 INPUT W(I,1),W(I,2),W(I,3)
9800 IF W(I,1) > W(I,2) THEN 9740
9810 FOR J=1 TO 3
9820 IF W(I,J) >= 1 THEN 9740
9830 IF W(I,J) <= 0 THEN 9740
9840 NEXT J
9850 NEXT I
9860 RETURN
9870 REM
9880 PRINT
9890 PRINT "********************"
9900 PRINT "* OPTIMAL SOLUTION *"
9910 PRINT "********************"
9920 FOR J=1 TO K9
9930 IF J <> 1 THEN 9950
9940 IF C$ = "ADAPT" THEN 9970
9950 PRINT "SEARCH ";W$(J);" FROM ";W(J,1);" TO ";W(J,2);
9960 PRINT " STEP ";W(J,3)
9970 NEXT J
9980 RETURN
9990 REM    PRINT FORECAST TABLE
10000 REM   ------------------
10010 IF K9 > 0 THEN 10050
10020 IF T0 < 0.5 THEN 10050
10030 PRINT
10040 PRINT "PERIOD    ACTUAL    FORECAST     ERROR  % ABS ERROR"
10050 Q0=0
10060 Q1=0
10070 Q2=0
10080 Q3=0
10090 N7=N-N8+1
10100 IF N7 <= 0 THEN 10280
10110 FOR T=N8 TO N
10120 Q5=0
10130 Q4=X(T)-F(T)
10140 Q2=Q2+Q4*Q4
10150 Q3=Q3+ABS(Q4)
10160 IF X(T) <=0 THEN 10230
10170 Q5=Q4/X(T)*100
10180 Q0=Q0+Q5
10190 Q1=Q1+ABS(Q5)
10200 IF K9 > 0 THEN 10230
10210 IF T0 < 0.5 THEN 10230
10220 PRINT USING G$(5),T,X(T),F(T),Q4,ABS(Q5)
10230 NEXT T
10240 Q0=Q0/N7
10250 Q1=Q1/N7
10260 Q2=Q2/N7
```

```
10270 Q3=Q3/N7
10280 IF K9 > 0 THEN 10400
10290 IF M2 <= 0 THEN 10340
10300 IF T0 < 0.5 THEN 10340
10310 FOR T=N+1 TO M2+N
10320 PRINT USING G$(4),T,F(T)
10330 NEXT T
10340 PRINT
10350 PRINT "NUMBER OF ERROR OBSNS     ";N7
10360 PRINT "MEAN % ERROR OR BIAS      ";Q0
10370 PRINT "MEAN ABSOLUTE % ERROR     ";Q1
10380 PRINT "MEAN SQUARED ERROR (MSE)  ";Q2
10390 PRINT "MEAN ABSOLUTE ERROR       ";Q3
10400 RETURN
10410 REM     DETERMINE SCALING AND PLOT OBSERVATIONS AND FORECASTS
10420 REM     --------------------------------------------------------
10430 A0=1.E20
10440 A1=-1.E20
10450 FOR I=1 TO N
10460 IF X(I) >= A0 THEN 10480
10470 A0=X(I)
10480 IF X(I) <= A1 THEN 10500
10490 A1=X(I)
10500 NEXT I
10510 FOR I=N8 TO N+M2
10520 IF F(I) >= A0 THEN 10540
10530 A0=F(I)
10540 IF F(I) <= A1 THEN 10560
10550 A1=F(I)
10560 NEXT I
10570 Y0=A0
10580 Y1=A1
10590 A2=(A1-A0)/4
10600 IF A2 <= 0 THEN 11150
10610 A3=LOG(A2)/LOG(10)
10620 REM     ROUND DOWN A3 TO A4
10630 A4=INT(A3)
10640 IF A4 <= A3 THEN 10660
10650 A4=A4-1
10660 A5=10**A4
10670 A6=A2/A5
10680 REM     ROUND UP A6 TO A7
10690 A7=INT(A6+1)
10700 H0=A5*A7
10710 REM     ROUND DOWN L0
10720 L0=INT(A0/H0)
10730 IF L0 <= A0/H0 THEN 10750
10740 L0=L0-1
10750 L0=L0*H0
10760 U0=L0+5*H0
10770 IF U0 >= A1 THEN 10800
10780 A1=A1+A2/2
10790 GOTO 10590
10800 IF L0 <= A0 THEN 10830
```

```
10810 A0=A0-A2/2
10820 GOTO 10590
10830 IF U0-H0 <= Y1 THEN 10860
10840 U0=U0-H0
10850 GOTO 10830
10860 IF L0+H0 >= Y0 THEN 10890
10870 L0=L0+H0
10880 GOTO 10860
10890 H0=(U0-L0)/4
10900 REM
10910 REM     PLOT ACTUAL VERSUS FORECAST
10920 PRINT
10930 GOSUB 11160
10940 GOSUB 11220
10950 H=H0/10
10960 IF N8=1 THEN 11020
10970 N2=-1
10980 FOR I=1 TO N8-1
10990 N1=INT((X(I)-L0)/H+0.5)
11000 GOSUB 11280
11010 NEXT I
11020 FOR I=N8 TO N
11030 N1=INT((X(I)-L0)/H+0.5)
11040 N2=INT((F(I)-L0)/H+0.5)
11050 GOSUB 11280
11060 NEXT I
11070 N1=-1
11080 IF M2 <= 0 THEN 11130
11090 FOR I=N+1 TO N+M2
11100 N2=INT((F(I)-L0)/H+0.5)
11110 GOSUB 11280
11120 NEXT I
11130 GOSUB 11220
11140 GOSUB 11160
11150 RETURN
11160 PRINT "                    ";
11170 FOR J=L0 TO U0 STEP H0
11180 PRINT USING G$(2),J;
11190 NEXT J
11200 PRINT
11210 RETURN
11220 PRINT "PERIOD ACTUAL  FORECAST  +";
11230 FOR J=1 TO 4
11240 PRINT "---------+";
11250 NEXT J
11260 PRINT
11270 RETURN
11280 FOR J=0 TO 40
11290 P$(J)=" "
11300 NEXT J
11310 J9=0
11320 IF N1 < 0 THEN 11360
11330 P$(N1)="O"
11340 IF J9 >= N1 THEN 11360
```

```
11350 J9=N1
11360 IF N2 < 0 THEN 11410
11370 IF P$(N2) = " " THEN 11400
11380 P$(N2)="+"
11390 GOTO 11410
11400 P$(N2)="*"
11410 IF J9 >= N2 THEN 11430
11420 J9=N2
11430 IF N1 >= 0 THEN 11460
11440 PRINT USING G$(6),I,F(I);
11450 GOTO 11480
11460 IF N2 < 0 THEN 11500
11470 PRINT USING G$(7),I,X(I),F(I);
11480 PRINT " -";
11490 GOTO 11520
11500 PRINT USING G$(7),I,X(I);
11510 PRINT "            -";
11520 FOR J=0 TO J9
11530 PRINT P$(J);
11540 NEXT J
11550 PRINT
11560 RETURN
11570 END
```

III. Dimensional Specifications

Program 8CURVS

Currently, the number of observations is limited to 50. If n is the maximum allowable number of observations, the following statements should be modified accordingly:

Line No.
---
70 DIM U(n),V(n),X(n),Y(n),P(n)
100 N0=n

Program FORCST

Let n be the maximum number of observations plus the number of future forecast periods. Also let p be the maximum number of periodicity in the Winters model. Currently, n=50 and p=12. To modify these values, the following statements should be changed as shown below:

Line No.
---
90 DIM X(n),Y(n),Z(n),F(n),W(3,3),R(p),S(p)
110 N0=n
120 S0=p

IV. References
----

1. Brown, R.G., SMOOTHING, FORECASTING AND PREDICTION OF DISCRETE TIME SERIES, Prentice-Hall, Englewood Cliffs, N.J., 1963.

2. Chambers, J.C., S.K. Mullick, and D.D. Smith, "How to Choose the Right Forecasting Technique," HARVARD BUSINESS REVIEW, July-August 1971, pp. 45-74.

3. Makridakis, S., and S.C. Wheelwright, INTERACTIVE FORECASTING, Scientific Press, Palo Alto, Calif., 1974.

4. Makridakis, S., and S.C. Wheelwright, INTERACTIVE FORECASTING - UNIVARIATE AND MULTIVARIATE METHODS, 2nd ed., Holden-Day, San Francisco, 1978.

5. Montgomery, D.C., and L.A. Johnson, FORECASTING AND TIME SERIES ANALYSIS, McGraw-Hill, New York, 1976.

6. Sullivan, W.G., and W.W. Claycombe, FUNDAMENTALS OF FORECASTING, Reston Publishing Company, Reston, Va., 1977.

7. Thomopoulos, N.T., APPLIED FORECASTING METHODS, Prentice-Hall, Englewood Cliffs, N.J., 1980.

8. Trigg, D.W., and D.H. Leach, "Exponentially Smoothing with an Adaptive Response Rate," OPERATIONS RESEARCH QUARTERLY, 18 (1967), 53-59.

9. Winters, P.R., "Forecasting Sales by Exponentially Weighted Moving Averages," MANAGEMENT SCIENCE, 6, No.3 (1960), 324-42

CHAPTER 4

LINEAR PROGRAMMING APPLICATIONS

I. General Linear Programming Problem

- The Program LP
- Input
- Output
- Example
- LP Example Terminal Session
- LP Program Listing

II. Aggregate Planning Model

- The Model
- The Program AGG
- Input
- Output
- Example
- AGG Example Terminal Session
- AGG Program Listing

III. Transportation-Type Problems

- The Classical Transportation Problem
- The Program TRANSP
- Input
- Output
- Example
- TRANSP Example Terminal Session
- TRANSP Program Listing

IV. Dimensional Specifications

V. References

This chapter contains three computer programs, two of which utilize the Simplex method to solve linear programming (LP) problems. The third program uses a special algorithm in LP.

The first program can be used to solve any LP formulation in which the variables are nonnegative. The program can handle both the maximization and minimization problems. We realize that there are many fine linear programming packages available, but the inclusion of this program in this chapter is meant to provide a source code for which more customized applications can be developed. This approach is demonstrated by the application of the second program.

The second program can be used as part of a capacity management planning function to produce aggregate plans for determining the optimal production and employment schedules over the planning horizon. The objective of the model is to minimize the sum of the production costs, inventory carrying costs, and shortage costs (if any), plus the hiring and laying-off costs. Two variants of the same model are imbedded into the program. One variant allows for production backorders while the other does not. In addition, the program contains a unique feature that allows the users to enter their own production and employment schedules and assess the impact of their own plans. Hence, not only can management use the program to develop the appropriate schedules, the program can also be used in developing skills to obtain good plans.

The third program solves transportation-type problems. It is well known that a large number of operations management problems fall into this category. This includes the classical transportation problem in which the objective is to minimize the sum of the transportation costs between various supply and demand points. Other well-known applications include the problem of assigning people to various tasks, transshipment problems, and certain formulations of production scheduling problems. The program is capable of handling the case where the sum of the requirements of various demand points is equal to the sum of the availabilities of the supply points as well as the case where they are not equal.

I. General Linear Programming Problem

This section deals with the program LP that can be used to solve the following formulation of a linear program:

(1) Maximize or Minimize $\sum_{j=1}^{n} c_j x_j$

subject to

(2) $\sum_{j=1}^{n} a_{ij} x_j \ \{<=, \ =, \ >=\} \ b_i \ , \ i=1,\ldots,m$

(3) and $x_j >= 0 \ , \ j=1,\ldots,n.$

The Program LP

The program accepts input data based on the formulation (1), (2), and (3) above. Note that condition (3) specifies that all variables are assumed to be nonnegative. It is furthur assumed that the right-hand side of constraints (2) will be nonnegative. The right-hand side can always be made nonnegative, if necessary, by multiplying the constraint by -1 which also reverses the inequality sign.

In order to use this program, the input data of the formulation must be entered through the use of DATA statements in BASIC. This may not be a very pleasant task when working with a large formulation. Many interactive linear programming packages can accept formulations in their natural form. They also utilize more sophisticated algorithms and computational schemes to avoid problems associated with numerical instability. However, for our purpose, it is not practical to provide such capabilities. Therefore the user should carefully review his or her own application to determine the appropriateness of using and/or adapting this program.

The program listing of LP is divided into three main sections - the main program section (starting at line number 10), the DATA statement section (starting at line number 2500), and the Simplex algorithm (starting at line number 5000). The algorithm utilizes the two-phase method for artificial variables [5]. It stores a tableau with only nonbasic vectors in order to reduce the amount of storage. Slack variables are automatically entered for inequality constraints. Their names coincide with those of the constraint names. Artificial variables are automatically added for greater-than-or-equal-to constraints, and equality constraints. These variables have no names because, if the problem is feasible, these variables will not be in the basis at the optimal solution.

The dimensionality of the program is currently set to handle problems with no more than 15 constraints. In addition, the number of variables plus the number of greater-than-or-equal-to constraints is limited to 20. See Section IV, "Dimensional Specifications", for possible modification.

INPUT

In BASIC language, the sequence of input data using DATA statements is dictated by their line numbers. We suggest that these statements be limited to between line 2500 and line 4999. The program LP requires that every constraint be given a name so that coefficients of variables can be entered properly. The sequence of input data is as follows:

1. Problem title - a one-line description of the problem
2. "MAX" or "MIN" - maximization or minimization problem
3. Number of variables (x_j's) - excluding slack and artificial variables
4. Objective function coefficients and their variable names. If a coefficient is zero, it must still be entered with its name.
5. Number of constraints - excluding nonnegativity condition
6. Name of each constraint, type of the constraint -i.e., <=, or =, or >=, and its right-hand side value
7. Number of nonzero elements on the left-hand side of all contraints.
8. Every nonzero left-hand side coefficient, its constraint name, and its variable name.

OUTPUT

If an optimal solution exists, the optimal value of the objective function together with the values of all basic and and nonbasic variables and their shadow prices is reported. If an optimal solution does not exist, an appropriate message is given as to whether the model has an infeasible or an unbounded solution.

EXAMPLE

Consider a much scaled-down version of a gasoline blending problem facing an oil refinery. On a daily basis the refinery produces superunleaded and regular unleaded gasoline by blending three different streams of output from its process units. These are Butane, Virgin Naptha, and Catalytic Cracked Gasoline. The properties of the streams and their availability per day are summarized below.

Stream	RVP*	Octane Number	Availability in 1000 Barrels/Day
Butane	12	120	25
Virgin Naptha	9	100	40
Catalytic Cracked Gasoline	6	80	100

* RVP stands for Reid Vapor Pressure

The desirable properties of the blended gasolines and their profit per thousand barrels are as follows:

Gasoline	Maximum RVP	Minimum Octane	Profit ($1000 per thousand barrels)
Superunleaded	<= 8	>= 92	3.0
Regular Unleaded	<= 8	>= 87	2.4

The blending process is illustrated in Figure 1. Note that we define x_{ij} as the quantity of the ith stream used in blending gasoline j. For example, x_{11} is the amount of Butane in thousands of barrel used in producing superunleaded gasoline, etc.

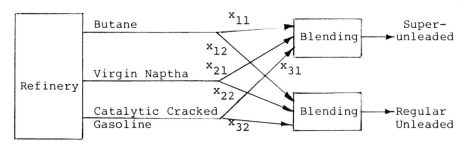

Figure 1. Gasoline Blending

In addition, each property of a blended gasoline is a volumetric average of the corresponding property of its components. This means that if p_i is a property (such as RVP) of the ith stream, x_{ij} is the quantity of the ith stream used in blending gasoline j, and q_j is the corresponding property of the blended gasoline, then we must have

111

$$\Sigma_i p_i x_{ij} = q_j \Sigma_i x_{ij} \quad , j=1,\ldots$$

For example, consider the RVP property of the superunleaded gasoline, the volumetric constraint is

$$12 x_{11} + 9 x_{21} + 6 x_{31} \le 8 (x_{11} + x_{21} + x_{31})$$

or

$$4 x_{11} + x_{21} - 2 x_{31} \le 0.$$

The objective is to determine the optimal blending quantities so as to maximize daily profit. The complete formulation is to

maximize $\quad 3x_{11} + 3x_{21} + 3x_{31} + 2.4x_{12} + 2.4x_{22} + 2.4x_{32}$

subject to

(RVPSUP) $\quad 4x_{11} + x_{21} - 2x_{31} \quad\quad\quad\quad\quad\quad\quad \le 0$

(OCTSUP) $\quad 28x_{11} + 8x_{21} - 12x_{31} \quad\quad\quad\quad\quad\quad \ge 0$

(RVPREG) $\quad\quad\quad\quad\quad\quad\quad\quad 4x_{12} + x_{22} - 2x_{32} \quad \le 0$

(OCTREG) $\quad\quad\quad\quad\quad\quad\quad\quad 33x_{12} + 13x_{22} - 7x_{32} \ge 0$

(SUPPLY-B) $\quad x_{11} \quad\quad\quad\quad + x_{12} \quad\quad\quad\quad \le 25$

(SUPPLY-V) $\quad\quad\quad x_{21} \quad\quad\quad\quad + x_{22} \quad\quad\quad \le 40$

(SUPPLY-C) $\quad\quad\quad\quad\quad x_{31} \quad\quad\quad\quad + x_{32} \le 100$

and all variables are nonnegative.

The DATA statements for this problem are shown below. They are merged with the source listing before the program is run.

```
2500 REM-----------I N P U T    D A T A--------------------
2510 REM---"PROBLEM TITLE" (ONE-LINE DESCRIPTION)
2520 DATA "GASOLINE BLENDING PROBLEM"
2530 REM---"MAX" OR "MIN"
2540 DATA "MAX"
2550 REM---NUMBER OF VARIABLES
2560 DATA 6
2570 REM---OBJECTIVE FUNCTION COEFFICIENT, "NAME OF VARIABLE"
2580 DATA 3, "X11"
2590 DATA 3, "X21"
2600 DATA 3, "X31"
2610 DATA 2.4, "X12"
2620 DATA 2.4, "X22"
2630 DATA 2.4, "X32"
2640 REM---NUMBER OF CONSTRAINTS
2650 DATA 7
2660 REM---"NAME", "TYPE", AND RHS VALUE OF EACH CONSTRAINT
2670 DATA "RVPSUP", "<=", 0
2680 DATA "OCTSUP", ">=", 0
2690 DATA "RVPREG", "<=", 0
2700 DATA "OCTREG", ">=", 0
2710 DATA "SUPPLY-B", "<=", 25
2720 DATA "SUPPLY-V", "<=", 40
2730 DATA "SUPPLY-C", "<=", 100
2740 REM---NO. OF NONZERO ELEMENTS ON LHS OF ALL CONSTRAINTS
2750 DATA 18
2760 REM---NONZERO COEFF., "CONSTRAINT NAME", "VARIABLE NAME"
2770 DATA 4, "RVPSUP", "X11"
2780 DATA 1, "RVPSUP", "X21"
2790 DATA -2, "RVPSUP", "X31"
2800 DATA 28, "OCTSUP", "X11"
2810 DATA 8, "OCTSUP", "X21"
2820 DATA -12, "OCTSUP", "X31"
2830 DATA 4, "RVPREG", "X12"
2840 DATA 1, "RVPREG", "X22"
2850 DATA -2, "RVPREG", "X32"
2860 DATA 33, "OCTREG", "X12"
2870 DATA 13, "OCTREG", "X22"
2880 DATA -7, "OCTREG", "X32"
2890 DATA 1, "SUPPLY-B", "X11"
2900 DATA 1, "SUPPLY-B", "X12"
2910 DATA 1, "SUPPLY-V", "X21"
2920 DATA 1, "SUPPLY-V", "X22"
2930 DATA 1, "SUPPLY-C", "X31"
2940 DATA 1, "SUPPLY-C", "X32"
```

The example terminal session for this example shows that the refinery can make a maximum daily profit of $473,400 by producing roughly 129,000 barrels of superunleaded gasoline, and roughly 36,000 barrels of regular unleaded gasoline. The compositions of the two types of gasoline are easily determined from the optimal solution output.

LP EXAMPLE TERMINAL SESSION
============================

G E N E R A L   L I N E A R   P R O G R A M
-------------   -----------   ---------------

NEED INTRODUCTION (Y OR N)...!N

HAVE ALL DATA BEEN ENTERED PROPERLY? (Y OR N)...!Y

R- E A D I N G   D A T A . . .
TITLE: GASOLINE BLENDING PROBLEM
MAX
NUMBER OF VARIABLES = 6
OBJECTIVE FUNCTION COEFFICIENT AND NAME OF EACH VARIABLE
3X11
3X21
3X31
2.4X12
2.4X22
2.4X32
NUMBER OF CONSTRAINTS = 7
NAME, TYPE, AND RHS VALUE OF EACH CONSTRAINT
RVPSUP<=0
OCTSUP>=0
RVPREG<=0
OCTREG>=0
SUPPLY-B<=25
SUPPLY-V<=40
SUPPLY-C<=100

TOTAL NUMBER OF NONZERO LHS COEFFICIENTS =18
NONZERO COEFFICIENT, CONSTRAINT NAME, VARIABLE NAME

4	RVPSUP	X11
1	RVPSUP	X21
-2	RVPSUP	X31
28	OCTSUP	X11
8	OCTSUP	X21
-12	OCTSUP	X31
4	RVPREG	X12
1	RVPREG	X22
-2	RVPREG	X32
33	OCTREG	X12
13	OCTREG	X22
-7	OCTREG	X32
1	SUPPLY-B	X11
1	SUPPLY-B	X12
1	SUPPLY-V	X21
1	SUPPLY-V	X22
1	SUPPLY-C	X31
1	SUPPLY-C	X32

TITLE : GASOLINE BLENDING PROBLEM

OPTIMAL   SOLUTION

MAX OBJECTIVE FUNCTION VALUE = 473.4

VARIABLE	VALUE	REDUCED COST
X11	18.700000	.000000
X21	40.000000	.000000
X31	70.300000	.000000
X12	6.300000	.000000
X22	.000000	.000000
X32	29.700000	.000000

CONSTRAINT	SLACK	DUAL PRICES
RVPSUP	25.800000	.000000
OCTSUP	.000000	-.120000
RVPREG	34.200000	.000000
OCTREG	.000000	-.120000
SUPPLY-B	.000000	6.360000
SUPPLY-V	.000000	3.960000
SUPPLY-C	.000000	1.560000

```
10 REM   G E N E R A L      L I N E A R     P R O G R A M
20 REM
30 DIM M$(15),N$(20)
40 DIM A(15),B(15,20),C(20),P(20),R(15),T(20),U(15),X(20),Z(20)
50 REM
60 M0=15
70 N0=20
80 PRINT "G E N E R A L     L I N E A R     P R O G R A M"
90 PRINT "-------------     -----------     -------------"
100 PRINT
110 PRINT "NEED INTRODUCTION (Y OR N)...";
120 INPUT Y$
130 IF Y$<>"Y"THEN 280
140 PRINT
150 PRINT "THIS PROGRAM DETERMINES THE OPTIMAL SOLUTION OF A GENERAL"
160 PRINT "LP PROBLEM.  THE PROBLEM MUST BE FORMULATED SUCH THAT THE"
170 PRINT "VARIABLES CAN ONLY ASSUME NONNEGATIVE VALUES.  ALL RIGHT"
180 PRINT "HAND SIDES MUST BE NONNEGATIVE."
190 PRINT
200 PRINT "THE TOTAL NUMBER OF CONSTRAINTS MUST NOT EXCEED ";M0
210 PRINT "THE NUMBER OF VARIABLES PLUS THE NUMBER OF >= CONSTRAINTS"
220 PRINT "MUST NOT EXCEED ";N0
230 PRINT
240 PRINT "INPUT REQUIREMENTS"
250 PRINT "------------------"
260 PRINT "DETAILS OF HOW TO PREPARE INPUT DATA ARE AVAILABLE"
270 PRINT "IN CHAPTER 4 OF THE BOOK."
280 PRINT
290 PRINT "HAVE ALL DATA BEEN ENTERED PROPERLY? (Y OR N)...";
300 INPUT Y$
310 IF Y$ <> "Y" THEN 9999
320 V7=0
330 PRINT
340 PRINT "R E A D I N G      D A T A  . . ."
350 READ T$
360 PRINT "TITLE: ";T$
370 REM    "MAX" OR "MIN"
380 READ Q$
390 PRINT Q$
400 READ N
410 PRINT "NUMBER OF VARIABLES = ";N
420 PRINT "OBJECTIVE FUNCTION COEFFICIENT AND NAME OF EACH VARIABLE"
430 FOR J=1 TO N
440 READ C(J),N$(J)
450 PRINT C(J);N$(J)
460 NEXT J
470 READ M
480 PRINT "NUMBER OF CONSTRAINTS = ";M
490 PRINT "NAME, TYPE, AND RHS VALUE OF EACH CONSTRAINT"
500 N3=0
510 FOR I=1 TO M
520 READ M$(I),C$,R(I)
530 PRINT M$(I);C$;R(I)
540 U(I)=0
```

```
550 IF C$="<=" THEN U(I)=1
560 IF C$="=" THEN U(I)=2
570 IF C$=">=" THEN U(I)=3
580 IF U(I) > 0 THEN 620
590 PRINT "*** ERROR IN THE ABOVE TYPE OF CONSTRAINT ! ***"
600 V7=V7+1
610 GOTO 630
620 IF U(I)=3 THEN N3=N3+1
630 NEXT I
640 IF N+N3<=N0 THEN 700
650 PRINT "NUMBER OF VARIABLES PLUS NUMBER OF >= CONSTRAINTS IS ";N+J
660 PRINT "THIS EXCEEDS THE LIMIT OF ";N0
670 PRINT "SEE DIMENSIONAL SPECIFICATIONS FOR MODIFICATIONS."
680 V7=V7+1
690 GOTO 860
700 REM    T(J)=CODED NAMES OF NONBASIC VARIABLES
710 FOR J=1 TO N
720 T(J)=J
730 NEXT J
740 L=0
750 FOR I=1 TO M
760 FOR J=1 TO N+N3
770 B(I,J)=0
780 NEXT J
790 IF U(I)<>3 THEN 850
800 L=L+1
810 B(I,N+L)=-1
820 C(N+L)=0
830 N$(N+L)=M$(I)
840 T(N+L)=N+I
850 NEXT I
860 PRINT
870 READ K1
880 PRINT "TOTAL NUMBER OF NONZERO LHS COEFFICIENTS =";K1
890 PRINT "NONZERO COEFFICIENT, CONSTRAINT NAME, VARIABLE NAME"
900 FOR L=1 TO K1
910 READ V0,C$,D$
920 PRINT V0,C$,D$
930 FOR I=1 TO M
940 IF C$=M$(I) THEN 980
950 NEXT I
960 V7=V7+1
970 PRINT C$;"  !!! ERROR !!!   NOT CONSISTENT !"
980 FOR J=1 TO N
990 IF D$=N$(J) THEN 1040
1000 NEXT J
1010 V7=V7+1
1020 PRINT D$;"  !!! ERROR !!!   NOT CONSISTENT !"
1030 GOTO 1050
1040 IF V7=0 THEN B(I,J)=V0
1050 NEXT L
1060 IF V7=0 THEN 1110
1070 PRINT
1080 PRINT V7;" ERRORS DETECTED !   EXECUTION TERMINATED !!!"
```

```
1090 STOP
1100 REM SET VECTOR Z( ) TO CONTAIN COEFFICIENTS OF OBJ FN.
1110 N=N+N3
1120 FOR J=1 TO N
1130 IF Q$="MAX" THEN 1160
1140 Z(J)=-C(J)
1150 GOTO 1170
1160 Z(J)=C(J)
1170 NEXT J
1180 REM
1190 REM    X(J) AND P(J) = SHADOW PRICES IN PHASES I AND II
1200 FOR J=1 TO N
1210 X(J)=0
1220 FOR I=1 TO M
1230 IF U(I)=1 THEN 1250
1240 X(J)=X(J)-B(I,J)
1250 NEXT I
1260 NEXT J
1270 REM    N9=INFEASIBILITY COUNT, A(I)=CODED NAMES OF BASIC VAR.
1280 N9=0
1290 FOR I=1 TO M
1300 A(I)=-I
1310 IF U(I)=1 THEN 1340
1320 N9=N9+1
1330 A(I)=-I-M
1340 U(I)=0
1350 NEXT I
1360 REM    C A L L    S I M P L E X    A L G O R I T H M
1370 GOSUB 5000
1380 REM    RETRIEVE THE OPTIMAL SOLUTION
1390 FOR I=1 TO 5
1400 PRINT
1410 NEXT I
1420 PRINT "TITLE : ";T$
1430 PRINT
1440 PRINT "O P T I M A L    S O L U T I O N"
1450 PRINT "------------    ---------------"
1460 C0=0
1470 FOR I = 1 TO M
1480 J=A(I)
1490 IF J <= 0 THEN 1510
1500 C0=C0+C(J)*R(I)
1510 NEXT I
1520 PRINT
1530 PRINT Q$;" OBJECTIVE FUNCTION VALUE = ";C0
1540 PRINT
1550 F$(1)="VARIABLE          VALUE           REDUCED COST"
1560 F$(2)="##########    -######.######    -######.######"
1570 F$(3)="CONSTRAINT        SLACK           DUAL PRICES"
1580 FOR J=1 TO N
1590 X(J)=0
1600 Z(J)=0
1610 NEXT J
1620 FOR I=1 TO M
```

```
1630 C(I)=0
1640 U(I)=0
1650 NEXT I
1660 REM    ASSIGN TABLE VALUES OF NONBASIC VARIABLES
1670 FOR J=1 TO N
1680 IF T(J) > N THEN 1770
1690 IF T(J)>0 THEN 1760
1700 IF T(J)>= -M THEN 1740
1710 K=ABS(T(J)+M)
1720 U(K)=P(J)
1730 GOTO 1770
1740 U(ABS(T(J)))=P(J)
1750 GOTO 1770
1760 Z(T(J))=P(J)
1770 NEXT J
1780 REM    ASSIGN TABLE VALUES OF BASIC VARIABLES
1790 FOR I=1 TO M
1800 K=A(I)
1810 IF K < 0 THEN 1870
1820 IF K > N-N3 THEN 1850
1830 X(K)=R(I)
1840 GOTO 1880
1850 C(K-N+N3)=R(I)
1860 GOTO 1880
1870 C(ABS(K))=R(I)
1880 NEXT I
1890 REM    PRINT REDUCED-COST AND DUAL-PRICE TABLES
1900 PRINT F$(1)
1910 FOR J=1 TO N-N3
1920 PRINT USING F$(2),N$(J),X(J),Z(J)
1930 NEXT J
1940 PRINT
1950 PRINT F$(3)
1960 FOR I=1 TO M
1970 PRINT USING F$(2),M$(I),C(I),U(I)
1980 NEXT I
1990 GOTO 9999
```

```
5000 REM
5010 REM     S I M P L E X      A L G O R I T H M
5020 I0=1
5030 FOR J = 1 TO N
5040 P(J)=-Z(J)
5050 FOR I = 1 TO M
5060 P(J)=P(J)+U(I)*B(I,J)
5070 NEXT I
5080 NEXT J
5090 REM
5100 REM    DETERMINE PIVOT COLUMN
5110 E9=-.0000001
5120 K9=0
5130 IF I0=2 THEN 5150
5140 IF N9 <= 0 THEN 5960
5150 FOR J=1 TO N
5160 IF T(J) < -M THEN 5250
5170 IF I0 = 2 THEN 5220
5180 IF X(J) >= E9 THEN 5250
5190 K9=J
5200 E9=X(J)
5210 GOTO 5250
5220 IF P(J) >= E9 THEN 5250
5230 K9=J
5240 E9=P(J)
5250 NEXT J
5260 IF K9 <= 0 THEN 5950
5270 REM
5280 REM    DETERMINE PIVOT ROW
5290 K8 = 0
5300 C9=E9
5310 E9=1E+20
5320 FOR I = 1 TO M
5330 IF B(I,K9) <= 0 THEN 5380
5340 R9=R(I)/B(I,K9)
5350 IF R9 >= E9 THEN 5380
5360 E9=R9
5370 K8=I
5380 NEXT I
5390 IF K8 > 0 THEN 5450
5400 PRINT
5410 PRINT " *** S O L U T I O N    U N B O U N D E D ***"
5420 PRINT "     ------------------------------------"
5430 STOP
5440 REM
5450 REM    TRANSFORM TABLEAU
5460 V9=B(K8,K9)
5470 FOR J = 1 TO N
5480 B(K8,J)=B(K8,J)/V9
5490 NEXT J
5500 R(K8)=R(K8)/V9
5510 FOR I = 1 TO M
5520 IF I=K8 THEN 5580
5530 R(I)=R(I)-R(K8)*B(I,K9)
```

```
5540 FOR J = 1 TO N
5550 IF J = K9 THEN 5570
5560 B(I,J)=B(I,J)-B(K8,J)*B(I,K9)
5570 NEXT J
5580 NEXT I
5590 FOR I = 1 TO M
5600 B(I,K9)=-B(I,K9)/V9
5610 B(K8,K9)=1/V9
5620 NEXT I
5630 REM
5640 REM    INTERCHANGE BASIC AND NONBASIC VARIABLES.
5650 R8=T(K9)
5660 T(K9)=A(K8)
5670 A(K8)=R8
5680 E9=Z(K9)
5690 Z(K9)=U(K8)
5700 U(K8)=E9
5710 IF T(K9) >= -M THEN 5730
5720 N9=N9-1
5730 IF I0 = 2 THEN 5820
5740 S9=P(K9)
5750 FOR J = 1 TO N
5760 P(J)=P(J)-S9*B(K8,J)
5770 X(J)=X(J)-C9*B(K8,J)
5780 NEXT J
5790 P(K9)=-S9/V9
5800 X(K9)=-C9/V9
5810 GOTO 5870
5820 FOR J = 1 TO N
5830 P(J)=P(J)-C9*B(K8,J)
5840 NEXT J
5850 P(K9)=-C9/V9
5860 REM    SET VERY SMALL ELEMENTS OF B(.,.) TO ZEROS.
5870 FOR I = 1 TO M
5880 FOR J = 1 TO N
5890 IF ABS(B(I,J)) > .0000001 THEN 5910
5900 B(I,J)=0
5910 NEXT J
5920 NEXT I
5930 GOTO 5110
5940 REM
5950 IF I0 = 2 THEN 6020
5960 I0=2
5970 IF N9 <= 0 THEN 5030
5980 PRINT
5990 PRINT " *** S O L U T I O N    I N F E A S I B L E ***"
6000 PRINT "      ---------------    -------------------"
6010 STOP
6020 RETURN
9999 END
```

II. Aggregate Planning Model

Aggregate plans are production and work force plans necessary for management to control the costs related to production. These costs include direct labor, overtime, inventory, and employment costs. The needs of management to control these costs are the obvious results of demand fluctuations. Various strategies are possible in trying to satisfy demands; for example, storing more inventory or using overtime or hiring additional people to cope with increasing demands, etc. The problem becomes that of finding the best strategy that will satisfy the demands and incurring the least cost. Aggregate planning is therefore only one facet of the overall capacity planning functions.

Most aggregate planning models in literature deal with an equivalent unit of production to overcome the massive amount of information that can be incorporated into one model if every product line is individually considered simultaneously. If several products are produced, the demands for each product may be forecasted and converted into an equivalent unit and aggregated to form the forecasts of the equivalent unit. This equivalent unit can be in terms of the dollar sales or an equivalent physical production unit. Also, historical information can provide data such as how many equivalent units a worker can produce in a period and what the inventory carrying cost is per period for the unit.

In this section, we discuss a mathematical model that can be formulated in terms of a linear program to solve for the optimal production schedule and the optimal work force level for a planning horizon of T periods. The original formulation of this model is due to Hanssmann and Hess [7]. This model and its variations are well known and used today. Some of the variations are the absence of the employment aspect of the model or imposing restrictions on the inventory capacity. See, for example, Schuermann and Kannan [10].

The Model

Let the planning horizon be T periods, and define the following quantities for period t, $t=1,\ldots,T$:

S_t = Forecast of demand (units),

P_t = Production rate (units per period),

I_t = Period ending inventory (units),

W_t = Work force level (number of workers),

H_t = Number of workers to be hired,

F_t = Number of workers to be laid off.

Let the following represent the relevant unit costs for the model:

c_h = Cost of hiring one worker,

c_f = Cost of laying off one worker,

c_s = Straight-time wage per worker per period,

c_o = Overtime wage per worker per period = fc_s
where f is the overtime factor,

c_1 = Unit inventory carrying cost per period,

c_2 = Unit backorder cost per period.

Also, let r be the number of units that a worker can produce in a period. Let $k=1/r$, I_0 be the initial inventory and W_0 be the initial number of workers. In order to express the model mathematically, we define for any real quantity, q,

(4a) $\quad q^+ = |q| \quad$ for $q >= 0$
$\quad\quad\quad\quad = 0, \quad$ otherwise

and

(4b) $\quad q^- = 0 \quad$ for $q >= 0$
$\quad\quad\quad\quad = |q|, \quad$ otherwise.

Note that $q^+ - q^- = q$.

Consider, for example, if $W_t < W_{t-1}$, this implies that some workers are laid off in period t; hence $(W_t - W_{t-1})^-$ is the number of workers laid off. Similarly, if $W_t > W_{t-1}$, $(W_t - W_{t-1})^+$ is the number of workers hired in period t, etc.

Hence the objective is to minimize the sum of the following cost components:

$$\text{Hiring cost} = c_h (W_t - W_{t-1})^+$$

123

$$\text{Layoff cost} = c_f (W_t - W_{t-1})^-$$

$$\text{Regular payroll} = c_s W_t$$

$$\text{Overtime cost} = c_o (kP_t - W_t)^+$$

$$\text{Inventory carrying cost} = c_1 I_t^+$$

$$\text{Backorder cost} = c_2 I_t^-.$$

Formally, the problem is to minimize

(5) $$\sum_{t=1}^{T} (c_h (W_t - W_{t-1})^+ + c_f (W_t - W_{t-1})^- + c_s W_t$$
$$+ c_o (kP_t - W_t)^+ + c_1 I_t^+ + c_2 I_t^-)$$

subject to

(6) $$W_t \geq 0$$

(7) $$P_t \geq 0$$

(8) $$I_t = I_{t-1} + P_t - S_t,$$

for $t=1,\ldots,T$.

In order to express (5) to (8) in the linear programming form, we define a new set of variables:

(9)
$$x_t = (W_t - W_{t-1})^+$$
$$y_t = (W_t - W_{t-1})^-$$
$$z_t = (kP_t - W_t)^+$$
$$w_t = (kP_t - W_t)^-$$

$$u_t = I_t^+$$
$$v_t = I_t^-$$

for t=1,...,T. These variables are substituted into (5) through (8) to obtain the complete LP formulation. For details of the algebraic manipulation, see Hillier and Lieberman [8]. The resulting model is to minimize

(10) $$\sum_{t=1}^{T} (c_h x_t + c_f y_t + (c_o - c_s)z_t + c_{st} w_t + c_1 u_t + c_2 v_t)$$

$$+ c_s k((u_T - v_T) - I_0 + \sum_{t=1}^{T} S_t)$$

subject to

(11) $$-(u_t - v_t) + (u_{t-1} - v_{t-1}) + (z_t - w_t)/k \le S_t,$$

(12) $$-(u_t - v_t) + (u_{t-1} - v_{t-1}) \le S_t,$$

(13) $$(x_t - y_t) + (z_t - w_t) - (z_{t-1} - w_{t-1})$$

$$- k(u_t - v_t) + 2k(u_{t-1} - v_{t-1}) - k(u_{t-2} - v_{t-2}) = k(S_t - S_{t-1}),$$

for t=1,...,T and all variables are nonnegative. When t=1, the constraints are as follows:

$$-k(u_1 - v_1) + (z_1 - w_1) \le k(S_1 - I_0),$$

$$-(u_1 - v_1) \le S_1 - I_0,$$

$$(x_1 - y_1) + (z_1 - w_1) - k(u_1 - v_1) = k(S_1 - I_0) - W_0.$$

When t=2, constraint (13) is simplified to

$$(x_2 - y_2) + (z_2 - w_2) - (z_1 - w_1) - k(u_2 - v_2) + 2k(u_1 - v_1) = k(S_2 - S_1 + I_0).$$

If ending inventory in period T is specified, we need one more constraint, that is, $u_T - v_T = I_T$. Hence we may have up to 6T variables and 3T+1 constraints.

The Program AGG

The program accepts the cost data, the forecasts, and other initial conditions in the form that the user can easily understand. It also allows the user to select either a backorder model or a no-backorder model. If the former is specified, the program formulates the problem according to (10) to (13). If the latter is specified, the variables v_t, $t=1,\ldots,T$ are not considered because they will have zero values. The user can also specify an ending inventory level at the end of the planning horizon. This requires another constraint if the value is nonzero.

The program solves for the optimal production schedule and the employment schedule only once and stores the solutions. The user can request the solution as many times as required. In general, the optimal values are not integers; therefore, the program provides a feature that allows the user to specify his or her own schedule which will be more realistic.

If a constant work force is desired throughout the planning horizon, the corresponding optimal program can be obtained by imposing very high cost values for the hiring and layoff unit costs.

Note also that the program AGG uses the same Simplex code as in the program LP. Currently, AGG can handle problems of up to 12 periods. See Section IV, "Dimensional Specifications," for possible expansion.

INPUT

1. Backorder/No-backorder model
2. Planning horizon
3. Cost of hiring one worker
4. Cost of laying off one worker
5. Straight-time wage per worker per period
6. Overtime factor
7. Number of units a worker can produce with no overtime in a period
8. Inventory carrying cost per unit per period
9. Unit backorder cost per period (required only if a backorder model is selected)
10. The initial number of workers
11. Initial inventory
12. Required ending inventory at the end of the planning horizon
13. Forecasts of the demands or sales

OUTPUT

Output consists of the optimal production schedule, the work force levels for all periods, and the minimum total cost. If the user specifies a plan, costs are calculated and broken down by components and percentages of the total cost of the plan.

EXAMPLE

Given a planning horizon of 12 months with the following forecasts: 7000, 9000, 10,000, 15,000, 20,000, 18,000, 10,000, 8000, 7000, 6000, 5000, and 5000 units. No backorder is allowed. The cost of hiring and laying off one worker is $500. The straight-time wage per month is $800. The overtime factor is 1.5 -i.e., if a worker-month of overtime is utilized, the firm will incur $1200. Each worker can produce on the average 60 units per month. Each unit costs $120 to produce on the average and if the annual inventory carrying charge is 10 percent, the inventory carrying cost per unit per period is $1. Assume also that the initial number of workers is 100. The firm currently has 700 units in inventory and desires an ending inventory of 500 units at the end of the planning horizon.

The optimal solution is shown in the example terminal session below with an optimal cost of $1,770,083. It indicates that the most cost-effective plan is to hire roughly 118 new workers at the beginning and maintain that work force for six months during which roughly 13,000 units can be produced per month. The plan also calls for a gradual reduction in the work force due to declining demands toward the second half of the planning horizon.

By roughly following the LP solution as done in the session, it can be seen that the total cost of the plan is approximately $1.8 million, of which almost 89 percent of the cost is due to the regular payroll. The ending inventory is 1700 units instead of the required 500 units. There is also a shortage in the sixth month of 300 units. Since the user selects a unit backorder cost per period of zero, the percentage of the shortage cost that appears in the summary table is also zero.

Note

The program AGG uses the same SIMPLEX algorithm as the program LP. In fact, the statements 5000 through 9999 must be appended to the end of the listing of AGG for the program to be complete. This is so because the program AGG calls the SIMPLEX algorithm at statement 2870 of the program.

```
AGG EXAMPLE TERMINAL SESSION
============================

A G G R E G A T E    P L A N N I N G    M O D E L
-----------------    ---------------    ---------

NEED INTRODUCTION (Y OR N)...!N

SELECT MODEL (1=BACKORDER, 2=NO BACKORDER)  !2

ENTER PLANNING HORIZON  !12

ENTER COST OF HIRING ONE WORKER  !500

ENTER COST OF LAYING OFF ONE WORKER  !500

ENTER STRAIGHT-TIME WAGE/WORKER/PERIOD  !800

ENTER OVERTIME FACTOR  !1.5

ENTER UNITS OF PRODUCTION/WORKER/PERIOD  !60

ENTER INVENTORY CARRYING COST/UNIT/PERIOD  !1

ENTER INITIAL NUMBER OF WORKERS  !100

ENTER INITIAL INVENTORY  !700

ENTER DESIRED ENDING INVENTORY (>=0 IF MODEL 2)...!500
FORECAST FOR PERIOD  1   !7000
FORECAST FOR PERIOD  2   !9000
FORECAST FOR PERIOD  3   !10000
FORECAST FOR PERIOD  4   !15000
FORECAST FOR PERIOD  5   !20000
FORECAST FOR PERIOD  6   !18000
FORECAST FOR PERIOD  7   !10000
FORECAST FOR PERIOD  8   !8000
FORECAST FOR PERIOD  9   !7000
FORECAST FOR PERIOD  10  !6000
FORECAST FOR PERIOD  11  !5000
FORECAST FOR PERIOD  12  !5000
```

```
OPTIMAL      PRODUCTION      SCHEDULE
-------------   --------------------   ----------------
            ENDING
PERIOD    INVENTORY        SALES       PRODUCTION
  0          700.00
  1         6750.00        7000.00      13050.00
  2        10800.00        9000.00      13050.00
  3        13850.00       10000.00      13050.00
  4        11900.00       15000.00      13050.00
  5         4950.00       20000.00      13050.00
  6             .00       18000.00      13050.00
  7             .00       10000.00      10000.00
  8             .00        8000.00       8000.00
  9             .00        7000.00       7000.00
 10             .00        6000.00       6000.00
 11          250.00        5000.00       5250.00
 12          500.00        5000.00       5250.00


OPTIMAL      WORK      FORCE     LEVEL
-------------   -------   ---------   ---------
PERIOD   WORK FORCE       HIRE         LAYOFF
  0         100.00
  1         217.50        117.50          .00
  2         217.50           .00          .00
  3         217.50           .00          .00
  4         217.50           .00          .00
  5         217.50           .00          .00
  6         217.50           .00          .00
  7         166.67           .00        50.83
  8         133.33           .00        33.33
  9         116.67           .00        16.67
 10         100.00           .00        16.67
 11          87.50           .00        12.50
 12          87.50           .00          .00


COST      COMPONENTS
-------   --------------------

$$$$$$$ HIRING COST       58750.00      3.319 %
$$$$$$$ LAYOFF COST       65000.00      3.672 %
$$$ REGULAR PAYROLL     1597333.33     90.241 %
$$$$ OVERTIME COST            .00       .000 %
$$$$ INVENTORY COST      49000.00      2.768 %
$$$$ SHORTAGE COST            .00       .000 %

TOTAL     COST        1770083.33
```

```
DO YOU WANT TO SPECIFY YOUR OWN PLAN (Y OR N)...!Y

ENTER UNIT BACKORDER COST PER PERIOD   !0

ENTER PRODUCTION PLAN
---------------------
PRODUCTION IN PERIOD 1    !13000
PRODUCTION IN PERIOD 2    !13000
PRODUCTION IN PERIOD 3    !13000
PRODUCTION IN PERIOD 4    !13000
PRODUCTION IN PERIOD 5    !13000
PRODUCTION IN PERIOD 6    !13000
PRODUCTION IN PERIOD 7    !11000
PRODUCTION IN PERIOD 8    !8000
PRODUCTION IN PERIOD 9    !7000
PRODUCTION IN PERIOD 10   !6000
PRODUCTION IN PERIOD 11   !5500
PRODUCTION IN PERIOD 12   !5500

ENTER WORK FORCE LEVEL
----------------------
NUMBER OF WORKERS IN PERIOD        1    !218
NUMBER OF WORKERS IN PERIOD        2    !218
NUMBER OF WORKERS IN PERIOD        3    !218
NUMBER OF WORKERS IN PERIOD        4    !218
NUMBER OF WORKERS IN PERIOD        5    !218
NUMBER OF WORKERS IN PERIOD        6    !218
NUMBER OF WORKERS IN PERIOD        7    !167
NUMBER OF WORKERS IN PERIOD        8    !134
NUMBER OF WORKERS IN PERIOD        9    !117
NUMBER OF WORKERS IN PERIOD       10    !100
NUMBER OF WORKERS IN PERIOD       11    !90
NUMBER OF WORKERS IN PERIOD       12    !90
```

```
P R O D U C T I O N    S C H E D U L E
                       ENDING
PERIOD      INVENTORY           SALES       PRODUCTION
  0            700.00
  1           6700.00          7000.00       13000.00
  2          10700.00          9000.00       13000.00
  3          13700.00         10000.00       13000.00
  4          11700.00         15000.00       13000.00
  5           4700.00         20000.00       13000.00
  6         -  300.00         18000.00       13000.00
  7            700.00         10000.00       11000.00
  8            700.00          8000.00        8000.00
  9            700.00          7000.00        7000.00
 10            700.00          6000.00        6000.00
 11           1200.00          5000.00        5500.00
 12           1700.00          5000.00        5500.00


W O R K     F O R C E     L E V E L

PERIOD      WORK FORCE         HIRE          LAYOFF
  0            100.00
  1            218.00         118.00            .00
  2            218.00            .00            .00
  3            218.00            .00            .00
  4            218.00            .00            .00
  5            218.00            .00            .00
  6            218.00            .00            .00
  7            167.00            .00          51.00
  8            134.00            .00          33.00
  9            117.00            .00          17.00
 10            100.00            .00          17.00
 11             90.00            .00          10.00
 12             90.00            .00            .00


C O S T     C O M P O N E N T S

$$$$$$$ HIRING COST         59000.00         3.269 %
$$$$$$$ LAYOFF COST         64000.00         3.546 %
$$$ REGULAR PAYROLL       1604800.00        88.928 %
$$$$$ OVERTIME COST         23600.00         1.308 %
$$$$ INVENTORY COST         53200.00         2.948 %
$$$$ SHORTAGE COST                .00         .000 %

T O T A L    C O S T      1804600.00

ENTER 1(SHOW LP SOLN), 2(TRY NEW PLAN), 3(RESTART), 4(EXIT)..!4
```

```
10  REM   A G G R E G A T E     P L A N N I N G     M O D E L
20  REM   VIA LINEAR PROGRAMMING
30  REM   REFERENCE : HILLIER,F.S. AND LIEBERMAN, G.J.,"OPERATIONS
40  REM   RESEARCH", HOLDEN DAY,INC., 1974, PP.166-171.
50  REM
60  DIM A(37),B(37,74),P(74),R(37),T(74),U(37),X(74),Z(74)
70  DIM S(12),W(12),D(12),E(12),H(12),F(12),FS(4)
80  GOSUB 4450
90  T9=12
100 PRINT "A G G R E G A T E     P L A N N I N G     M O D E L"
110 PRINT "-----------------     ---------------     ---------"
120 PRINT
130 PRINT "NEED INTRODUCTION (Y OR N)...";
140 INPUT Y$
150 PRINT
160 IF Y$="N" THEN 500
170 PRINT "THIS PROGRAM FINDS THE OPTIMAL PRODUCTION AND EMPLOYMENT"
180 PRINT "SCHEDULES BY MINIMIZING THE SUM OF HIRING, LAYING OFF"
190 PRINT "EMPLOYEES, INVENTORY CARRYING AND SHORTAGE COSTS"
200 PRINT "(OPTIONAL) OVER A PLANNING HORIZON OF T PERIODS."
210 PRINT
220 PRINT "TWO BASIC MODELS ARE ALLOWED :-"
230 PRINT "   1. MODEL WHICH ALLOWS BACKORDERS, I.E. SHORTAGE"
240 PRINT "   2. MODEL WHICH DOES NOT ALLOW ANY SHORTAGE."
250 PRINT
260 PRINT "YOU ALSO HAVE THE OPTION OF SPECIFYING YOUR OWN"
270 PRINT "PRODUCTION AND EMPLOYMENT SCHEDULES AND COMPARE THE"
280 PRINT "COSTS WITH THE LINEAR PROGRAMMING SOLUTION."
290 PRINT
300 PRINT
310 PRINT "INPUT REQUIREMENTS"
320 PRINT "------------------"
330 PRINT "   1. SELECTION OF A MODEL"
340 PRINT "   2. LENGTH OF THE PLANNING HORIZON"
350 PRINT "   3. COST OF HIRING ONE WORKER"
360 PRINT "   4. COST OF LAYING OFF ONE WORKER"
370 PRINT "   5. STRAIGHT-TIME WAGE OF ONE WORKER"
380 PRINT "   6. OVERTIME FACTOR"
390 PRINT "   7. NUMBER OF UNITS ONE WORKER CAN PRODUCE PER PERIOD"
400 PRINT "      WITH NO OVERTIME"
410 PRINT "   8. COST OF CARRYING A UNIT IN INVENTORY PER PERIOD"
420 PRINT "   9. UNIT BACKORDER COST PER PERIOD (OPTIONAL)"
430 PRINT "  10. INITIAL NUMBER OF WORKERS AVAILABLE"
440 PRINT "  11. INITIAL INVENTORY"
450 PRINT "  12. INVENTORY AT END OF PLANNING HORIZON"
460 PRINT "  13. FORECASTS OF DEMANDS FOR T PERIODS."
470 FOR I=1 TO 5
480 PRINT
490 NEXT I
500 PRINT "SELECT MODEL (1=BACKORDER, 2=NO BACKORDER)   ";
510 INPUT B0
520 PRINT
530 PRINT "ENTER PLANNING HORIZON   ";
540 INPUT T0
```

```
550 IF T0 > T9 THEN 520
560 IF T0 <= 1 THEN 520
570 PRINT
580 PRINT "ENTER COST OF HIRING ONE WORKER   ";
590 INPUT H0
600 PRINT
610 PRINT "ENTER COST OF LAYING OFF ONE WORKER   ";
620 INPUT L0
630 PRINT
640 PRINT "ENTER STRAIGHT-TIME WAGE/WORKER/PERIOD   ";
650 INPUT R0
660 PRINT
670 PRINT "ENTER OVERTIME FACTOR   ";
680 INPUT F0
690 V0=R0*F0
700 PRINT
710 PRINT "ENTER UNITS OF PRODUCTION/WORKER/PERIOD   ";
720 INPUT K0
730 IF K0<=0 THEN 700
740 K0=1/K0
750 PRINT
760 PRINT "ENTER INVENTORY CARRYING COST/UNIT/PERIOD   ";
770 INPUT C1
780 IF B0=2 THEN 820
790 PRINT
800 PRINT "ENTER PER UNIT BACKORDER COST PER PERIOD   ";
810 INPUT C2
820 PRINT
830 PRINT "ENTER INITIAL NUMBER OF WORKERS   ";
840 INPUT W0
850 PRINT
860 PRINT "ENTER INITIAL INVENTORY   ";
870 INPUT I1
880 PRINT
890 PRINT "ENTER DESIRED ENDING INVENTORY (>=0 IF MODEL 2)...";
900 INPUT I2
910 IF I2 >= 0 THEN 960
920 IF B0=1 THEN 960
930 PRINT
940 PRINT "  ***** NO BACKORDER MODEL WAS SELECTED *****";
950 GOTO 890
960 PRINT
970 FOR I=1 TO T0
980 PRINT "FORECAST FOR PERIOD",I;"    ";
990 INPUT S(I)
1000 NEXT I
1010 REM    F O R M U L A T E      L P     M O D E L
1020 S0=0
1030 FOR I=1 TO T0
1040 S0=S0+S(I)
1050 NEXT I
1060 N=5*T0
1070 IF B0=2 THEN 1090
1080 N=N+T0
```

```
1090 T1=2*T0
1100 T2=3*T0
1110 T3=4*T0
1120 T4=5*T0
1130 M=T2
1140 FOR I=1 TO T2
1150 FOR J=1 TO N
1160 B(I,J)=0
1170 NEXT J
1180 NEXT I
1190 REM ASSIGN FIRST SET OF CONSTRAINT COEFFICIENTS TO MATRIX B
1200 J1=2*T0
1210 FOR I=1 TO T0
1220 J1=J1+1
1230 J2=J1+T0
1240 J3=J1+T1
1250 IF B0=2 THEN 1270
1260 J4=J1+T2
1270 B(I,J1)=1
1280 B(I,J2)=-1
1290 B(I,J3)=-K0
1300 IF B0=1 THEN B(I,J4)=K0
1310 NEXT I
1320 J1=T3
1330 FOR I=2 TO T0
1340 J1=J1+1
1350 B(I,J1)=K0
1360 IF B0=2 THEN 1390
1370 J2=J1+T0
1380 B(I,J2)=-K0
1390 NEXT I
1400 REM ASSIGN SECOND SET OF CONSTRAINT COEFFICIENTS TO MATRIX B
1410 J=T3
1420 FOR I6=1 TO T0
1430 I=T0+I6
1440 J=J+1
1450 B(I,J)=-1
1460 IF B0=2 THEN 1480
1470 B(I,J+T0)=1
1480 NEXT I6
1490 J=T3
1500 FOR I6=2 TO T0
1510 I=T0+I6
1520 J=J+1
1530 B(I,J)=1
1540 IF B0=1 THEN B(I,J+T0)=-1
1550 NEXT I6
1560 REM ASSIGN THIRD SET OF CONSTRAINT COEFFICIENTS TO MATRIX B
1570 FOR I6=1 TO T0
1580 I=T1+I6
1590 J1=T0+I6
1600 J2=I
1610 J3=I+T0
1620 J4=I+T1
```

```
1630 B(I,I6)=1
1640 B(I,J1)=-1
1650 B(I,J2)=1
1660 B(I,J3)=-1
1670 B(I,J4)=-K0
1680 IF B0=2 THEN 1710
1690 J5=I+T2
1700 B(I,J5)=K0
1710 NEXT I6
1720 K2=2*K0
1730 J=T1
1740 FOR I6=2 TO T0
1750 I=T1+I6
1760 J=J+1
1770 J1=J+T0
1780 J2=J1+T0
1790 B(I,J)=-1
1800 B(I,J1)=1
1810 B(I,J2)=K2
1820 IF B0=2 THEN 1850
1830 J3=J2+T0
1840 B(I,J3)=-K2
1850 NEXT I6
1860 IF T0=2 THEN 2030
1870 J=T3
1880 FOR I6=3 TO T0
1890 I=T1+I6
1900 J=J+1
1910 B(I,J)=-K0
1920 IF B0=2 THEN 1940
1930 B(I,J+T0)=K0
1940 NEXT I6
1950 REM SET ANOTHER CONSTRAINT IF NECESSARY
1960 IF I2<>0 THEN 1980
1970 IF B0=2 THEN 2060
1980 M=M+1
1990 FOR J=1 TO N
2000 B(M,J)=0
2010 NEXT J
2020 IF B0=2 THEN 2040
2030 B(M,N)=-1
2040 B(M,T4)=1
2050 R(M)=I2
2060 REM    SET VECTOR R(.) TO CONTAIN RHS OF CONSTRAINTS
2070 R(1)=K0*(S(1)-I1)
2080 R(T0+1)=S(1)-I1
2090 FOR I=2 TO T0
2100 R(I)=K0*S(I)
2110 R(T0+I)=S(I)
2120 NEXT I
2130 R(T1+1)=R(1)-W0
2140 R(T1+2)=K0*(S(2)-S(1)+I1)
2150 IF T0<=2 THEN 2200
2160 FOR I6=3 TO T0
```

```
2170 I=T1+I6
2180 R(I)=K0*(S(I6)-S(I6-1))
2190 NEXT I6
2200 REM SET VECTOR Z(.) TO CONTAIN COEFFICIENTS OF OBJ FN.
2210 C4=V0-R0
2220 C5=R0*K0
2230 FOR J=1 TO T0
2240 Z(J)=-H0
2250 Z(T0+J)=-L0
2260 Z(T1+J)=-C4
2270 Z(T2+J)=-R0
2280 Z(T3+J)=-C1
2290 IF B0=2 THEN Z(T4+J)=-C2
2300 NEXT J
2310 Z(T4)=Z(T4)-C5
2320 IF B0=1 THEN Z(T1+T3)=Z(T1+T3)+C5
2330 REM    ASSIGN STRUCTURAL VARIABLES NONBASIC FOR A START
2340 FOR J=1 TO N
2350 T(J)=J
2360 NEXT J
2370 REM    MAKE SURE RHS IS NONNEGATIVE
2380 REM    ALSO NAME VARIABLES TO BE CONSISTENT WITH PROGRAM LP
2390 FOR I=1 TO T1
2400 U(I)=1
2410 NEXT I
2420 J=T1+1
2430 FOR I=J TO M
2440 U(I)=2
2450 NEXT I
2460 N3=0
2470 N9=0
2480 FOR I=1 TO M
2490 IF R(I) >= 0 THEN 2670
2500 IF U(I) = 2 THEN 2640
2510 N3=N3+1
2520 FOR L=1 TO M
2530 B(L,N+N3)=0
2540 NEXT L
2550 B(I,N+N3)=-1
2560 T(N+N3)=N+I
2570 N9=N9+1
2580 REM    MULTIPLY THIS CONSTRAINT WITH -1
2590 R(I)=-R(I)
2600 FOR J=1 TO N
2610 B(I,J)=-B(I,J)
2620 NEXT J
2630 GOTO 2720
2640 N9=N9+1
2650 A(I)=-I-M
2660 GOTO 2590
2670 IF U(I)=1 THEN 2710
2680 N9=N9+1
2690 A(I)=-I-M
2700 GOTO 2720
```

```
2710 A(I)=-I
2720 NEXT I
2730 REM     N3 = NO. OF SURPLUS VARIABLES
2740 N=N+N3
2750 REM     GET READY FOR PHASE I
2760 FOR J=1 TO N
2770 X(J)=0
2780 FOR I=1 TO M
2790 IF U(I) = 1 THEN 2810
2800 X(J)=X(J)-B(I,J)
2810 NEXT I
2820 NEXT J
2830 FOR I=1 TO M
2840 U(I)=0
2850 NEXT I
2860 REM     C A L L    S I M P L E X    A L G O R I T H M
2870 GOSUB 5000
2880 N=N-N3
2890 REM     RETRIEVE THE OPTIMAL SOLUTION
2900 PRINT
2910 FOR J=1 TO N
2920 T(J)=0
2930 NEXT J
2940 FOR I=1 TO M
2950 IF A(I)<0 THEN 2980
2960 IF A(I)>N THEN 2980
2970 T(A(I))=R(I)
2980 NEXT I
2990 REM CALCULATE ENDING INVENTORY, E(.)
3000 J=T3
3010 FOR I=1 TO T0
3020 J=J+1
3030 IF B0=2 THEN 3060
3040 E(I)=T(J)-T(J+T0)
3050 GOTO 3070
3060 E(I)=T(J)
3070 NEXT I
3080 REM CALCULATE PRODUCTION, D(.)  AND  WORK FORCE, W(.)
3090 D(1)=E(1)-I1+S(1)
3100 W(1)=W0+T(1)-T(T0+1)
3110 FOR I=2 TO T0
3120 D(I)=E(I)-E(I-1)+S(I)
3130 W(I)=W(I-1)+T(I)-T(T0+I)
3140 NEXT I
3150 REM CALCULATE NO. OF WORKERS HIRED/LAID OFF
3160 FOR I=1 TO T0
3170 H(I)=T(I)
3180 F(I)=T(T0+I)
3190 NEXT I
3200 PRINT
3210 PRINT
3220 PRINT "O P T I M A L    P R O D U C T I O N    S C H E D U L E"
3230 PRINT "-------------    --------------------    ---------------"
3240 GOSUB 4250
```

```
3250 PRINT
3260 PRINT "O P T I M A L    W O R K    F O R C E    L E V E L"
3270 PRINT "-------------    -------    ---------    ---------"
3280 GOSUB 4350
3290 GOSUB 3840
3300 PRINT
3310 PRINT "DO YOU WANT TO SPECIFY YOUR OWN PLAN (Y OR N)...";
3320 INPUT Y$
3330 IF Y$ = "N" THEN 4190
3340 IF B0=2 THEN 3390
3350 PRINT
3360 PRINT "DO YOU WANT TO USE THE SAME BACKORDER COST (Y OR N)...";
3370 INPUT Y$
3380 IF Y$ = "Y" THEN 3420
3390 PRINT
3400 PRINT "ENTER UNIT BACKORDER COST PER PERIOD   ";
3410 INPUT C2
3420 PRINT
3430 PRINT "ENTER PRODUCTION PLAN"
3440 PRINT "---------------------"
3450 FOR I=1 TO T0
3460 PRINT "PRODUCTION IN PERIOD",I;"   ";
3470 INPUT D(I)
3480 NEXT I
3490 PRINT
3500 PRINT "ENTER WORK FORCE LEVEL"
3510 PRINT "----------------------"
3520 FOR I=1 TO T0
3530 PRINT "NUMBER OF WORKERS IN PERIOD",I;"   ";
3540 INPUT W(I)
3550 NEXT I
3560 E(1)=I1+D(1)-S(1)
3570 FOR I=2 TO T0
3580 E(I)=E(I-1)+D(I)-S(I)
3590 NEXT I
3600 PRINT
3610 PRINT "P R O D U C T I O N    S C H E D U L E"
3620 PRINT "-------------------    ---------------"
3630 GOSUB 4250
3640 IF W(1) > W0 THEN 3680
3650 F(1)=W0-W(1)
3660 H(1)=0
3670 GOTO 3700
3680 H(1)=W(1)-W0
3690 F(1)=0
3700 FOR I=2 TO T0
3710 IF W(I) > W(I-1) THEN 3750
3720 F(I)=W(I-1)-W(I)
3730 H(I)=0
3740 GOTO 3770
3750 H(I)=W(I)-W(I-1)
3760 F(I)=0
3770 NEXT I
3780 PRINT
```

```
3790 PRINT "W O R K     F O R C E     L E V E L"
3800 PRINT "-------     ---------     ---------"
3810 GOSUB 4350
3820 GOSUB 3840
3830 GOTO 4160
3840 FOR I=1 TO 6
3850 T(I)=0
3860 NEXT I
3870 FOR I=1 TO T0
3880 T(1)=T(1)+H0*H(I)
3890 T(2)=T(2)+L0*F(I)
3900 T(3)=T(3)+R0*W(I)
3910 A1=K0*D(I)
3920 IF A1 <= W(I) THEN 3940
3930 T(4)=T(4)+V0*(A1-W(I))
3940 IF E(I) < 0 THEN 3970
3950 T(5)=T(5)+C1*E(I)
3960 GOTO 3980
3970 T(6)=T(6)-C2*E(I)
3980 NEXT I
3990 T(13)=T(1)+T(2)+T(3)+T(4)+T(5)+T(6)
4000 FOR I=1 TO 6
4010 T(I+6)=T(I)/T(13)*100
4020 NEXT I
4030 PRINT
4040 PRINT "C O S T     C O M P O N E N T S"
4050 PRINT "-------     ------------------"
4060 PRINT
4070 PRINT USING F$(4),"$$$$$$$ HIRING COST",T(1),T(7)
4080 PRINT USING F$(4),"$$$$$$$ LAYOFF COST",T(2),T(8)
4090 PRINT USING F$(4),"$$$ REGULAR PAYROLL",T(3),T(9)
4100 PRINT USING F$(4),"$$$$$ OVERTIME COST",T(4),T(10)
4110 PRINT USING F$(4),"$$$$ INVENTORY COST",T(5),T(11)
4120 PRINT USING F$(4),"$$$$$ SHORTAGE COST",T(6),T(12)
4130 PRINT
4140 PRINT USING F$(4),"T O T A L     C O S T",T(13)
4150 RETURN
4160 FOR I=1 TO 5
4170 PRINT
4180 NEXT I
4190 PRINT "ENTER 1(SHOW LP SOLN), 2(TRY NEW PLAN), 3(RESTART), 4(EXIT).."
4200 INPUT L2
4210 IF L2=1 THEN 2890
4220 IF L2=2 THEN 3340
4230 IF L2=3 THEN 470
4240 GOTO 9999
4250 REM     PRINT PRODUCTION SCHEDULE
4260 I=0
4270 PRINT F$(0)
4280 PRINT F$(1)
4290 PRINT USING F$(2),I,I1
4300 FOR I=1 TO T0
4310 PRINT USING F$(2),I,E(I),S(I),D(I)
4320 NEXT I
```

```
4330 PRINT
4340 RETURN
4350 REM     PRINT WORK FORCE LEVEL/HIRING/LAYOFF
4360 I=0
4370 PRINT F$(3)
4380 PRINT USING F$(2),I,W0
4390 FOR I=1 TO T0
4400 PRINT USING F$(2),I,W(I),H(I),F(I)
4410 NEXT I
4420 PRINT
4430 PRINT
4440 RETURN
4450 REM INITIALIZE FORMAT VARIABLES
4460 F$(0)="              ENDING"
4470 F$(1)="PERIOD     INVENTORY      SALES      PRODUCTION"
4480 F$(2)="###     -#######.##   -#######.##   -#######.##"
4490 F$(3)="PERIOD    WORK FORCE       HIRE         LAYOFF"
4500 F$(4)="##################  #########.##   ####.### %"
4510 RETURN
```

III. Transportation-Type Problems

Many LP problems can be formulated as a transportation problem, which is to minimize the sum of the transportation costs between the supply points and the demand points. Applications that are frequently cited in operations research literature include the Transshipment problem, Assignment problem (i.e., problem of assigning people to various tasks), Caterer problem, Warehouse problem, Equipment Replacement problem, Shortest Route problem, Longest Route problem, and Bowman production scheduling problem [2, 3]. All of these models have network structures and therefore can be solved very efficiently using special techniques in LP rather than using the Standard Simplex method.

The Classical Transportation Problem

Let m be the number of supply points; n be the number of demand points; S_i, $i=1,\ldots,m$ be the supply amounts for the m supply points; and D_j, $j=1,\ldots,n$ be the corresponding demand amounts for the n demand points. Also let c_{ij} be the unit transportation cost from supply point i to demand point j, and x_{ij} be the amount to be shipped from i to j. Then the problem is to

(14) $\quad$ minimize $\quad \sum_{i=1}^{m} \sum_{j=1}^{n} c_{ij} x_{ij}$

subject to

(15) $\quad \sum_{j=1}^{n} x_{ij} = S_i$, $i=1,\ldots,m$

(16) $\quad \sum_{i=1}^{m} x_{ij} = D_j$, $j=1,\ldots,n$

(17) and $\quad x_{ij} \geq 0$ for all i and j.

This requires that the sum of the demands at various locations be equal to the sum of all supplies. It implies that one of the above constraints is redundant, and therefore can be dropped. As a result, the problem can be solved using the Simplex method with no more than m+n-1 values of x_{ij} being positive. However, because of the special structure of the problem, it can be solved much more efficiently using the Transportation Algorithm [4, 5, 8, 12].

The Program TRANSP

For problems in which the sum of the supplies is greater than the sum of the demands, a dummy demand point called "NEW-DEMAND" is created by TRANSP to handle the excess supply. If the sum of the demands is greater than the sum of the supplies, a new dummy supply point called "NEW-SUPPLY" is created by TRANSP to handle the excess demand. A unit cost of zero is assigned to the route between the dummy demand point or the dummy supply point and various locations. When the total demand equals the total supply, no dummy location is added. The program is command-driven. After all data are entered, the command "GO" will solve for the optimal solution. Commands are also provided for editing the supply and demand amounts and the unit transportation costs.

The program determines the initial basic solution using the row-minimum method with standard perturbation technique to avoid degeneracy [5]. It utilizes the Predecessor Index Method [6] to locate the stepping-stone paths during each iteration of the algorithm. The current setting of the maximum number of supply points (m) and the demand points (n) is 10 but can easily be expanded by merely changing the sizes of the arrays and certain parameters. See Section IV, "Dimensional Specifications," for more information.

INPUT
-----
1. Number of supply points (>=2)
2. Number of demand points (>=2)
3. Names of the supply and demand points
4. The supply and demand amounts (>0)
5. The unit transportation costs from the supply points to the demand points

OUTPUT

The default optimal solution includes the amounts to be shipped from the supply points to the demand points and the minimum total cost. The shadow prices at the optimal solution can be obtained by entering the command "SHADOWON" before issuing the "GO" command.

EXAMPLE

Consider the transportation problem below with three supply points and four demand points in which the total demand equals the total supply of 2200 units. The following table shows the unit transportation costs together with the supply and demand amounts.

To From	Dallas	Portland	Milwaukee	St. Louis	Supply
Chicago	1.0	1.2	0.4	0.7	500
New York	1.4	1.8	1.1	1.2	1000
Los Angeles	1.3	0.6	1.2	1.1	700
Demand	600	700	600	300	2200

The optimal solution is to ship 500 units from Chicago to Milwaukee; 100, 300, and 600 units from New York to Milwaukee, St. Louis, and Dallas, respectively; and 700 units from Los Angeles to Portland. The optimal cost is 1930. If St. Louis demand increases from 300 units to 500 units and at the same time the unit cost from New York to Dallas is increased from 1.4 to 1.9, then the optimal cost becomes 2010. The example terminal session shows that the total demand exceeds the total supply by 200 units. Hence a new dummy supply point called "NEW-SUPPLY" is introduced because Dallas will be short of supply by 200 units.

As a final note, the program TRANSP was designed to solve transportation-type problems of relatively small sizes. For readers who are interested in solving large-scale problems, reference 1 refers to one of the fastest transportation codes known in existence today.

```
TRANSP EXAMPLE TERMINAL SESSION
================================

T R A N S P O R T A T I O N    P R O B L E M
---------------------------     -------------

NEED INTRODUCTION (Y OR N)...!N

ENTER NUMBER OF SUPPLY POINTS (>=2)  !3

ENTER NUMBER OF DEMAND POINTS (>=2)  !4

ENTER THE NAME OF EACH SUPPLY POINT
-----------------------------------
SUPPLY POINT          1       !CHICAGO
SUPPLY POINT          2       !NEW YORK
SUPPLY POINT          3       !LOS ANGELES

ENTER THE NAME OF EACH DEMAND POINT
-----------------------------------
DEMAND POINT          1       !DALLAS
DEMAND POINT          2       !PORTLAND
DEMAND POINT          3       !MILWAUKEE
DEMAND POINT          4       !ST.LOUIS

ENTER AVAILABLE SUPPLY FOR EACH SUPPLY POINT
--------------------------------------------
CHICAGO                       !500
NEW YORK                      !1000
LOS ANGELES                   !700

ENTER DEMAND FOR EACH DEMAND POINT
----------------------------------
DALLAS                        !600
PORTLAND                      !700
MILWAUKEE                     !600
ST.LOUIS                      !300

ENTER UNIT TRANSPORTATION COST FOR ALL ROUTES
---------------------------------------------
FROM CHICAGO          TO DALLAS!1
FROM CHICAGO          TO PORTLAND!1.2
FROM CHICAGO          TO MILWAUKEE!.4
FROM CHICAGO          TO ST.LOUIS!.7
FROM NEW YORK         TO DALLAS!1.4
FROM NEW YORK         TO PORTLAND!1.8
FROM NEW YORK         TO MILWAUKEE!1.1
FROM NEW YORK         TO ST.LOUIS!1.2
FROM LOS ANGELES      TO DALLAS!1.3
FROM LOS ANGELES      TO PORTLAND!.6
FROM LOS ANGELES      TO MILWAUKEE!1.2
FROM LOS ANGELES      TO ST.LOUIS!1.1
```

```
PLEASE ENTER COMMAND OR TYPE HELP

COMMAND --> !HELP

AVAILABLE COMMANDS
==================
HELP       - PRINT THIS MESSAGE
LDEMAND    - LIST ALL DEMANDS
LSUPPLY    - LIST ALL SUPPLIES
LCOST      - LIST ALL UNIT TRANSPORTATION COSTS
CDEMAND    - CHANGE A DEMAND AMOUNT
CSUPPLY    - CHANGE A SUPPLY AMOUNT
CCOST      - CHANGE A UNIT COST
SHADOWON   - PRINT SHADOW PRICES AT SOLUTION
SHADOWOFF  - SUPPRESS PRINTING OF SHADOW PRICES
GO         - FIND THE OPTIMAL SOLUTION
RESTART    - START A NEW PROBLEM
EXIT       - EXIT FROM THE PROGRAM

COMMAND --> !GO


O P T I M A L    S O L U T I O N
-------------    ---------------
FROM CHICAGO            TO MILWAUKEE     500
FROM NEW YORK           TO MILWAUKEE     100
FROM NEW YORK           TO ST.LOUIS      300
FROM NEW YORK           TO DALLAS        600
FROM LOS ANGELES        TO PORTLAND      700
FROM LOS ANGELES        TO DALLAS          0

OPTIMAL COST =      1930
************

COMMAND --> !SHADOWON


COMMAND --> !CDEMAND

ENTER DEMAND POINT    !ST.LOUIS
ENTER DEMAND AMOUNT   !500

COMMAND --> !CCOST

ENTER SUPPLY POINT    !NEW YORK
ENTER DEMAND POINT    !DALLAS
ENTER UNIT TRANSPORTATION COST   !1.9

COMMAND --> !LDEMAND

DALLAS         600
PORTLAND       700
MILWAUKEE      600
```

```
ST.LOUIS                500

TOTAL DEMAND =  2400

COMMAND --> !GO


OPTIMAL    SOLUTION
-------------      ----------------
FROM CHICAGO            TO MILWAUKEE      100
FROM NEW YORK           TO MILWAUKEE      500
FROM NEW YORK           TO ST.LOUIS       500
FROM NEW-SUPPLY         TO PORTLAND       3.637978807092E-12
FROM LOS ANGELES        TO PORTLAND       700
FROM CHICAGO            TO DALLAS         400
FROM NEW-SUPPLY         TO DALLAS         200

OPTIMAL COST =      2010
*************

SHADOW    PRICES
--------------------------

FROM CHICAGO      TO  DALLAS    0
FROM CHICAGO      TO  PORTLAND  -.2
FROM CHICAGO      TO  MILWAUKEE 0
FROM CHICAGO      TO  ST.LOUIS  -.2
FROM NEW YORK     TO  DALLAS    -.2
FROM NEW YORK     TO  PORTLAND  -.1
FROM NEW YORK     TO  MILWAUKEE 0
FROM NEW YORK     TO  ST.LOUIS  0
FROM LOS ANGELES  TO  DALLAS    -.7
FROM LOS ANGELES  TO  PORTLAND  0
FROM LOS ANGELES  TO  MILWAUKEE -1.2
FROM LOS ANGELES  TO  ST.LOUIS  -1
FROM NEW-SUPPLY   TO  DALLAS    0
FROM NEW-SUPPLY   TO  PORTLAND  0
FROM NEW-SUPPLY   TO  MILWAUKEE -.6
FROM NEW-SUPPLY   TO  ST.LOUIS  -.5

COMMAND --> !EXIT
```

```
10 REM    T R A N S P O R T A T I O N     P R O B L E M
20 REM    VIA THE MODI METHOD(ROW-COLUMN SUM METHOD)
30 REM
40 DIM D$(11),S$(11)
50 DIM A(22),B(22),C(11,11),D(11),S(11),K(11),R(11)
60 DIM U(11),V(11),E(11),F(11),H(22),T(22),X(22)
70 REM
80 REM   M0 = MAXIMUM NUMBER OF SUPPLY POINTS.
90 REM   N0 = MAXIMUM NUMBER OF DEMAND POINTS.
100 M0=10
110 N0=10
120 PRINT
130 PRINT
140 PRINT "T R A N S P O R T A T I O N    P R O B L E M"
150 PRINT "--------------------------    -------------"
160 PRINT
170 PRINT "NEED INTRODUCTION (Y OR N)...";
180 INPUT Y$
190 IF Y$<>"Y" THEN 410
200 PRINT
210 PRINT "    THIS PROGRAM DETERMINES THE OPTIMAL SOLUTION OF"
220 PRINT "    THE TRANSPORTATION TYPE PROBLEMS USING THE"
230 PRINT "    WELL KNOWN MODI METHOD.   THE TOTAL DEMAND NEEDS"
240 PRINT "    NOT EQUAL THE TOTAL SUPPLY."
250 PRINT
260 PRINT "    MAXIMUM NUMBER OF SUPPLY POINTS =",M0
270 PRINT "    MAXIMUM NUMBER OF DEMAND POINTS =",N0
280 PRINT
290 PRINT "INPUT REQUIREMENTS"
300 PRINT "------------------"
310 PRINT "    1. NUMBER OF SUPPLY POINTS"
320 PRINT "    2. NUMBER OF DEMAND POINTS"
330 PRINT "    3. NAMES OF SUPPLY POINTS"
340 PRINT "    4. NAMES OF DEMAND POINTS"
350 PRINT "    5. QUANTITIES OF SUPPLIES AND DEMANDS"
360 PRINT "    6. UNIT TRANSPORTATION COST FROM SUPPLY POINTS"
370 PRINT "       TO DEMAND POINTS."
380 REM
390 REM   I N P U T     S E C T I O N
400 Q0=0
410 PRINT
420 PRINT "ENTER NUMBER OF SUPPLY POINTS (>=2)  ";
430 INPUT M9
440 IF M9<=1 THEN 410
450 IF M9<=M0 THEN 480
460 PRINT "*** ERROR *** MAX ALLOWED",M0
470 GOTO 410
480 PRINT
490 PRINT "ENTER NUMBER OF DEMAND POINTS (>=2)  ";
500 INPUT N9
510 IF N9<=1 THEN 480
520 IF N9<=N0 THEN 550
530 PRINT "*** ERROR *** MAX ALLOWED",N0
540 GOTO 480
```

```
550 PRINT
560 PRINT "ENTER THE NAME OF EACH SUPPLY POINT"
570 PRINT "---------------------------------"
580 FOR I=1 TO M9
590 PRINT "SUPPLY POINT ",I;"     ";
600 INPUT S$(I)
610 NEXT I
620 PRINT
630 PRINT "ENTER THE NAME OF EACH DEMAND POINT"
640 PRINT "---------------------------------"
650 FOR J=1 TO N9
660 PRINT "DEMAND POINT ",J;"     ";
670 INPUT D$(J)
680 NEXT J
690 PRINT
700 PRINT "ENTER AVAILABLE SUPPLY FOR EACH SUPPLY POINT"
710 PRINT "---------------------------------------------"
720 FOR I=1 TO M9
730 PRINT S$(I),"     ";
740 INPUT S(I)
750 IF S(I) > 0 THEN 780
760 PRINT "*** ERROR, PLEASE RE-ENTER ***"
770 GOTO 730
780 NEXT I
790 PRINT
800 PRINT "ENTER DEMAND FOR EACH DEMAND POINT"
810 PRINT "----------------------------------"
820 FOR J=1 TO N9
830 PRINT D$(J),"     ";
840 INPUT D(J)
850 IF D(J) > 0 THEN 880
860 PRINT "*** ERROR, PLEASE RE-ENTER ***"
870 GOTO 830
880 NEXT J
890 REM
900 PRINT
910 PRINT "ENTER UNIT TRANSPORTATION COST FOR ALL ROUTES"
920 PRINT "---------------------------------------------"
930 FOR I=1 TO M9
940 FOR J=1 TO N9
950 PRINT "FROM ";S$(I),"   TO ";D$(J);
960 INPUT C(I,J)
970 NEXT J
980 NEXT I
990 PRINT
1000 PRINT "PLEASE ENTER COMMAND OR TYPE HELP"
1010 PRINT
1020 PRINT "COMMAND --> ";
1030 INPUT C$
1040 PRINT
1050 IF C$ <> "HELP" THEN 1080
1060 GOSUB 2030
1070 GOTO 1010
1080 IF C$ <> "LDEMAND" THEN 1110
```

```
1090 GOSUB 2180
1100 GOTO 1010
1110 IF C$ <> "LSUPPLY" THEN 1140
1120 GOSUB 2250
1130 GOTO 1010
1140 IF C$ <> "LCOST" THEN 1170
1150 GOSUB 2320
1160 GOTO 1010
1170 IF C$ <> "CDEMAND" THEN 1200
1180 GOSUB 2480
1190 GOTO 1010
1200 IF C$ <> "CSUPPLY" THEN 1230
1210 GOSUB 2540
1220 GOTO 1010
1230 IF C$ <> "CCOST" THEN 1260
1240 GOSUB 2600
1250 GOTO 1010
1260 IF C$ <> "SHADOWON" THEN 1290
1270 Q0=1
1280 GOTO 1010
1290 IF C$ <> "SHADOWOFF" THEN 1320
1300 Q0=0
1310 GOTO 1010
1320 IF C$ <> "GO" THEN 1350
1330 GOSUB 1400
1340 GOTO 1010
1350 IF C$ <> "RESTART" THEN 1370
1360 GOTO 120
1370 IF C$ = "EXIT" THEN 5300
1380 PRINT "*** ERROR *** PLEASE TRY AGAIN !"
1390 GOTO 1010
1400 REM
1410 REM    G O      C O M M A N D
1420 GOSUB 2430
1430 GOSUB 2380
1440 REM
1450 REM  CHECK WHICH IS GREATER, SUPPLY OR DEMAND
1460 IF S0=D0 THEN 1640
1470 IF S0>D0 THEN 1560
1480 M=M9+1
1490 N=N9
1500 S(M)=D0-S0
1510 S$(M)="NEW-SUPPLY"
1520 FOR J=1 TO N
1530 C(M,J)=0
1540 NEXT J
1550 GOTO 1660
1560 M=M9
1570 N=N9+1
1580 D(N)=S0-D0
1590 D$(N)="NEW-DEMAND"
1600 FOR I=1 TO M
1610 C(I,N)=0
1620 NEXT I
```

```
1630 GOTO 1660
1640 M=M9
1650 N=N9
1660 M1=M+N-1
1670 REM
1680 REM   CALL THE TRANSPORTATION ALGORITHM
1690 REM   ---------------------------------
1700 GOSUB 2880
1710 GOSUB 3460
1720 IF I9=0 THEN 1760
1730 GOSUB 4220
1740 GOSUB 5000
1750 GOTO 1710
1760 PRINT
1770 PRINT
1780 PRINT "O P T I M A L     S O L U T I O N"
1790 PRINT "-------------    ----------------"
1800 Z=0
1810 FOR L=1 TO M1
1820 I=A(L)
1830 J=B(L)
1840 Z=Z+X(L)*C(I,J)
1850 PRINT "FROM ";S$(I),"   TO ";D$(J),X(L)
1860 NEXT L
1870 PRINT
1880 PRINT "OPTIMAL COST =",Z
1890 PRINT "*************"
1900 IF Q0 = 0 THEN 2010
1910 PRINT
1920 PRINT "S H A D O W     P R I C E S"
1930 PRINT "--------------------------"
1940 PRINT
1950 FOR I=1 TO M
1960 FOR J=1 TO N
1970 G1=U(I)+V(J)-C(I,J)
1980 PRINT "FROM ";S$(I);"   TO ";D$(J);"   ";G1
1990 NEXT J
2000 NEXT I
2010 RETURN
2020 REM
2030 PRINT "AVAILABLE COMMANDS"
2040 PRINT "=================="
2050 PRINT "HELP     - PRINT THIS MESSAGE"
2060 PRINT "LDEMAND  - LIST ALL DEMANDS"
2070 PRINT "LSUPPLY  - LIST ALL SUPPLIES"
2080 PRINT "LCOST    - LIST ALL UNIT TRANSPORTATION COSTS"
2090 PRINT "CDEMAND  - CHANGE A DEMAND AMOUNT"
2100 PRINT "CSUPPLY  - CHANGE A SUPPLY AMOUNT"
2110 PRINT "CCOST    - CHANGE A UNIT COST"
2120 PRINT "SHADOWON - PRINT SHADOW PRICES AT SOLUTION"
2130 PRINT "SHADOWOFF- SUPPRESS PRINTING OF SHADOW PRICES"
2140 PRINT "GO       - FIND THE OPTIMAL SOLUTION"
2150 PRINT "RESTART  - START A NEW PROBLEM"
2160 PRINT "EXIT     - EXIT FROM THE PROGRAM"
```

```
2170 RETURN
2180 FOR J=1 TO N9
2190 PRINT D$(J),D(J)
2200 NEXT J
2210 GOSUB 2380
2220 PRINT
2230 PRINT "TOTAL DEMAND =   ";D0
2240 RETURN
2250 FOR I=1 TO M9
2260 PRINT S$(I),S(I)
2270 NEXT I
2280 GOSUB 2430
2290 PRINT
2300 PRINT "TOTAL SUPPLY =   ";S0
2310 RETURN
2320 FOR I=1 TO M9
2330 FOR J=1 TO N9
2340 PRINT "FROM ";S$(I);"   TO   ";D$(J);"   ";C(I,J)
2350 NEXT J
2360 NEXT I
2370 RETURN
2380 D0=0
2390 FOR J=1 TO N9
2400 D0=D0+D(J)
2410 NEXT J
2420 RETURN
2430 S0=0
2440 FOR I=1 TO M9
2450 S0=S0+S(I)
2460 NEXT I
2470 RETURN
2480 REM CDEMAND   COMMAND
2490 GOSUB 2770
2500 IF J <= 0 THEN 2530
2510 PRINT "ENTER DEMAND AMOUNT   ";
2520 INPUT D(J)
2530 RETURN
2540 REM   CSUPPLY   COMMAND
2550 GOSUB 2690
2560 IF I <= 0 THEN 2590
2570 PRINT "ENTER SUPPLY AMOUNT   ";
2580 INPUT S(I)
2590 RETURN
2600 REM   CCOST   COMMAND
2610 GOSUB 2690
2620 IF I > 0 THEN 2640
2630 GOTO 2680
2640 GOSUB 2770
2650 IF J <= 0 THEN 2680
2660 PRINT "ENTER UNIT TRANSPORTATION COST   ";
2670 INPUT C(I,J)
2680 RETURN
2690 PRINT "ENTER SUPPLY POINT   ";
2700 INPUT Y$
```

```
2710 FOR I=1 TO M9
2720 IF Y$ = S$(I) THEN 2760
2730 NEXT I
2740 I=0
2750 GOSUB 2850
2760 RETURN
2770 PRINT "ENTER DEMAND POINT   ";
2780 INPUT Y$
2790 FOR J=1 TO N9
2800 IF Y$ = D$(J) THEN 2840
2810 NEXT J
2820 J=0
2830 GOSUB 2850
2840 RETURN
2850 PRINT
2860 PRINT "*** ERROR ***    COMMAND ABORTED !!!"
2870 RETURN
2880 REM
2890 REM  DETERMINE INITIAL BASIC SOLUTION USING ROW-MINIMUM
2900 REM  METHOD WITH STANDARD PERTURBATION TECHNIQUE TO
2910 REM  ELIMINATE DEGENERACY.
2920 E7=0.001
2930 FOR I=1 TO M
2940 R(I)=E7
2950 U(I)=S(I)+E7
2960 NEXT I
2970 FOR J=1 TO N-1
2980 K(J)=0
2990 V(J)=D(J)
3000 NEXT J
3010 K(N)=M*E7
3020 V(N)=D(N)+K(N)
3030 REM
3040 REM  DETERMINE MINIMUM COST IN THE REMAINING COLUMNS OF THE ROW
3050 L1=0
3060 FOR L=1 TO M
3070 IF U(L)<>0 THEN 3100
3080 NEXT L
3090 GOTO 3350
3100 I=L
3110 C0=1.E20
3120 FOR J=1 TO N
3130 IF V(J)=0 THEN 3170
3140 IF C0<=C(I,J) THEN 3170
3150 C0=C(I,J)
3160 L=J
3170 NEXT J
3180 J=L
3190 L1=L1+1
3200 A(L1)=I
3210 B(L1)=J
3220 IF U(I) > V(J) THEN 3290
3230 X(L1)=U(I)
3240 V(J)=V(J)-U(I)
```

```
3250 U(I)=0
3260 K(J)=K(J)-R(I)
3270 T(L1)=R(I)
3280 GOTO 3060
3290 X(L1)=V(J)
3300 U(I)=U(I)-V(J)
3310 V(J)=0
3320 R(I)=R(I)-K(J)
3330 T(L1)=K(J)
3340 GOTO 3110
3350 IF L1=M1 THEN 3390
3360 PRINT
3370 PRINT "* ERROR * # OF BASIC CELLS NOT EQUAL TO M+N-1"
3380 STOP
3390 REM UPDATE THE BASIC CELL VALUES
3400 FOR L=1 TO M1
3410 X(L)=X(L)-T(L)
3420 IF X(L) >= E7 THEN 3440
3430 X(L)=0
3440 NEXT L
3450 RETURN
3460 REM
3470 REM DETERMINE DUAL VARIABLES AND PREDECESSOR INDEXES
3480 REM SIMULTANEOUSLY
3490 REM
3500 REM INITIALIZE SOME INDEXES
3510 FOR I=1 TO M
3520 R(I)=-1
3530 NEXT I
3540 FOR J=1 TO N
3550 K(J)=-1
3560 NEXT J
3570 M3=1
3580 N3=0
3590 REM
3600 REM INITIALIZE THE FIRST ROW INDEX FROM A(1)
3610 I=A(1)
3620 U(I)=0
3630 R(I)=0
3640 FOR L=1 TO M1
3650 IF A(L) <> I THEN 3700
3660 J=B(L)
3670 K(J)=I
3680 V(J)=C(I,J)
3690 N3=N3+1
3700 NEXT L
3710 REM
3720 REM ASSIGN MORE INDEXES TO THE ROWS
3730 FOR J=1 TO N
3740 IF K(J) < 0 THEN 3830
3750 FOR L=1 TO M1
3760 IF B(L)<>J THEN 3820
3770 IF R(A(L)) >= 0 THEN 3820
3780 I=A(L)
```

```
3790 R(I)=J
3800 U(I)=C(I,J)-V(J)
3810 M3=M3+1
3820 NEXT L
3830 NEXT J
3840 L0=1
3850 GOTO 4030
3860 REM
3870 REM
3880 REM ASSIGN MORE INDEXES TO THE COLUMNS
3890 FOR I=1 TO M
3900 IF R(I) <= 0 THEN 3990
3910 FOR L=1 TO M1
3920 IF A(L) <> I THEN 3980
3930 IF K(B(L)) > 0 THEN 3980
3940 J=B(L)
3950 K(J)=I
3960 V(J)=C(I,J)-U(I)
3970 N3=N3+1
3980 NEXT L
3990 NEXT I
4000 L0=2
4010 REM
4020 REM CHECK IF WE HAVE FOUND ALL VALUES OR NOT
4030 IF N3 <> N THEN 4050
4040 IF M3 = M THEN 4070
4050 IF L0=1 THEN 3870
4060 GOTO 3710
4070 REM
4080 REM   DETERMINE ENTERING BASIC VARIABLE I9,J9
4090 G0=0.0
4100 I9=0
4110 FOR I=1 TO M
4120 FOR J=1 TO N
4130 G1=U(I)+V(J)-C(I,J)
4140 IF G1<=1.E-8 THEN 4190
4150 IF G1<=G0 THEN 4190
4160 G0=G1
4170 I9=I
4180 J9=J
4190 NEXT J
4200 NEXT I
4210 RETURN
4220 REM
4230 REM   F I N D    S T E P P I N G - S T O N E    P A T H S
4240 REM
4250 REM
4260 REM   TRACE FLAGGED BACKWARD PATH, H(.)
4270 H(1)=I9
4280 L0=I9
4290 L=0
4300 L=L+1
4310 IF H(L) < 0 THEN 4370
4320 IF R(L0)=0 THEN 4410
```

```
4330 H(L+1)=-R(L0)
4340 R(L0)=-R(L0)
4350 L0=-R(L0)
4360 GOTO 4300
4370 H(L+1)=K(L0)
4380 K(L0)=-K(L0)
4390 L0=-K(L0)
4400 GOTO 4300
4410 H0=L
4420 REM
4430 REM    TRACE BACKWARD PATH, T(.)
4440 T(1)=-J9
4450 L0=J9
4460 L=0
4470 L=L+1
4480 IF T(L)<0 THEN 4550
4490 IF R(L0)<=0 THEN 4600
4500 T(L+1)=-R(L0)
4510 R(L0)=-R(L0)
4520 L0=-R(L0)
4530 GOTO 4470
4540 REM    ........
4550 IF K(L0)<=0 THEN 4600
4560 T(L+1)=K(L0)
4570 K(L0)=-K(L0)
4580 L0=-K(L0)
4590 GOTO 4470
4600 T0=L
4610 REM
4620 REM    ELIMINATE PORTION OF PATHS WHICH COINCIDE.
4630 L0=T(T0)
4640 FOR L=1 TO H0
4650 IF H(L)=L0 THEN 4670
4660 NEXT L
4670 H0=L
4680 REM
4690 REM    TRACE PATH H(.) FOR STONES
4700 REM    E(.),F(.) STORE ROW AND COLUMN INDEXES, RESPECTIVELY
4710 E(1)=I9
4720 F(1)=J9
4730 L1=1
4740 H0=H0-1
4750 IF H0<=0 THEN 4870
4760 FOR L=1 TO H0
4770 L1=L1+1
4780 IF H(L)>0 THEN 4820
4790 E(L1)=H(L+1)
4800 F(L1)=-H(L)
4810 GOTO 4840
4820 E(L1)=H(L)
4830 F(L1)=-H(L+1)
4840 NEXT L
4850 REM
4860 REM    TRACE PATH T(.) FOR STONES
```

```
4870 T0=T0-1
4880 IF T0<=0 THEN 4990
4890 FOR L=1 TO T0
4900 L1=L1+1
4910 J=T0-L+2
4920 IF T(J)<0 THEN 4960
4930 E(L1)=T(J)
4940 F(L1)=-T(J-1)
4950 GOTO 4980
4960 E(L1)=T(J-1)
4970 F(L1)=-T(J)
4980 NEXT L
4990 RETURN
5000 REM
5010 REM   IMPROVE THE BASIC SOLUTION
5020 REM
5030 FOR L=2 TO L1
5040 FOR J=1 TO M1
5050 IF E(L)<>A(J) THEN 5090
5060 IF F(L)<>B(J) THEN 5090
5070 H(L)=J
5080 GOTO 5100
5090 NEXT J
5100 NEXT L
5110 X0=1.E20
5120 L2=L1/2
5130 FOR L=1 TO L2
5140 L3=L*2
5150 IF X(H(L3))>=X0 THEN 5180
5160 X0=X(H(L3))
5170 L4=H(L3)
5180 NEXT L
5190 X(L4)=X(L4)+X0
5200 A(L4)=I9
5210 B(L4)=J9
5220 FOR L=1 TO L2
5230 I=L*2
5240 J=I-1
5250 IF J=1 THEN 5270
5260 X(H(J))=X(H(J))+X0
5270 X(H(I))=X(H(I))-X0
5280 NEXT L
5290 RETURN
5300 END
```

IV. DIMENSIONAL SPECIFICATIONS

Program LP

Let m be the allowable number of constraints and n be the number of structural variables plus the number of >= constraints. Currently, m is set to 15, and n is set to 20. To accommodate other sizes, the following statements must be adjusted:

Line No.
----
30 DIM M$(m),N$(n)
40 A(m),B(m,n),C(n),P(n),R(m),T(n),U(m),X(n),Z(n)
60 M0=m
70 N0=n

Program AGG

Let t be the maximum planning horizon length, which is currently set at 12. It can be increased or decreased easily by chaning a few statements in AGG. Define f=3t+1 and g=6t+2; then the following statements should be adjusted accordingly:

Line No.
----
60 DIM A(f),B(f,g),P(g),R(f),T(g),U(f),X(g),Z(g)
70 S(t),W(t),D(t),E(t),H(t),F(t),F$(4)
90 T9=t

Program TRANSP

The maximum number of supply points, m, and the maximum number of demand points, n, are currently set to 10. To accommodate larger problems, define s=m+1, d=n+1, and r=d+s-1; then the following statements in TRANSP should be modified as follows:

Line No.
----
40 DIM D$(d),S$(s)
50 DIM A(r),B(r),C(s,d),D(d),S(s),K(d),R(s)
60 DIM U(s),V(d),E(s),F(d),H(r),T(r),X(r)
100 M0=m
110 N0=n

V. REFERENCES

1. Barr, R.S., F. Glover, and D. Klingman, "A New Optimization Method for Large Scale Fixed Charge Transportation Problem, OPERATIONS RESEARCH, 29, No.3 (1981), 448-63.

2. Bowman, E.H., "Production Scheduling by the Transportation Method of Linear Programming," JOURNAL OF THE OPERATIONS RESEARCH SOCIETY, 1956, pp. 100-103.

3. Chase, R.B., and N.J. Aquilano, PRODUCTION AND OPERATIONS MANAGEMENT, Richard D. Irwin, Homewood, Ill., 1973.

4. Dantzig, G.B., LINEAR PROGRAMMING AND EXTENSIONS, Princeton University Press, Princeton, N.J., 1963.

5. Hadley, G., LINEAR PROGRAMMING, Addison-Wesley, Reading, Mass., 1962.

6. Glover, F., and D. Klingman, "Locating Stepping-Stone Paths in Distribution Problems via the Predecessor Index Method," TRANSPORTATION SCIENCE, 1970, pp. 220-25.

7. Hanssmann, F., and S.W. Hess, "A Linear Programming Approach to Production and Employment Scheduling," MANAGEMENT TECHNOLOGY, 1960, pp. 46-52.

8. Hillier, F.S., and G.J. Lieberman, OPERATIONS RESEARCH, Holden-Day, San Francisco, 1974.

9. Loomba, N.P., and E. Turban, APPLIED PROGRAMMING FOR MANAGEMENT, Holt, Reinhart & Winston, New York, 1974.

10. Schuermann, A.C., and N.P. Kannan, "A Production Forecasting and Planning System for Dairy Processing," COMPUTERS & INDUSTRIAL ENGINEERING, 1978, pp. 153-58.

11. Srinivasan, V., and G.L. Thompson, "Benefit-Cost Analysis of Coding Techniques for the Primal Transportation Algorithm," Management Sciences Research Report No. 229, Graduate School of Industrial Administration, Carnegie-Mellon University, Pittsburgh, 1970.

12. Wagner, H.M., PRINCIPLES OF OPERATIONS RESEARCH, 2nd ed., Prentice-Hall, Englewood Cliffs, N.J., 1975.

CHAPTER 5

PROJECT MANAGEMENT (CPM/PERT)

I. The Concept of Project Management

  ○ Notation Used
  ○ Definitions

II. The Program CPM

  ○ Input
  ○ Output
  ○ Example
  ○ CPM Example Terminal Session
  ○ CPM Program Listing

III. The Program PERT

  ○ Theory
  ○ Input
  ○ Output
  ○ Example
  ○ PERT Example Terminal Session
  ○ PERT Program Listing

IV. Dimensional Specifications

V. References

During the past three decades, management has seen an increasing usage of project management techniques such as CPM (Critical Path Method) and PERT (Program Evaluation and Review Technique) for controlling all kinds of projects and also for planning, scheduling, and budgeting purposes.

In this chapter, we present two computer programs dealing with both CPM and PERT in their most fundamental forms. In section I, we discuss in general the concept of project management in terms of how to construct a project network. We also give various definitions and state the difference between CPM and PERT. Numerous textbooks on this subject are available (see, for example, [2, 3, 6, 8]). Section II describes the usage of the program CPM in details with an example. Section III discusses briefly the underlying theory of PERT and the assumptions used in the program. It also describes the operating procedure of the program with one more example terminal session. All terminal sessions discussed in this chapter refer to only one network with varying assumptions.

I. The Concept of Project Management

A project is a set of activities to be executed in some logical sequence to accomplish a goal or several goals. Each activity usually requires some type of resources and time. The resources may be manpower, machine-hours, materials, etc. The control of the various activities to meet the target dates and to keep track of a variety of resources and cost can be a full-time occupation and an important task.

The logical sequencing of various activities forms a project network or networks. Two traditional methods of constructing a project network are still widely used today. The first is called the Activity-on-Arrow network (or i-j notation) in which the activity is represented by an arrow drawn from the start event node to the finish event node of the activity. Therefore a network is a collection of such representation where activities are interrelated. Figure 1 shows an activity-on-arrow network with six activities (A1, A2, A3, A4, A5, A6) and five events (1, 2, 3, 4, 5). By virtue of the nature of the precedence relationship, the network must be acyclic. This means that no path within the network can form a loop.

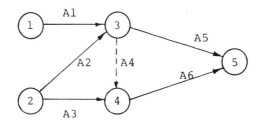

Figure 1. Activity-on-Arrow Network

The basic rule in constructing a network is that any pair of nodes may be connected by at most one arc. This is to facilitate machine computations of the schedule, since activities are referred to by their start and finish events and confusion can arise when we have parallel activities, as shown in Figure 2a. The alternative representation is to add a dummy event (k) and a dummy activity (i,k) with zero duration, as shown in Figure 2b.

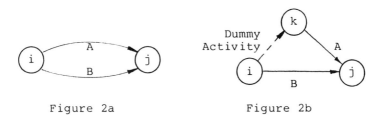

Figure 2a   Figure 2b

The second method of constructing the project network is called Precedence Diagramming (Precedence notation) or Activity-on-Node network. In this method, an activity is represented by a node and the arrow connecting the nodes represents the precedence relationship between them. Figure 3 shows an equivalent network of Figure 1 in Precedence notation. It should be noted that activity A4 in Figure 1 is a dummy activity and is eliminated by this method of constructing the network.

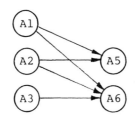

Figure 3. Activity-on-Node Network

Generally, the Precedence notation is more powerful, since it can also accommodate lead/lag relationships of the activities in a more natural manner than the i-j notation. However, it is not convenient when we want to refer to certain milestone events in the network. In the latter instance, the i-j notation is more natural to observe. Most commercial packages nowadays can accept data in both notations.

Notation Used

We will, however, adopt the i-j notation for both programs because of its simplicity. We also adopt the convention that all event names must be positive integers and that the labeling of the event numbers must be such that the start event number is always less than the finish event number for all activities. Since the network must be acyclic, it is always possible to label them that way. It is suggested that the events be numbered sequentially starting with 1,2,..., and so on, because the program is limited not only by the number of activities but also by the highest node number in the network. However, it is trivial to vary the capacity of the program. See Section IV for more information.

We also adopt a now widely accepted distinction between CPM and PERT. That is, CPM refers to project scheduling when the activity duration is assumed to be fixed or deterministic, whereas PERT refers to the method used when some of the activity durations are variable or probabilistic.

Definitions

In order to determine the schedule of each activity, the program performs the forward pass and the backward pass calculations as shown in standard textbooks. The forward pass gives the earliest start (ES) and earliest finish (EF) times of each activity. The backward pass, on the other hand, yields the latest start (LS) and the latest finish (LF) times. Three measurements of floats are given by the CPM program -i.e., total float (TF), free float (FF), and interfering float (IF).

We use the following formulas to determine the floats. For activity (i,j):

(1) $$TF_{ij} = LF_{ij} - EF_{ij} = LS_{ij} - ES_{ij}$$

(2) $$FF_{ij} = ES_{jk} - EF_{ij}$$

where ES_{jk} is the earliest start time of activity (j,k) which immediately follows activity (i,j), and

(3) $\quad IF_{ij} = TF_{ij} - FF_{ij}$.

The critical activities forming the critical path are defined as those activities that have nonpositive total floats. Note that an activity can have negative float. This is due to the fact that if an unreasonable target due date is set for the project (i.e., too tight a target), an impossible condition will occur.

The normal interpretation of the activity total float is the maximum delay an activity can have without delaying the entire project. Free float of an activity is the amount of delay an activity can have assuming that all of its preceding activities start and finish as early as possible. If the free float of an activity is used up, activities that follow it can still start and finish at their earliest scheduled time. On the other hand, interfering float is defined as the difference between the total float and the free float, and it is the amount of time that an activity can be delayed without delaying the project. However, any interfering float that is used up by an activity will affect the earliest start time of activities that follow it. If all interfering float of an activity is used up, its following activities will become critical.

II. The Program CPM

The program CPM computes the schedule of the activities utilizing the project start date as the earliest start time of all independent start activities. It uses the project due date (if known) to be the latest finish time of all independent finish activities. This implies that the program can handle multiple independent subprojects and projects with multiple starts and finishes. It must, however, obey the node labeling rule as described earlier in Section I. If the project due date is unknown, the latest finish of all independent finish activities will be set equal to the finish date of the last activity on the critical path.

INPUT
-----
1. The project title
2. The number of activities (>=2)
3. The highest node number in the network (which is equal to the number of event nodes if events are numbered sequentially starting from 1)
4. The project start date (>=1)
5. The project due date (if known)

6. The activity data, which consist of
   I-node, J-node, duration
   where I-node is the start event, J-node is the finish event, duration is the number of time units in integral number (days, weeks, months, etc.) required to complete the activity

OUTPUT

There are basically two types of output from CPM. The first output is a table showing 10 columns of the input data plus calculated results. For example, column 3 refers to the activity duration (DU), and column 4 refers to the activity earliest start date (ES), etc. The second output is a bar chart showing the early start schedule of all activities together with a graph showing the schedule. A critical activity is represented by a bar of C's. The X's represent noncritical activities, and the dots indicate the amount of available total floats. Since the program is command-driven, the TABLE and the BAR commands generate the two types of output, respectively.

In addition, the SORT command will allow the user to sort the output reports from low to high by selecting any two columns of the table in sort order. For example, if the SORT command is entered and the user selects to sort on column 8 (total float, TF) and column 4 (earliest start date, ES), subsequent outputs will be reported in that sort order.

EXAMPLE

Consider a project network called SAMPLE NETWORK with seven events and eleven activities, as shown in Figure 4. Table 1 shows the activity durations. Assume that the project starts on day 1 and we wish to use the program CPM to find the critical path and the activity schedules.

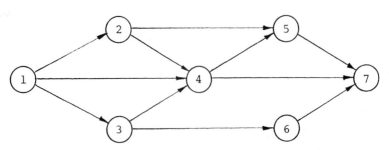

Figure 4. Project SAMPLE NETWORK

Table 1. Activity Durations

i	j	Duration
1	2	5
1	4	6
1	3	9
2	4	6
3	4	10
2	5	10
4	5	12
3	6	5
4	7	6
5	7	8
6	7	10

The example terminal session is shown below. Note that we assume that the project due date is not known. After all of the activity data are entered, the HELP command is entered to show all available commands. Next the TABLE command is entered to obtain the schedules. The project duration is 39 days, which is also equal to the earliest finish date because the project start date is day 1. The SORT command is entered next, and we sort the activities by columns 8 and 4 of the table. The BAR command produces a bar chart whose activities are sorted by the total floats and the earliest start dates as required. Note that the bar chart does indeed display the critical activities before the noncritical ones. The critical path consists of activities (1,3), (3,4), (4,5), and (5,7).

```
CPM EXAMPLE TERMINAL SESSION
============================

C R I T I C A L     P A T H     M E T H O D
---------------     -------     -----------

NEED INTRODUCTION (Y OR N) ... !N

ENTER PROJECT TITLE : !SAMPLE NETWORK

ENTER NUMBER OF ACTIVITIES (>=2) !11

ENTER THE HIGHEST NODE NUMBER IN YOUR NETWORK  !7

ENTER PROJECT START DATE (>=1) !1

DO YOU KNOW THE PROJECT DUE DATE (Y OR N)  !N

ENTER ACTIVITY DATA (I NODE, J NODE, DURATION)
----------------------------------------------
ENTER DATA FOR ACTIVITY  1   !1,2,5
ENTER DATA FOR ACTIVITY  2   !1,4,6
ENTER DATA FOR ACTIVITY  3   !1,3,9
ENTER DATA FOR ACTIVITY  4   !2,4,6
ENTER DATA FOR ACTIVITY  5   !3,4,10
ENTER DATA FOR ACTIVITY  6   !2,5,10
ENTER DATA FOR ACTIVITY  7   !4,5,12
ENTER DATA FOR ACTIVITY  8   !3,6,5
ENTER DATA FOR ACTIVITY  9   !4,7,6
ENTER DATA FOR ACTIVITY  10  !5,7,8
ENTER DATA FOR ACTIVITY  11  !6,7,10

PLEASE ENTER COMMAND OR TYPE HELP

COMMAND --> !HELP

AVAILABLE COMMANDS
==================
HELP    - PRINT THIS MESSAGE
TABLE   - GENERATE TABLE OF SCHEDULE IN SORT ORDER
SORT    - SORT THE ACTIVITIES ACCORDING TO THE TABLE
          TWO COLUMNS AT A TIME
BAR     - GENERATE BAR CHART IN SORT ORDER
RESTART - START A NEW PROBLEM
EXIT    - EXIT FROM THE PROGRAM

COMMAND --> !TABLE
```

```
PROJECT : SAMPLE NETWORK
START DATE = 1
EARLIEST FINISH DATE = 39
PROJECT DURATION = 39

---------------------------------------------------------
  1   2    3    4    5    6    7    8    9   10
  I   J   DU   ES   EF   LS   LF   TF   FF   IF
---------------------------------------------------------
  1   2    5    1    5    9   13    8    0    8
  1   4    6    1    6   14   19   13   13    0
  1   3    9    1    9    1    9    0    0    0  CRITICAL
  2   4    6    6   11   14   19    8    8    0
  3   4   10   10   19   10   19    0    0    0  CRITICAL
  2   5   10    6   15   22   31   16   16    0
  4   5   12   20   31   20   31    0    0    0  CRITICAL
  3   6    5   10   14   25   29   15    0   15
  4   7    6   20   25   34   39   14   14    0
  5   7    8   32   39   32   39    0    0    0  CRITICAL
  6   7   10   15   24   30   39   15   15    0


COMMAND --> !SORT

SELECT TWO COLUMNS IN SORT ORDER   !8,4
TABLE SORTED !!!

COMMAND --> !BAR

PROJECT : SAMPLE NETWORK
START DATE = 1
EARLIEST FINISH DATE = 39
PROJECT DURATION = 39

   I    J   DU   ES   EF           EACH * =     1
----------------------------I------------------------------------I
   1    3    9    1    9   ICCCCCCCCC                            I
   3    4   10   10   19   I         CCCCCCCCCC                  I
   4    5   12   20   31   I                   CCCCCCCCCCCC      I
   5    7    8   32   39   I                               CCCCCCCC I
   1    2    5    1    5   IXXXXX........                       I
   2    4    6    6   11   I    XXXXXX........                  I
   1    4    6    1    6   IXXXXXX.............                 I
   4    7    6   20   25   I                   XXXXXX...............  I
   3    6    5   10   14   I        XXXXX................       I
   6    7   10   15   24   I             XXXXXXXXXX...............    I
   2    5   10    6   15   I      XXXXXXXXXX................    I
----------------------------I------------------------------------I

COMMAND --> !EXIT
```

167

```
10 REM      C R I T I C A L      P A T H      M E T H O D
20 REM      USING I-J NOTATION (ACTIVITY ON ARROW NETWORK).
30 DIM A(50),P(10,50),E(60),F(60),P$(40),F$(3)
40 N9=60
50 M9=50
60 GOSUB 2940
70 PRINT
80 PRINT "C R I T I C A L      P A T H      M E T H O D"
90 PRINT "---------------      -------      -----------"
100 PRINT
110 PRINT "NEED INTRODUCTION (Y OR N) ... ";
120 INPUT Y$
130 IF Y$ <> "Y" THEN 380
140 PRINT
150 PRINT "THIS PROGRAM PERFORMS CRITICAL PATH ANALYSIS TO"
160 PRINT "DETERMINE SCHEDULES OF ACTIVITIES IN MULTIPLE "
170 PRINT "INDEPENDENT PROJECTS.  LABELING OF NODES OF EVERY"
180 PRINT "ACTIVITY MUST BE SUCH THAT THE START EVENT NODE"
190 PRINT "IS SMALLER THAN THE FINISH EVENT NODE."
200 PRINT
210 PRINT "IT IS SUGGESTED THAT EACH NODE BE NUMBERED SEQUENTIALLY"
220 PRINT "STARTING WITH 1 AS THE START EVENT FOR THE FIRST"
230 PRINT "ACTIVITY.  ALL NODES MUST BE LABELED AS POSITIVE"
240 PRINT "INTEGERS."
250 PRINT
260 PRINT "INPUT REQUIREMENTS"
270 PRINT "------------------"
280 PRINT "   1. PROJECT TITLE"
290 PRINT "   2. NUMBER OF ACTIVITIES"
300 PRINT "   3. THE HIGHEST NODE NUMBER"
310 PRINT "   4. PROJECT START DATE"
320 PRINT "   5. PROJECT DUE DATE (IF ANY)"
330 PRINT "   6. ACTIVITY DATA WHICH CONSIST OF"
340 PRINT "       I NODE, J NODE, DURATION"
350 PRINT
360 PRINT "MAXIMUM NUMBER OF ACTIVITIES = ";M9
370 PRINT "HIGHEST NODE NUMBER = ";N9
380 PRINT
390 PRINT "ENTER PROJECT TITLE : ";
400 INPUT T$
410 PRINT
420 PRINT "ENTER NUMBER OF ACTIVITIES (>=2) ";
430 INPUT M
440 IF M <= 0 THEN 380
450 IF M > M9 THEN 380
460 PRINT
470 PRINT "ENTER THE HIGHEST NODE NUMBER IN YOUR NETWORK   ";
480 INPUT N
490 IF N <= 0 THEN 460
500 IF N > N9 THEN 460
510 PRINT
520 PRINT "ENTER PROJECT START DATE (>=1) ";
530 INPUT S7
540 IF S7 <= 0 THEN 510
```

```
550 REM     ADJUST START DATE FOR FORWARD AND BACKWARD PASS CALC.
560 S=S7-1
570 PRINT
580 PRINT "DO YOU KNOW THE PROJECT DUE DATE (Y OR N)    ";
590 INPUT D$
600 IF D$ = "Y" THEN 630
610 IF D$ = "N" THEN 660
620 GOTO 570
630 PRINT
640 PRINT "ENTER PROJECT DUE DATE   ";
650 INPUT D
660 PRINT
670 PRINT "ENTER ACTIVITY DATA (I NODE, J NODE, DURATION)"
680 PRINT "------------------------------------------"
690 FOR I=1 TO M
700 PRINT "ENTER DATA FOR ACTIVITY   ";I;"   ";
710 INPUT P(1,I),P(2,I),P(3,I)
720 IF P(1,I) > N THEN 760
730 IF P(2,I) > N THEN 760
740 A(I)=I
750 GOTO 820
760 IF P(2,I) > N9 THEN 790
770 N=P(2,I)
780 GOTO 740
790 PRINT
800 PRINT "*** ERROR *** HIGHEST NODE LABEL EXCEEDS ";N9
810 GOTO 700
820 NEXT I
830 REM     SORT ACTIVITIES IN TOPOLOGICAL SEQUENCE
840 FOR I=1 TO M-1
850 FOR J=I+1 TO M
860 IF P(1,A(I)) <> P(2,A(J)) THEN 930
870 A1=A(J)
880 FOR K=J TO I+1 STEP -1
890 A(K)=A(K-1)
900 NEXT K
910 A(I)=A1
920 GOTO 850
930 NEXT J
940 NEXT I
950 REM     T = CRITICAL PATH DURATION
960 REM     E(.) = EARLIEST START TIME OF EVENTS (NODES)
970 T=0
980 FOR I=1 TO N
990 E(I)=S
1000 NEXT I
1010 REM    F O R W A R D    P A S S
1020 REM    ------------------------
1030 REM    DETERMINE ES, EF OF EACH ACTIVITY
1040 FOR I=1 TO M
1050 K=A(I)
1060 I1=P(1,K)
1070 J1=P(2,K)
1080 P(4,K)=E(I1)
```

```
1090 P(5,K)=E(I1)+P(3,K)
1100 IF P(5,K) <= E(J1) THEN 1120
1110 E(J1)=P(5,K)
1120 IF P(5,K) <= T THEN 1140
1130 T=P(5,K)
1140 NEXT I
1150 REM    IF PROJECT DUE DATE IS KNOWN -- SET THE LATEST FINISH
1160 REM    DATE OF THE INDEPENDENT FINISH ACTIVITIES TO THE DUE
1170 REM    DATE. OTHERWISE, ASSIGN THE LATEST FINISH DATE AS THE
1180 REM    LONGEST PATH OBTAINED FROM THE FORWARD PASS.
1190 D9=D
1200 IF D$="Y" THEN 1220
1210 D9=T
1220 FOR I=1 TO N
1230 F(I)=D9
1240 NEXT I
1250 REM    B A C K W A R D    P A S S
1260 REM    ------------------------
1270 REM    DETERMINE LS AND LF OF EACH ACTIVITY
1280 FOR J=1 TO M
1290 I=M-J+1
1300 K=A(I)
1310 I1=P(1,K)
1320 J1=P(2,K)
1330 P(7,K)=F(J1)
1340 P(6,K)=F(J1)-P(3,K)
1350 IF P(6,K) >= F(I1) THEN 1370
1360 F(I1)=P(6,K)
1370 NEXT J
1380 REM    DETERMINE TOTAL FLOAT, FREE FLOAT AND INTERFERENCE FLOAT
1390 FOR I=1 TO M
1400 P(8,I)=P(6,I)-P(4,I)
1410 J1=P(2,I)
1420 P(9,I)=E(J1)-P(5,I)
1430 P(10,I)=P(8,I)-P(9,I)
1440 P(4,I)=P(4,I)+1
1450 P(6,I)=P(6,I)+1
1460 NEXT I
1470 PRINT
1480 PRINT "PLEASE ENTER COMMAND OR TYPE HELP"
1490 PRINT
1500 PRINT "COMMAND --> ";
1510 INPUT C$
1520 PRINT
1530 IF C$ <> "HELP" THEN 1560
1540 GOSUB 1690
1550 GOTO 1490
1560 IF C$ <> "TABLE" THEN 1590
1570 GOSUB 1800
1580 GOTO 1490
1590 IF C$ <> "SORT" THEN 1620
1600 GOSUB 2050
1610 GOTO 1490
1620 IF C$ <> "BAR" THEN 1650
```

```
1630 GOSUB 2340
1640 GOTO 1490
1650 IF C$="RESTART" THEN 380
1660 IF C$ = "EXIT" THEN 3130
1670 PRINT "***ERROR ***   PLEASE TRY AGAIN !!!"
1680 GOTO 1490
1690 REM    *** HELP COMMAND ***
1700 PRINT "AVAILABLE COMMANDS"
1710 PRINT "=================="
1720 PRINT "HELP    - PRINT THIS MESSAGE"
1730 PRINT "TABLE   - GENERATE TABLE OF SCHEDULE IN SORT ORDER"
1740 PRINT "SORT    - SORT THE ACTIVITIES ACCORDING TO THE TABLE"
1750 PRINT "          TWO COLUMNS AT A TIME"
1760 PRINT "BAR     - GENERATE BAR CHART IN SORT ORDER"
1770 PRINT "RESTART - START A NEW PROBLEM"
1780 PRINT "EXIT    - EXIT FROM THE PROGRAM"
1790 RETURN
1800 REM    *** TABLE COMMAND ***
1810 GOSUB 3040
1820 L3=50
1830 GOSUB 2990
1840 PRINT
1850 FOR J=1 TO 10
1860 PRINT USING F$(1),J;
1870 NEXT J
1880 PRINT
1890 PRINT F$(2);F$(3)
1900 GOSUB 2990
1910 PRINT
1920 FOR I=1 TO M
1930 K=A(I)
1940 C$="  "
1950 IF P(8,K) > 0.00001 THEN 1970
1960 C$=" CRITICAL"
1970 FOR J=1 TO 10
1980 PRINT USING F$(1),P(J,K);
1990 NEXT J
2000 PRINT C$
2010 NEXT I
2020 PRINT
2030 PRINT
2040 RETURN
2050 REM    *** SORT COMMAND ***
2060 PRINT
2070 PRINT "SELECT TWO COLUMNS IN SORT ORDER  ";
2080 INPUT L,L9
2090 IF L <= 0 THEN 2060
2100 IF L > 10 THEN 2060
2110 IF L9 <= 0 THEN 2060
2120 IF L9 > 10 THEN 2060
2130 FOR I=1 TO M
2140 A(I)=I
2150 NEXT I
2160 FOR I=1 TO M-1
```

```
2170 K=A(I)
2180 B0=P(L,K)
2190 B1=P(L9,K)
2200 FOR J=I+1 TO M
2210 L1=A(J)
2220 IF B0 < P(L,L1) THEN 2300
2230 IF B0 > P(L,L1) THEN 2250
2240 IF B1 <= P(L9,L1) THEN 2300
2250 B0=P(L,L1)
2260 A0=A(I)
2270 A(I)=A(J)
2280 A(J)=A0
2290 B1=P(L9,L1)
2300 NEXT J
2310 NEXT I
2320 PRINT "TABLE SORTED !!!"
2330 RETURN
2340 REM    *** BAR COMMAND ***
2350 V=INT((T-S)/40)+1
2360 GOSUB 3040
2370 PRINT F$(2);"           EACH * = ";
2380 PRINT USING F$(1),V
2390 L3=27
2400 GOSUB 2860
2410 FOR I=1 TO M
2420 K=A(I)
2430 FOR J=1 TO 40
2440 P$(J)=" "
2450 NEXT J
2460 IF P(3,K)=0 THEN 2730
2470 I1=(P(4,K)-S)/V
2480 I2=INT(I1)
2490 IF I1=I2 THEN 2510
2500 I1=I2+1
2510 I2=(P(5,K)-S)/V
2520 I3=INT(I2)
2530 IF I2=I3 THEN 2550
2540 I2=I3+1
2550 G$="X"
2560 IF P(8,K) > 0 THEN 2580
2570 G$="C"
2580 IF (I2-I1+1)*V-P(3,K) < V THEN 2600
2590 I2=I2-1
2600 FOR J=I1 TO I2
2610 P$(J)=G$
2620 NEXT J
2630 IF P(8,K) <= 0 THEN 2730
2640 I3=(P(7,K)-S)/V
2650 I4=INT(I3)
2660 IF I3=I4 THEN 2680
2670 I3=I4+1
2680 IF (I3-I2+1)*V-P(8,K) <= V THEN 2700
2690 I3=I3-1
2700 FOR J=I2+1 TO I3
```

```
2710 P$(J)="."
2720 NEXT J
2730 FOR J=1 TO 4
2740 PRINT USING F$(1),P(J,K);
2750 NEXT J
2760 PRINT USING F$(1),P(5,K);
2770 PRINT "  I";
2780 FOR J=1 TO 40
2790 PRINT P$(J);
2800 NEXT J
2810 PRINT "I"
2820 NEXT I
2830 GOSUB 2870
2840 PRINT
2850 RETURN
2860 REM    PRINT A LINE BEFORE OR AFTER THE BAR CHART
2870 GOSUB 2990
2880 PRINT "I";
2890 FOR J=1 TO 40
2900 PRINT "-";
2910 NEXT J
2920 PRINT "I"
2930 RETURN
2940 REM    I N I T I A L I Z A T I O N
2950 F$(1)="#####"
2960 F$(2)="    I      J     DU    ES    EF"
2970 F$(3)="    LS    LF    TF    FF    IF"
2980 RETURN
2990 REM    PRINT A LINE ACROSS A PAGE AND HOLD CR/LF
3000 FOR J=1 TO L3
3010 PRINT "-";
3020 NEXT J
3030 RETURN
3040 PRINT
3050 PRINT "PROJECT : ";T$
3060 PRINT "START DATE = ";S7
3070 IF D$ <> "Y" THEN 3090
3080 PRINT "DUE DATE = ";D
3090 PRINT "EARLIEST FINISH DATE = ";T+S7-1
3100 PRINT "PROJECT DURATION = ";T
3110 PRINT
3120 RETURN
3130 END
```

III. The Program PERT

Before we describe the input and output for the PERT program, we will briefly review and discuss some of the findings in this area and the underlying assumptions used in this program.

THEORY

The development of PERT evolved naturally from the fact that the time it takes to accomplish an activity is rarely known to be fixed. There are many factors affecting an activity duration, some of which are beyond our control -e.g., weather may affect construction projects, etc. Sometimes managers require additional information to help them make better decisions. Some information cannot be readily obtained through the use of the regular CPM technique. Perhaps the question with the most-sought-after answer is the probability that the project will finish on or before a target date. Another question of interest is in estimating the probability that an activity will lie on a critical path. This is also called a "Criticality Index" of an activity [13].

The answer to the first question can be obtained through the use of the traditional PERT method, although there are several objections to the technique itself - for example, the merge event bias problem [8, p.300], which is the problem of neglecting the noncritical activities when determining the activity earliest start times. The traditional method of determining the PERT critical path always leads to an underestimate of the expected project completion time due to this problem. In addition, there are other controversies concerning the definitions of the activity time estimates used in the technique itself.

In the original PERT development, the activity duration is assumed to be a random variable from a beta distribution. The beta distribution was chosen because of its flexibility in fitting empirical distributions. The mean and variance of the activity duration can be estimated from three time estimates A, M, and B where

- A = Optimistic duration -i.e., the shortest time possible to finish the activity,

- M = Most likely duration -i.e., if the activity is repeated time and time again, the most frequent duration will be M, and

- B = Pessimistic duration -i.e., the longest possible duration required to finish the activity.

The mean, $E(t)$, and the variance, $V(t)$, of an activity

duration can be estimated by

(4) $\qquad E(t) = (A+4M+B)/6$,

(5) $\qquad V(t) = (B-A)^2/36$.

We will refer to the use of equations (4) and (5) and the above definitions of A, M, and B as the BETA MODEL. This is to distinguish it from another set of definitions given by Moder and Phillips [8] where

- A = Optimistic duration, which is defined as the duration for which there is only a 5 percent chance that it will be smaller,

- M = Most likely duration, and

- B = Pessimistic duration, which is defined as the duration for which the chance of exceeding it is 5 percent.

Based on these definitions, they propose that V(t) can be calculated using equation (6):

(6) $\qquad V(t) = ((B-A)/3.2)^2$.

The calculation for E(t) remains the same. Moder and Phillips [8] suggest that equation (6) is a good approximation for a large number of activity time distributions. We will, however, assume that the distribution is normal. We will also refer to the use of equations (4) and (6) as the NORMAL model in our program. Figure 5 illustrates the difference between the two models.

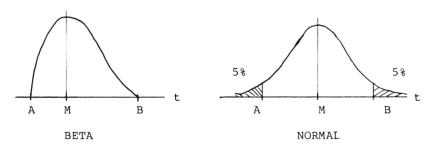

Figure 5. Difference in the definition of A, M, and B in the two models.

Obviously, there is an inconsistency in the NORMAL model because the estimation of the activity mean duration and its variance is based on the three time estimates (A, M, and B) which implies the beta-type distribution. On the other hand, the normally distributed random duration would imply that $(B-M)=(M-A)$. However, since the true form of the distribution is rarely known, the use of the normally distributed time as the model may not be a serious drawback in gaining insight into the behavior of the project duration distribution.

In the simulation of the NORMAL model, the random duration is generated using a standard normal deviate, z, which is based on the Central Limit Theorem, whereby

$$(7) \quad z = \sum_{i=1}^{12} r_i - 6$$

where r_i is a uniform random number between zero and one [9].

For the BETA model, we use Don T. Phillips's algorithm [11] as described in Shannon [12] to generate gamma variates as the basis for generating beta variates. This is because a normalized beta variate is defined as the ratio of two gamma variates.

Consider a gamma random variable x with probability density $f(x)$:

$$(8) \quad f(x) = \begin{cases} \dfrac{r^p x^{p-1} e^{-rx}}{\Gamma(p)} & , x > 0 \\ 0 & , \text{otherwise} \end{cases}$$

with $r>0$ and $p>0$ as the two parameters. Also let x_1 and x_2 be two independent gamma variables with parameters (p, r), and (q, r), respectively. If we define $y = x_1/(x_1+x_2)$, then y is a normalized beta variable with the density function

$$(9) \quad g(y) = \begin{cases} \dfrac{\Gamma(p+q)}{\Gamma(p)\Gamma(q)} y^{p-1} (1-y)^{q-1} & , 0 \leq y \leq 1 \\ 0 & , \text{otherwise.} \end{cases}$$

The expected value and the variance of y are as follows:

$$(10) \quad E(y) = p/(p+q)$$

(11) $$V(y) = [pq/(p+q+1)]/[(p+q)^2] .$$

Let t be the activity duration, which is beta distributed. We can write

(12) $$t = A + (B-A)y .$$

Hence, $E(t) = A+(B-A)E(y)$ and $V(t) = V(y)(B-A)^2$. By equating $E(t)$ and $V(t)$ to equations (4) and (5), respectively, and solving for p and q, we obtain

(13) $$p = 36 S (S - S^2 - 1/36) ,$$

and

(14) $$q = (p-pS)/S ,$$

where $$S = \frac{4M+B-5A}{6(B-A)} .$$

To generate the activity random time t, we generate x_1 and x_2 with parameters (p, r=1) and (q, r=1), respectively. Then the random variate y is determined by its definition, and t is is obtained using the transformation (12).

The program PERT uses the above-mentioned procedures to generate random activity durations. A simulation trial consists of generating random durations, performing forward and backward pass calculations, and collecting the various statistics.

INPUT
-----
1. The project title
2. The number of activities
3. The highest node number in the network
4. The project start date (>=1)
5. Activity data, which consist of
   I node, J node, A, M, B

OUTPUT

Two types of output are available from PERT. First, the TABLE-command output, which consists of the schedule of activities based on expected activity durations and the

probability that the project will be completed on or before a specified date (PROB command). The second type of output is the result of simulating the network a number of times. The distribution of project duration together with the probability that an activity lies on a critical path (i.e., criticality index), is reported every time a SIMULATE command is activated.

In the TABLE-command output, if the network contains more than one critical path, the standard deviation of the project duration will not be correct. This is because the program does not trace the paths through the network in order to find the unique path with the largest variance. It merely sums all critical activity variances. Thus the resulting standard deviation may be larger than expected. This implies that, for this case, the calculation of the probability that the project will finish on or before a target date will be incorrect according to standard textbook procedure. One way to overcome this problem is to slightly perturb some critical activities so that a unique critical path results.

It should be noted that the activity time estimates A, M, B are entered only once. However, the user has access to both the BETA and the NORMAL models through the use of the BETA and the NORMAL commands. The default model is BETA. The program will remain in a model unless it is changed. This is useful when performing simulation because the SIMULATE command can be executed as many times as desired and the program will accumulate the simulation results until the desired accuracy of the estimates is achieved. If a change in the model is made, all previous simulation statistics will be destroyed because the program will reinitialize variables ready for the new model simulation.

The user should also be cautioned of the amount of time needed to perform simulation. For example, we simulated for 10,000 times the fourteen-activity network (BETA model) as described in [13] on the PRIME-400 at Illinois Institute of Technology. Despite the fact that we used one of the fastest variate generators [11, 12], it took (on one Sunday afternoon) approximately 55 minutes before obtaining the simulation output.

We also experienced the same behavior as described in Van Slyke [13] -i.e., the running time appears to be a linear function of the number of simulation trials. It should be expected that the program would have performed much faster if it had been written in a language other than BASIC (e.g., FORTRAN).

EXAMPLE

We will use the same network logic as in Figure 4, but with different activity time estimates. The data are listed in Table 2.

Table 2. PERT Activity Data for SAMPLE NETWORK

i	j	A	M	B
1	2	6	12	30
1	4	8	15	20
1	3	5	15	30
2	4	3	5	7
3	4	4	6	8
2	5	6	9	15
4	5	3	6	10
3	6	10	18	25
4	7	15	30	50
5	7	12	25	30
6	7	6	12	20

We assume that the project start date is the start of day 1. After the data of all 11 activities are entered, the HELP command is entered to display all available commands. We first investigate the BETA model. Since this is the default model when the program is first executed, the TABLE command generates the schedule of the activities based on the traditional PERT method. Activities (1,3), (3,4), and (4,7) are identified as being critical. The expected project duration is 52.67 days, with a standard deviation of 7.2 days.

Next the command PROB is entered and we specify a target finish date of 60. The probability of the project being completed on or before day 60 is 0.8458. This is based on the assumption that the project duration is normally distributed with a mean of 52.67 days and a standard deviation of 7.2 days.

The next command entered is the COUNT command. It allows the user to specify the frequency of simulation count, that the program will report to the user, as to the progress of the simulation. The number 250 is entered for this example. The default value is 100.

The SIMULATE command is then entered, and we select to simulate 500 trials. The mean critical path duration is 56.74 days, which is longer than the 52.67 days obtained from the TABLE command. This confirms that the theoretical PERT method always underestimates the expected project duration. The standard deviation of the mean project duration is 0.2673, which implies that our estimate of the true mean is indeed very close. The histogram of the project duration is slightly skewed to the right. This is typical of the project duration distribution. The table immediately following the histogram gives the activity criticality indexes. For example, activity (1,3) is on a critical path 320 out of the 500 simulation trials. Thus its criticality index is 0.64. The columns E(T) and V(T) are the sample means and variances of the activity durations, respectively. They compare very well with the earlier results in the table generated from the TABLE command. It should be noted, however, that the sample variances contain slight positive bias for every activity. This is due to the approximating procedure used in generating GAMMA variates [11].

We will now consider using the same data but interpret the values of A, M, and B as being from the NORMAL model. The NORMAL command is entered next followed by the TABLE command. The result is almost identical to that of the BETA model except column V(t) whose calculation is based on equation (6) instead of (5). The standard deviation of the project duration becomes 13.50 days. When the PROB command is entered, the probability that the project will finish on or before day 60 is now 0.7065.

Next we simulate 500 trials and observe essentially the same behavior that we have observed under the BETA model in terms of the expected project duration and the skewness of the distribution.

```
PERT EXAMPLE TERMINAL SESSION
=============================

P E R T
-------

NEED INTRODUCTION (Y OR N) ... !N

ENTER PROJECT TITLE  :  !SAMPLE NETWORK

ENTER NUMBER OF ACTIVITIES -- !11

ENTER THE HIGHEST NODE NUMBER IN THE NETWORK -- !7

ENTER PROJECT START DATE -- !1

ENTER ACTIVITY DATA (I NODE, J NODE, A, M, B)
---------------------------------------------
ENTER DATA FOR ACTIVITY 1     !1,2,6,12,30
ENTER DATA FOR ACTIVITY 2     !1,4,8,15,20
ENTER DATA FOR ACTIVITY 3     !1,3,5,15,30
ENTER DATA FOR ACTIVITY 4     !2,4,3,5,7
ENTER DATA FOR ACTIVITY 5     !3,4,4,6,8
ENTER DATA FOR ACTIVITY 6     !2,5,6,9,15
ENTER DATA FOR ACTIVITY 7     !4,5,3,6,10
ENTER DATA FOR ACTIVITY 8     !3,6,10,18,25
ENTER DATA FOR ACTIVITY 9     !4,7,15,30,50
ENTER DATA FOR ACTIVITY 10    !5,7,12,25,30
ENTER DATA FOR ACTIVITY 11    !6,7,6,12,20

PLEASE ENTER COMMAND OR TYPE HELP

COMMAND --> !HELP

AVAILABLE COMMAND
=================
HELP     - PRINT THIS MESSAGE
BETA     - SELECT BETA MODEL
NORMAL   - SELECT NORMAL MODEL
TABLE    - PRINT TABLE OF PERT OUTPUT
COUNT    - SET SIMULATION COUNTER
SIMULATE - SIMULATE THE NETWORK BASED ON SELECTED MODEL
RESTART  - START A NEW PROBLEM
EXIT     - EXIT FROM THE PROGRAM

COMMAND --> !TABLE
```

P E R T TABLE OUTPUT

P R O J E C T : SAMPLE NETWORK

M O D E L   : BETA

EXPECTED PROJECT DURATION   52.66666666667
STD.DEV. OF PROJECT DURATION   7.199537022152

PROJECT START DATE   : 1
EXPECTED COMPLETION DATE : 52.66666666667

```
------------------------------------------------------------------
 I   J   A   M   B   E(T)   V(T)   E(ES)  E(EF)  E(LS)  E(LF)  E(TF)
------------------------------------------------------------------
 1   2   6  12  30  14.0   16.0   1.0    14.0   3.8    16.8   2.8
 1   4   8  15  20  14.7    4.0   1.0    14.7   8.2    21.8   7.2
 1   3   5  15  30  15.8   17.4   1.0    15.8   1.0    15.8    .0  CRITICAL
 2   4   3   5   7   5.0     .4  15.0    19.0  17.8    21.8   2.8
 3   4   4   6   8   6.0     .4  16.8    21.8  16.8    21.8    .0  CRITICAL
 2   5   6   9  15   9.5    2.3  15.0    23.5  20.5    29.0   5.5
 4   5   3   6  10   6.2    1.4  22.8    28.0  23.8    29.0   1.0
 3   6  10  18  25  17.8    6.3  16.8    33.7  23.5    40.3   6.7
 4   7  15  30  50  30.8   34.0  22.8    52.7  22.8    52.7    .0  CRITICAL
 5   7  12  25  30  23.7    9.0  29.0    51.7  30.0    52.7   1.0
 6   7   6  12  20  12.3    5.4  34.7    46.0  41.3    52.7   6.7
```

COMMAND --> !PROB

PROB(PROJECT WILL FINISH ON OR BEFORE TARGET DATE)

P R O J E C T : SAMPLE NETWORK

M O D E L   : BETA

ENTER TARGET FINISH DATE (OR -1 TO RETURN)
TARGET FINISH DATE   : !60
PROBABILITY   =   .8457997699282
TARGET FINISH DATE   : !-1

COMMAND --> !COUNT

ENTER THE SIMULATION COUNTER (INTEGER > 0)   !250

COMMAND --> !SIMULATE

HOW MANY TRIALS !500

S I M U L A T I N G, PLEASE STAND BY ... !!!
SIMULATED   250
SIMULATED   500

```
P E R T    SIMULATION OUTPUT
---------------------------

P R O J E C T  :  SAMPLE NETWORK

M O D E L      :  BETA

NUMBER OF TRIALS       500
MEAN CRITICAL PATH DURATION          56.74315669081
STD.DEV. OF CRITICAL PATH DURATION    5.976632766961
STD.DEV. OF MEAN CRITICAL PATH DURATION  .2672831428695

PROBABILITY DISTRIBUTION OF CRITICAL PATH DURATION
--------------------------------------------------
    DURATION           CUM.
 FROM    TO    PROB   PROB.   I .0000                                      .1200I
  .0    36.5  .0000  .0000   I                                                  I
 36.5   38.3  .0000  .0000   I                                                  I
 38.3   40.1  .0020  .0020   I*                                                 I
 40.1   41.9  .0000  .0020   I                                                  I
 41.9   43.7  .0080  .0100   I***                                               I
 43.7   45.5  .0060  .0160   I**                                                I
 45.5   47.3  .0300  .0460   I**********                                        I
 47.3   49.1  .0620  .1080   I*********************                             I
 49.1   50.9  .0660  .1740   I**********************                            I
 50.9   52.7  .0920  .2660   I*******************************                   I
 52.7   54.5  .1060  .3720   I************************************              I
 54.5   56.3  .1020  .4740   I***********************************               I
 56.3   58.1  .1120  .5860   I**************************************            I
 58.1   59.9  .1020  .6880   I***********************************               I
 59.9   61.7  .0880  .7760   I******************************                    I
 61.7   63.5  .0800  .8560   I***************************                       I
 63.5   65.3  .0480  .9040   I****************                                  I
 65.3   67.1  .0480  .9520   I****************                                  I
 67.1   68.9  .0320  .9840   I***********                                       I
 68.9   70.7  .0120  .9960   I****                                              I
 70.7   72.5  .0040 1.0000   I*                                                 I
 72.5   74.3  .0000 1.0000   I                                                  I
 74.3    UP   .0000 1.0000   I                                                  I


PROBABILITY THAT AN ACTIVITY LIES ON A CRITICAL PATH
----------------------------------------------------

  I    J    E(T)     V(T)    FREQUENCY  PROBABILITY
  1    2    14.0     17.9        168       .3360
  1    4    14.7      4.4         12       .0240
  1    3    15.8     21.3        320       .6400
  2    4     5.0       .5        150       .3000
  3    4     6.0       .5        307       .6140
  2    5     9.5      2.6         18       .0360
  4    5     6.2      1.7        187       .3740
  3    6    17.8      7.4         13       .0260
  4    7    31.0     41.5        282       .5640
  5    7    23.6     10.0        205       .4100
  6    7    12.3      6.7         13       .0260
```

```
COMMAND --> !NORMAL


COMMAND --> !TABLE



P E R T   TABLE OUTPUT
---------------------

P R O J E C T  :  SAMPLE NETWORK

M O D E L      :  NORMAL

EXPECTED PROJECT DURATION   52.66666666667
STD.DEV. OF PROJECT DURATION  13.49913191654

PROJECT START DATE    :  1
EXPECTED COMPLETION DATE  :  52.66666666667

-----------------------------------------------------------------
 I  J   A   M   B   E(T)   V(T)   E(ES)  E(EF)  E(LS)  E(LF)  E(TF)
-----------------------------------------------------------------
 1  2   6  12  30  14.0   56.3    1.0   14.0    3.8   16.8   2.8
 1  4   8  15  20  14.7   14.1    1.0   14.7    8.2   21.8   7.2
 1  3   5  15  30  15.8   61.0    1.0   15.8    1.0   15.8    .0  CRITICAL
 2  4   3   5   7   5.0    1.6   15.0   19.0   17.8   21.8   2.8
 3  4   4   6   8   6.0    1.6   16.8   21.8   16.8   21.8    .0  CRITICAL
 2  5   6   9  15   9.5    7.9   15.0   23.5   20.5   29.0   5.5
 4  5   3   6  10   6.2    4.8   22.8   28.0   23.8   29.0   1.0
 3  6  10  18  25  17.8   22.0   16.8   33.7   23.5   40.3   6.7
 4  7  15  30  50  30.8  119.6   22.8   52.7   22.8   52.7    .0  CRITICAL
 5  7  12  25  30  23.7   31.6   29.0   51.7   30.0   52.7   1.0
 6  7   6  12  20  12.3   19.1   34.7   46.0   41.3   52.7   6.7



COMMAND --> !PROB


PROB(PROJECT WILL FINISH ON OR BEFORE TARGET DATE)
--------------------------------------------------

P R O J E C T  :  SAMPLE NETWORK

M O D E L      :  NORMAL


ENTER TARGET FINISH DATE (OR -1 TO RETURN)
TARGET FINISH DATE   :  !60
PROBABILITY    =       .7065194702534
TARGET FINISH DATE   :  !-1
```

```
COMMAND --> !SIMULATE

HOW MANY TRIALS  !500

S I M U L A T I N G, PLEASE STAND BY ... !!!
SIMULATED   250
SIMULATED   500



P E R T    SIMULATION OUTPUT
---------------------------

P R O J E C T  :  SAMPLE NETWORK

M O D E L      :  NORMAL

NUMBER OF TRIALS        500
MEAN CRITICAL PATH DURATION              61.02424434399
STD.DEV. OF CRITICAL PATH DURATION        9.429423832669
STD.DEV. OF MEAN CRITICAL PATH DURATION    .4216966535701

PROBABILITY DISTRIBUTION OF CRITICAL PATH DURATION
--------------------------------------------------
     DURATION           CUM.
  FROM     TO    PROB   PROB.  I .0000                                    .2000I
    .0   29.0   .0000   .0000  I                                               I
  29.0   32.4   .0000   .0000  I                                               I
  32.4   35.8   .0020   .0020  I                                               I
  35.8   39.2   .0000   .0020  I                                               I
  39.2   42.5   .0100   .0120  I**                                             I
  42.5   45.9   .0300   .0420  I******                                         I
  45.9   49.3   .0620   .1040  I************                                   I
  49.3   52.7   .0780   .1820  I****************                               I
  52.7   56.0   .1080   .2900  I**********************                         I
  56.0   59.4   .1760   .4660  I************************************           I
  59.4   62.8   .1400   .6060  I****************************                   I
  62.8   66.2   .1060   .7120  I*********************                          I
  66.2   69.5   .1180   .8300  I************************                       I
  69.5   72.9   .0580   .8880  I************                                   I
  72.9   76.3   .0520   .9400  I**********                                     I
  76.3   79.7   .0260   .9660  I*****                                          I
  79.7   83.0   .0140   .9800  I***                                            I
  83.0   86.4   .0080   .9880  I**                                             I
  86.4   89.8   .0060   .9940  I*                                              I
  89.8   93.2   .0060  1.0000  I*                                              I
  93.2   96.5   .0000  1.0000  I                                               I
  96.5     UP   .0000  1.0000  I                                               I
```

PROBABILITY THAT AN ACTIVITY LIES ON A CRITICAL PATH

I	J	E(T)	V(T)	FREQUENCY	PROBABILITY
1	2	14.0	50.5	189	.3780
1	4	14.6	12.9	32	.0640
1	3	15.9	58.5	279	.5580
2	4	5.0	1.4	152	.3040
3	4	6.0	1.6	243	.4860
2	5	9.6	8.1	37	.0740
4	5	6.2	5.2	180	.3600
3	6	17.6	21.8	36	.0720
4	7	30.7	116.9	247	.4940
5	7	23.8	31.5	217	.4340
6	7	12.2	18.9	36	.0720

COMMAND --> !EXIT

```
10  REM     P E R T
20  REM     PROGRAM EVALUATION AND REVIEW TECHNIQUE
30  REM     USING I-J NOTATION (ACTIVITY ON ARROW NETWORK).
40  REM
50  DIM A(50),P(16,50),E(60),L(60),Q(50),H(50),G(50),U(50),V(50)
60  DIM Y(49),Z(50),P$(40),F$(10)
70  N9=60
80  M9=50
90  M4=49
100 M5=50
110 GOSUB 5740
120 PRINT
130 PRINT "P E R T"
140 PRINT "-------"
150 PRINT
160 PRINT "NEED INTRODUCTION (Y OR N) ... ";
170 INPUT Y$
180 IF Y$ <> "Y" THEN 480
190 PRINT
200 PRINT "THIS PROGRAM DETERMINES THE EXPECTED COMPLETION TIME"
210 PRINT "OF A SINGLE INDEPENDENT PROJECT USING THE PERT"
220 PRINT "TECHNIQUE.  IT USES THE SAME NODE LABELING"
230 PRINT "CONVENTION AS IN THE PROGRAM CPM."
240 PRINT "IN ADDITION TO THE STANDARD PERT OUTPUT, THE"
250 PRINT "PROGRAM CAN BE USED TO SIMULATE THE ACTIVITIES"
260 PRINT "IN THE NETWORK TO DETERMINE THE PROBABILITY"
270 PRINT "THAT AN ACTIVITY WILL LIE ON A CRITICAL PATH."
280 PRINT
290 PRINT "TWO DISTINCT MODELS BASED ON THE FORM OF THE"
300 PRINT "DISTRIBUTION OF THE ACTIVITY DURATION ARE ALLOWED"
310 PRINT "   1. BETA DISTRIBUTION"
320 PRINT "   2. NORMAL DISTRIBUTION."
330 PRINT
340 PRINT "INPUT REQUIREMENTS"
350 PRINT "------------------"
360 PRINT "   1. PROJECT TITLE"
370 PRINT "   2. NUMBER OF ACTIVITIES"
380 PRINT "   3. THE HIGHEST NODE NUMBER IN THE NETWORK"
390 PRINT "   4. PROJECT START DATE"
400 PRINT "   5. ACTIVITY DATA WHICH CONSIST OF -"
410 PRINT "         I NODE, J NODE, A, M, B"
420 PRINT "       WHERE   A = OPTIMISTIC DURATION"
430 PRINT "               M = MOST LIKELY DURATION"
440 PRINT "               B = PESSIMISTIC DURATION."
450 PRINT
460 PRINT "MAXIMUM NUMBER OF ACTIVITIES =",M9
470 PRINT "HIGHEST NODE NUMBER = ",N9
480 PRINT
490 PRINT "ENTER PROJECT TITLE  :   ";
500 INPUT T$
510 PRINT
520 PRINT "ENTER NUMBER OF ACTIVITIES -- ";
530 INPUT M
540 IF M <= 0 THEN 510
```

```
550 IF M > M9 THEN 510
560 PRINT
570 PRINT "ENTER THE HIGHEST NODE NUMBER IN THE NETWORK -- ";
580 INPUT N
590 IF N <= 0 THEN 560
600 IF N > N9 THEN 560
610 PRINT
620 PRINT "ENTER PROJECT START DATE -- ";
630 INPUT S
640 PRINT
650 PRINT "ENTER ACTIVITY DATA (I NODE, J NODE, A, M, B)"
660 PRINT "-------------------------------------------"
670 FOR I=1 TO M
680 PRINT "ENTER DATA FOR ACTIVITY ";I;"   ";
690 INPUT P(1,I),P(2,I),P(3,I),P(4,I),P(5,I)
700 IF P(1,I) > N THEN 740
710 IF P(2,I) > N THEN 740
720 A(I)=I
730 GOTO 800
740 IF P(2,I) > N9 THEN 770
750 N=P(2,I)
760 GOTO 720
770 PRINT
780 PRINT "*** ERROR *** HIGHEST NODE LABEL EXCEEDS",N9
790 GOTO 680
800 NEXT I
810 REM    SORT THE ACTIVITIES IN TOPOLOGICAL ORDER
820 FOR I=1 TO M
830 P(6,I)=(P(3,I)+4*P(4,I)+P(5,I))/6
840 FOR J=I+1 TO M
850 IF P(1,A(I)) <> P(2,A(J)) THEN 920
860 A1=A(J)
870 FOR K=J TO I+1 STEP -1
880 A(K)=A(K-1)
890 NEXT K
900 A(I)=A1
910 GOTO 840
920 NEXT J
930 NEXT I
940 GOSUB 1290
950 GOSUB 4420
960 PRINT
970 PRINT "PLEASE ENTER COMMAND OR TYPE HELP"
980 PRINT
990 PRINT "COMMAND --> ";
1000 INPUT C$
1010 PRINT
1020 IF C$ <> "HELP" THEN 1050
1030 GOSUB 5890
1040 GOTO 980
1050 IF C$ <> "BETA" THEN 1080
1060 GOSUB 6010
1070 GOTO 980
1080 IF C$ <> "NORMAL" THEN 1110
```

```
1090 GOSUB 6010
1100 GOTO 980
1110 IF C$ <> "TABLE" THEN 1140
1120 GOSUB 1560
1130 GOTO 980
1140 IF C$ <> "COUNT" THEN 1170
1150 GOSUB 1880
1160 GOTO 980
1170 IF C$ <> "SIMULATE" THEN 1200
1180 GOSUB 1940
1190 GOTO 980
1200 IF C$ <> "PROB" THEN 1230
1210 GOSUB 3130
1220 GOTO 980
1230 IF C$ <> "RESTART" THEN 1260
1240 M$="BETA"
1250 GOTO 480
1260 IF C$ = "EXIT" THEN 6070
1270 PRINT "*** ERROR *** PLEASE TRY AGAIN !!!"
1280 GOTO 980
1290 REM    CALCULATE ACTIVITY VARIANCE AND DETERMINE SCHEDULE
1300 B=6.0
1310 IF M$="BETA" THEN 1330
1320 B=3.2
1330 FOR I=1 TO M
1340 IF P(6,I) > 0.00001 THEN 1390
1350 P(6,I)=0.
1360 P(7,I)=0.
1370 U(I)=0.
1380 GOTO 1420
1390 P(7,I)=(P(5,I)-P(3,I))/B
1400 U(I)=P(7,I)
1410 P(7,I)=U(I)*U(I)
1420 NEXT I
1430 IF M$<>"BETA" THEN 1450
1440 GOSUB 5130
1450 K0=0
1460 GOSUB 3440
1470 GOSUB 3500
1480 V2=0.
1490 FOR I=1 TO M
1500 IF P(12,I) > 0.00001 THEN 1520
1510 V2=V2+P(7,I)
1520 NEXT I
1530 T4=T
1540 V4=SQR(V2)
1550 RETURN
1560 REM    *** TABLE COMMAND ***
1570 IF N1 <= 0 THEN 1590
1580 GOSUB 1290
1590 GOSUB 3960
1600 PRINT "P E R T  TABLE OUTPUT"
1610 PRINT "--------------------"
1620 GOSUB 3370
```

```
1630 PRINT "EXPECTED PROJECT DURATION   ";T4
1640 PRINT "STD.DEV. OF PROJECT DURATION   ";V4
1650 PRINT
1660 PRINT "PROJECT START DATE  :  ";S
1670 S1=S+T4-1
1680 PRINT "EXPECTED COMPLETION DATE  :  ";S1
1690 PRINT
1700 L8=66
1710 GOSUB 4020
1720 PRINT "   I  J    A     M     B    E(T)";
1730 PRINT "    V(T)   E(ES)   E(EF)   E(LS)   E(LF)   E(TF)"
1740 GOSUB 4020
1750 FOR I=1 TO M
1760 K=A(I)
1770 G$=" "
1780 IF P(12,K) > 0 THEN 1800
1790 G$=" CRITICAL"
1800 PRINT USING F$(1),P(1,K),P(2,K),P(3,K),P(4,K),P(5,K),P(6,K);
1810 FOR J= 7 TO 12
1820 PRINT USING F$(10),P(J,K);
1830 NEXT J
1840 PRINT G$
1850 NEXT I
1860 GOSUB 3960
1870 RETURN
1880 REM    *** COUNT COMMAND ***
1890 PRINT "ENTER THE SIMULATION COUNTER (INTEGER > 0)  ";
1900 INPUT U9
1910 PRINT
1920 IF U9 <= 0 THEN 1890
1930 RETURN
1940 REM    *** SIMULATE COMMAND ***
1950 IF N1 > 0 THEN 2000
1960 PRINT "HOW MANY TRIALS   ";
1970 INPUT N0
1980 IF N0 <= 0 THEN 980
1990 GOTO 2030
2000 PRINT "HOW MANY MORE TRIALS   ";
2010 INPUT N0
2020 IF N0 <= 0 THEN 3120
2030 N1=N1+N0
2040 REM    SIMULATE N0 TIMES
2050 K0=1
2060 PRINT
2070 PRINT "S I M U L A T I N G, PLEASE STAND BY ... !!!"
2080 Z9=0
2090 FOR K1=1 TO N0
2100 Z9=Z9+1
2110 W9=INT(Z9/U9)
2120 IF W9*U9 <> Z9 THEN 2140
2130 PRINT "SIMULATED   ";Z9
2140 GOSUB 4090
2150 GOSUB 3500
2160 T0=T0+T
```

```
2170 V0=V0+T*T
2180 FOR I=1 TO M
2190 H(I)=H(I)+G(I)
2200 V(I)=V(I)+G(I)*G(I)
2210 IF P(12,I) > 0.00001 THEN 2230
2220 Q(I)=Q(I)+1
2230 NEXT I
2240 FOR J=I5 TO M4
2250 IF T > Y(J) THEN 2280
2260 Z(J)=Z(J)+1
2270 GOTO 2310
2280 NEXT J
2290 J=M5
2300 Z(J)=Z(J)+1
2310 IF J >= N4 THEN 2330
2320 N4=J
2330 IF J <= N5 THEN 2350
2340 N5=J
2350 NEXT K1
2360 PRINT
2370 GOSUB 3960
2380 PRINT "P E R T   SIMULATION OUTPUT"
2390 PRINT "--------------------------"
2400 GOSUB 3370
2410 PRINT "NUMBER OF TRIALS ",N1
2420 T7=T0/N1
2430 IF N1 <= 1 THEN 2480
2440 V7=V0-T0*T0/N1
2450 IF V7 <= 0 THEN 2480
2460 V7=SQR(V7/(N1-1))
2470 GOTO 2490
2480 V7=0
2490 PRINT "MEAN CRITICAL PATH DURATION",T7
2500 PRINT "STD.DEV. OF CRITICAL PATH DURATION",V7
2510 V7=V7/SQR(N1)
2520 PRINT "STD.DEV. OF MEAN CRITICAL PATH DURATION",V7
2530 PRINT
2540 PRINT "PROBABILITY DISTRIBUTION OF CRITICAL PATH DURATION"
2550 L8=50
2560 GOSUB 4020
2570 PRINT F$(3)
2580 PRINT F$(4);
2590 P9=-1
2600 FOR I=I5 TO M5
2610 IF Z(I) <= P9 THEN 2630
2620 P9=Z(I)
2630 NEXT I
2640 P9=P9/N1
2650 GOSUB 4740
2660 PRINT USING F$(9),L0;
2670 FOR I=1 TO 28
2680 PRINT " ";
2690 NEXT I
2700 PRINT USING F$(9),U0;
```

```
2710 PRINT "I"
2720 K4=N4-2
2730 IF K4 >= I5 THEN 2750
2740 K4=I5
2750 K5=N5+2
2760 IF K5 <= M5 THEN 2780
2770 K5=M5
2780 Z7=Z(K4)/N1
2790 Z8=Z7
2800 I=0
2810 PRINT USING F$(5),I,Y(K4),Z7,Z8;
2820 GOSUB 4980
2830 FOR I=K4+1 TO K5-1
2840 Z7=Z(I)/N1
2850 Z8=Z8+Z7
2860 PRINT USING F$(5),Y(I-1),Y(I),Z7,Z8;
2870 GOSUB 4980
2880 NEXT I
2890 Z7=Z(K5)/N1
2900 Z8=Z8+Z7
2910 PRINT USING F$(5),Y(K5-1);
2920 PRINT "        UP";
2930 PRINT USING F$(6),Z7,Z8;
2940 GOSUB 4980
2950 PRINT
2960 PRINT
2970 PRINT "PROBABILITY THAT AN ACTIVITY LIES ON A CRITICAL PATH"
2980 L8=52
2990 GOSUB 4020
3000 PRINT
3010 PRINT F$(7)
3020 FOR I=1 TO M
3030 K=A(I)
3040 Z7=H(K)/N1
3050 Z8=Q(K)/N1
3060 Z6=0.
3070 IF N1 <= 1 THEN 3090
3080 Z6=(V(K)-H(K)*H(K)/N1)/(N1-1)
3090 PRINT USING F$(8),P(1,K),P(2,K),Z7,Z6,Q(K),Z8
3100 NEXT I
3110 GOSUB 3960
3120 RETURN
3130 REM *** PROB COMMAND ***
3140 REM CALCULATE PROB(PROJECT DURATION <= SPECIFIED DATE)
3150 PRINT
3160 PRINT "PROB(PROJECT WILL FINISH ON OR BEFORE TARGET DATE)"
3170 L8=50
3180 GOSUB 4020
3190 GOSUB 3370
3200 PRINT
3210 PRINT "ENTER TARGET FINISH DATE (OR -1 TO RETURN)"
3220 PRINT "TARGET FINISH DATE  :  ";
3230 INPUT B0
3240 IF B0 < 0 THEN 3360
```

```
3250 G0=B0-S+1
3260 IF V4 <= 0.00001 THEN 3300
3270 G0=(G0-T4)/V4
3280 GOSUB 5630
3290 GOTO 3340
3300 IF B0 < S1 THEN 3330
3310 W1=1.0
3320 GOTO 3340
3330 W1=0.0
3340 PRINT "PROBABILITY  =  ",W1
3350 GOTO 3220
3360 RETURN
3370 REM
3380 PRINT
3390 PRINT "P R O J E C T   :   ";T$
3400 PRINT
3410 PRINT "M O D E L      :   ";M$
3420 PRINT
3430 RETURN
3440 REM
3450 REM    COPY E(T) TO G(.) AND WORK WITH G(.)
3460 FOR I=1 TO M
3470 G(I)=P(6,I)
3480 NEXT I
3490 RETURN
3500 T=0.
3510 FOR I=1 TO N
3520 E(I)=0
3530 NEXT I
3540 REM
3550 REM    F O R W A R D    P A S S
3560 REM    ------------------------
3570 FOR I=1 TO M
3580 K=A(I)
3590 I1=P(1,K)
3600 J1=P(2,K)
3610 P(8,K)=E(I1)
3620 P(9,K)=E(I1)+G(K)
3630 IF P(9,K) <= E(J1) THEN 3650
3640 E(J1)=P(9,K)
3650 IF P(9,K) <= T THEN 3670
3660 T=P(9,K)
3670 NEXT I
3680 REM
3690 FOR I=1 TO N
3700 L(I)=T
3710 NEXT I
3720 REM
3730 REM    B A C K W A R D    P A S S
3740 REM    --------------------------
3750 FOR J=1 TO M
3760 I=M-J+1
3770 K=A(I)
3780 I1=P(1,K)
```

```
3790 J1=P(2,K)
3800 P(11,K)=L(J1)
3810 P(10,K)=L(J1)-G(K)
3820 IF P(10,K) >= L(I1) THEN 3840
3830 L(I1)=P(10,K)
3840 NEXT J
3850 REM
3860 REM    DETERMINE TOTAL FLOAT FOR EACH ACTIVITY
3870 FOR I= 1 TO M
3880 P(12,I)=P(10,I)-P(8,I)
3890 IF K0 > 0 THEN 3940
3900 P(8,I)=P(8,I)+S
3910 P(9,I)=P(9,I)+S-1
3920 P(10,I)=P(10,I)+S
3930 P(11,I)=P(11,I)+S-1
3940 NEXT I
3950 RETURN
3960 REM
3970 REM    *** SKIP L9 LINES ***
3980 FOR I=1 TO L9
3990 PRINT
4000 NEXT I
4010 RETURN
4020 REM    *** PRINT A LINE ACROSS THE PAGE OF LENGTH L8 ***
4030 FOR I=1 TO L8
4040 PRINT "-";
4050 NEXT I
4060 PRINT " "
4070 RETURN
4080 REM    *** GENERATE RANDOM ACTIVITY TIMES BETA/NORMAL ***
4090 IF M$="BETA" THEN 4260
4100 REM    GENERATE NORMAL DEVIATES
4110 FOR I=1 TO M
4120 IF P(6,I) <= 0.00001 THEN 4240
4130 IF P(7,I) > 0.00001 THEN 4160
4140 G(I)=P(6,I)
4150 GOTO 4240
4160 T5=0.
4170 FOR J=1 TO 12
4180 R=RND(X9)
4190 T5=T5+R
4200 NEXT J
4210 G(I)=U(I)*(T5-6)+P(6,I)
4220 IF G(I) >= 0 THEN 4240
4230 G(I)=0
4240 NEXT I
4250 GOTO 4410
4260 REM    GENERATE BETA VARIATE AS THE RATIO OF TWO GAMMAS.
4270 FOR I=1 TO M
4280 IF P(6,I) <= 0.00001 THEN 4400
4290 IF P(7,I) > 0.00001 THEN 4320
4300 G(I)=P(6,I)
4310 GOTO 4400
4320 R=RND(X9)
```

```
4330 IF R <=0 THEN 4320
4340 G0=(-P(13,I)*LOG(R))**P(14,I)
4350 R=RND(X9)
4360 IF R <= 0 THEN 4350
4370 G1=(-P(15,I)*LOG(R))**P(16,I)
4380 G(I)=G0/(G0+G1)
4390 G(I)=P(3,I)+G(I)*(P(5,I)-P(3,I))
4400 NEXT I
4410 RETURN
4420 REM    INITIALIZING COUNTERS READY FOR SIMULATION
4430 N1=0
4440 FOR I=1 TO M
4450 Q(I)=0.
4460 H(I)=0.
4470 V(I)=0.
4480 NEXT I
4490 REM    GET CLASS LIMITS FOR PROJECT DURATION FREQ. DISTRIB.
4500 V3=V4/4
4510 Y(21)=T
4520 FOR J=1 TO 20
4530 J6=21-J
4540 J7=21-J+1
4550 J8=21+J
4560 Y(J6)=Y(J7)-V3
4570 Y(J8)=Y(J8-1)+V3
4580 NEXT J
4590 FOR J=J8+1 TO M4
4600 Y(J)=Y(J-1)+V3
4610 NEXT J
4620 FOR I=1 TO M4
4630 IF Y(I) > 0 THEN 4650
4640 NEXT I
4650 I5=I
4660 T0=0.
4670 V0=0.
4680 FOR J=1 TO M5
4690 Z(J)=0.
4700 NEXT J
4710 N4=M5
4720 N5=1
4730 RETURN
4740 REM    SCALING ROUTINE FOR VALUES >= 0.
4750 L0=0
4760 A1=P9
4770 Y1=A1
4780 A2=A1/4
4790 A3=LOG(A2)/LOG(10)
4800 REM    ROUND DOWN A3 TO A4
4810 A4=INT(A3)
4820 IF A4 <= A3 THEN 4870
4830 A4=A4-1
4840 A5=10**A4
4850 A6=A2/A5
4860 REM    ROUND UP A6 TO A7
```

```
4870 A7=INT(A6+1)
4880 H0=A5*A7
4890 U0=L0+5*H0
4900 IF U0 >= A1 THEN 4930
4910 A1=A1+A2/2
4920 GOTO 4780
4930 IF U0-H0 <= Y1 THEN 4960
4940 U0=U0-H0
4950 GOTO 4930
4960 H0=(U0-L0)/40
4970 RETURN
4980 REM
4990 REM    PLOT THE PROBABILITY, Z7
5000 K9=INT(Z7/H0+0.5)
5010 P$(0)="  I"
5020 FOR J=1 TO K9
5030 P$(J)="*"
5040 NEXT J
5050 FOR J=K9+1 TO 40
5060 P$(J)=" "
5070 NEXT J
5080 FOR J=0 TO 40
5090 PRINT P$(J);
5100 NEXT J
5110 PRINT "I"
5120 RETURN
5130 REM
5140 REM    CALCULATE PARAMETERS OF BETA DISTRIBUTION
5150 FOR I=1 TO M
5160 IF P(7,I) <= 0.00001 THEN 5210
5170 A1=(4*P(4,I)+P(5,I)-5*P(3,I))
5180 A1=A1/(6*(P(5,I)-P(3,I)))
5190 P(13,I)=36*A1*(A1-A1*A1-1/36)
5200 P(15,I)=(P(13,I)-P(13,I)*A1)/A1
5210 NEXT I
5220 REM
5230 REM    CALCULATE PARAMETERS OF DON T. PHILLIPS'S ALGORITHM
5240 REM    --------------------------------------------------
5250 FOR J=13 TO 15 STEP 2
5260 FOR I=1 TO M
5270 IF P(7,I) <= 0.00001 THEN 5600
5280 W1=P(J,I)
5290 W2=W1*W1
5300 X3=1.0
5310 IF W1 <= 1.5 THEN 5340
5320 IF W1 <= 19.0 THEN 5380
5330 GOTO 5420
5340 B0=0.24797 + 1.34735740*W1 - 1.00004204*W2
5350 B0=B0 + 0.53203176*W1*W2 - 0.13671536*W2*W2
5360 B0=B0 + 0.01320864*W1*W2*W2
5370 GOTO 5440
5380 B0=0.64350 + 0.45839602*W1 - 0.02952801*W2
5390 B0=B0 + 0.00172718*W1*W2 - 0.00005810*W2*W2
5400 B0=B0 + 0.00000082*W1*W2*W2
```

```
5410 GOTO 5440
5420 B0=1.33408 + 0.22499991*W1 - 0.00230695*W2
5430 B0=B0 + 0.00001623*W1*W2 - 0.00000006*W2*W2
5440 Y0=1.0 + 1.0/B0
5450 IF Y0-1. < 0. THEN 5500
5460 IF Y0-1. = 0. THEN 5550
5470 Y0=Y0-1.0
5480 X3=X3*Y0
5490 GOTO 5450
5500 G9=Y0*( 0.2548205 - 0.05149930*Y0)
5510 G9=Y0*( 0.8328212 + Y0*( -0.5684729 + G9))
5520 G9=Y0*( 0.985854  + Y0*( -0.8764218 + G9))
5530 G9=1.0 + Y0*( -0.5771017 + G9)
5540 X3=X3*G9/Y0
5550 A1=(X3/W1)**B0
5560 B0=1.0/B0
5570 A1=1.0/A1
5580 P(J,I)=A1
5590 P(J+1,I)=B0
5600 NEXT I
5610 NEXT J
5620 RETURN
5630 REM
5640 REM    CALC. CUMULATIVE NORMAL PROB VIA HASTINGS'S APPROX.
5650 REM    --------------------------------------------------
5660 W2=ABS(G0)
5670 Y0=1.0/(1.0+0.2316419*W2)
5680 A1=0.3989423*EXP(-G0*G0/2.0)
5690 W1=((1.330274*Y0-1.821256)*Y0+1.781478)*Y0
5700 W1=1.0-A1*Y0*((W1-0.3565638)*Y0+0.3193815)
5710 IF G0 >= 0. THEN 5730
5720 W1=1.0-W1
5730 RETURN
5740 REM    I N I T I A L I Z A T I O N
5750 M$="BETA"
5760 F$(1)=" ## ## ### ### ### ###.#"
5770 F$(2)="####.#"
5780 F$(3)="      DURATION          CUM."
5790 F$(4)="    FROM      TO    PROB  PROB.   I"
5800 F$(5)="#####.# #####.# #.#### #.####"
5810 F$(6)="#.#### #.####"
5820 F$(7)="   I   J       E(T)     V(T)   FREQUENCY PROBABILITY"
5830 F$(8)="##### #### ######.# ######.# ##########   #.####"
5840 F$(9)="#.####"
5850 F$(10)="#####.#"
5860 L9=3
5870 U9=100
5880 RETURN
5890 REM   *** HELP COMMAND ***
5900 PRINT "AVAILABLE COMMAND"
5910 PRINT "================="
5920 PRINT "HELP     - PRINT THIS MESSAGE"
5930 PRINT "BETA     - SELECT BETA MODEL"
5940 PRINT "NORMAL   - SELECT NORMAL MODEL"
```

```
5950 PRINT "TABLE    - PRINT TABLE OF PERT OUTPUT"
5960 PRINT "COUNT    - SET SIMULATION COUNTER"
5970 PRINT "SIMULATE - SIMULATE THE NETWORK BASED ON SELECTED MODEL"
5980 PRINT "RESTART  - START A NEW PROBLEM"
5990 PRINT "EXIT     - EXIT FROM THE PROGRAM"
6000 RETURN
6010 REM    *** BETA/NORMAL COMMANDS ***
6020 IF M$=C$ THEN 6060
6030 M$=C$
6040 GOSUB 1290
6050 GOSUB 4420
6060 RETURN
6070 END
```

IV. DIMENSIONAL SPECIFICATIONS

Program CPM

Currently, the maximum number of activities, m, and the highest node number, n, in the networks are set at 50 and 60, respectively. They can be altered by replacing the m and n in the following statements with appropriate values:

Line No.
```
30 DIM A(m),P(10,m),E(n),F(n),P$(40),F$(3)
40 N9=n
50 M9=m
```

Program PERT

Similar to the CPM program, the limits on m and n for the PERT program are set at 50 and 60. They can be changed by modifying the following statements:

Line No.
```
50 DIM A(m),P(16,m),E(n),L(n),Q(m),H(m),G(m),U(m),V(m)
70 N9=n
80 M9=m
```

V. REFERENCES

1. Abramowitz, M., and I.A. Stegun, HANDBOOK OF MATHEMATICAL FUNCTIONS, Dover Publications, New York, 1970.

2. Antill, J.M., and R.W. Woodhead, CRITICAL PATH METHODS IN CONSTRUCTION PRACTICE, 2nd ed., Wiley-Interscience, New York, 1970.

3. Archibald, R.D., and R.L. Villoria, NETWORK-BASED MANAGEMENT SYSTEMS (PERT/CPM), John Wiley, New York, 1967.

4. Elmaghraby, S.E., "The Theory of Networks and Management Science: Part II," MANAGEMENT SCIENCE, 17, No. 2 (October 1970), B54-71.

5. King, W.R., "Network Simulation Using Historical Estimating Behavior," AIIE TRANSACTIONS, 3, No. 2 (June 1971), 150-55.

6. Kerzner, H., PROJECT MANAGEMENT - A SYSTEMS APPROACH TO PLANNING, SCHEDULING AND CONTROLLING, Van Nostrand Reinhold, New York, 1979.

7. Klingel, A.R.,Jr., "Bias in PERT Project Completion Time Calculations for a Real Network," MANAGEMENT SCIENCE, 13, No. 4 (1966), B194-201.

8. Moder, J.J., and C.R. Phillips, PROJECT MANAGEMENT WITH CPM AND PERT, 2nd ed., Van Nostrand Reinhold, New York, 1970.

9. Naylor, T.H., J.L. Balintfy, D.S. Burdick, and K. Chu, COMPUTER SIMULATION TECHNIQUES, John Wiley, New York, 1966.

10. Phillips, C.R., "Fifteen Key Features of Computer Programs for CPM and PERT," JOURNAL OF INDUSTRIAL ENGINEERING, 15, No. 1 (January-February 1964), 14-20.

11. Phillips, D.T., "Generation of Random Gamma Variates from the Two-Parameter Gamma," AIIE TRANSACTIONS, 3, No. 3 (September 1971), 191-98.

12. Shannon, R.E., SYSTEMS SIMULATION - THE ART AND SCIENCE, Prentice-Hall, Englewood Cliffs, N.J., 1975.

13. Van Slyke, R.M., "Monte Carlo Methods and the PERT Problem," OPERATIONS RESEARCH, 11, No. 5 (1963), 839-60.

CHAPTER 6

WORK MEASUREMENT

I. Time Study

　　The Program TIME.STY
　　○ Input
　　○ Output
　　○ Example
　　○ TIME.STY Example Terminal Session 1
　　○ TIME.STY Example Terminal Session 2
　　○ TIME.STY Program Listing

II. Work Sampling

　　The Program WORK.SMPLG
　　○ Input
　　○ Output
　　○ Example
　　○ WORK.SMPLG Example Terminal Session 1
　　○ WORK.SMPLG Example Terminal Session 2
　　○ WORK.SMPLG Program Listing

III. Dimensional Specifications

IV. References

Work measurement is concerned with setting labor standards, that is, the amount of time that should be reasonably incurred in performing a specified task by an average operator (operator of average skill and aptitude) under normal working conditions. The three formal methods of determining task time upon which standards are based are

- Time Study
- Work Sampling
- Predetermined Motion Time Systems

This chapter contains two computer programs dealing with those portions of the Time Study and Work Sampling procedures in which computers can be useful. Proprietary packages of the two most often used Predetermined Motion Time Systems, namely, Methods Time Measurement (MTM) [3] and Work Factor (WF) [6], have been developed and are available. Section I deals with the Time Study Method, and Section II discusses the Work Sampling Procedure.

I. Time Study

The stopwatch time study technique of work measurement was introduced by Frederick W. Taylor in 1881. The labor standards for an average operator working under normal working conditions are estimated by observing a single worker performing the task according to the standard method. It is applied to repetitive existing tasks that are of short duration. The labor standards are usually expressed in minutes per unit of output. The following procedure is generally employed in a stopwatch time study:

1. Select the task and define the standard method.
2. Divide the task into basic work elements.
3. Select an appropriate operator (operator of good skill and aptitude).
4. Observe and record the actual time, X_{ij}, required for each work element i, over j cycles (at least 5-10 cycles), in 0.01 minutes.
5. Rate the operator performance for each work element with respect to normal pace.
6. Calculate the appropriate number of cycles for each element i, for the desired accuracy and level of confidence, N_i:

$$N_i = [100 Z S_i / (A \bar{X}_i)]^2$$

202

where Z = Confidence level factor - i.e., value from standard normal distribution,
A = Desired accuracy, expressed as a percentage of the true value,
S_i = Estimated standard deviation of the distribution of performance time for element i

$$= \sqrt{\sum_{j=1}^{n} (X_{ij} - \overline{X}_i)^2 / (n-1)}$$

where $\overline{X}_i = \sum_{j=1}^{n} X_{ij} / n$

n = Number of cycles observed before computing the desired number of cycles.

The largest value of N_i, denoted by N, will be the appropriate number of cycles for the task.

7. Observe and record the actual time for additional cycles (N-n) if required.

8. Compute the average performance time for each work element:

$$\overline{X}_i = \sum_{j=1}^{N} X_{ij} / N \ .$$

9. Compute the normal time for each element in minutes:

$$\{NT_i\} = (\overline{X}_i / 100) \cdot PR_i$$

where PR_i = Performance rating of element i.

10. Compute normal time for the task in minutes:

$$\{NT\} = \sum_{i=1}^{m} NT_i$$

where m = Number of elements in a task.

11. Compute standard time for the task in minutes:

$$\{ST\} = \{NT\}(1 + \text{Allowance Fraction}).$$

Allowance Fraction represents allowances for personal needs, unavoidable delays, and worker fatigue.

12. Compute standard unit of output per hour:

$$\{SU\} = 60 / \{ST\} \ .$$

The Program TIME.STY

TIME.STY has been designed such that it can accommodate virtually an unlimited number of observed cycles. Unlike most of the programs in this book whose input data are prompted from the terminal, we ask the users to input data through the use of the DATA statements in BASIC.

After the analysis of the first set of observed data, the procedure may call for an additional set of data to be collected in order to satisfy the desired percent accuracy and confidence level. In our context, a set of data refers to a group of observed cycles. On most computer systems, separate data files can be created for each set of data and appended to the program before running. The line numbers in BASIC provide a means to divide the data into blocks logically, with each block representing a data set.

The program TIME.STY assumes that we have at least one data set. Additional data sets, when appended to the program, can enter the calculations via the user's command -i.e., the ADD command. Each time the ADD command is invoked, the program will read in another set of data. Hence, if it is executed the first time, we will be combining the second data set with the first one. Normally, only one additional data set is required if the first set does not provide sufficient accuracy.

INPUT
-----
1. The company's name
2. The department's name
3. The task description
4. The operator's name
5. The analyst's name
6. The study date
7. Percent personal allowance
8. Percent unavoidable delay allowance
9. Percent fatigue allowance
10. Number of work elements in the task
11. Descriptions of all work elements in the task; each description is limited to 24 characters in the output
12. The performance ratings of all work elements
13. Type of time data -i.e., "SNAPBACK" or "CONTINUOUS"
14. First time data set
15. Second time data set
16. Third time data set, etc.

Each time data set consists of the number of observed cycles in the set, and the observed times of every work element arranged in order of the work elements and grouped in one cycle at a time. It is important to note that the observed times must be entered in units of 0.01 minutes.

OUTPUT

The main output from the program is a report from the ANALYZE command. The report is a summary of the average times, the normal times, the standard times, and the desired number of cycles by work elements. It also gives the additional number of observations required to obtain the desired accuracy and confidence level, as well as the standard number of units of output per hour. The program also contains commands for changing the desired accuracy (ACC command), the confidence level (CL command), and the three percent allowances (ALLOW command). The STATUS command displays the values of these parameters.

EXAMPLE

Suppose John Doe, a time study analyst at XYZ company, is setting a time standard for an assembly operation that consists of five work elements. An initial set of 10 cycles of observations has been collected using the "continuous method" of timing. The data are given below in terms of 0.01 minutes and the performance ratings of the operator named J. Dooer.

	Perform.	Cycle Number									
Element	Rating	1	2	3	4	5	6	7	8	9	10
1	0.98	23	94	171	243	320	393	469	547	625	704
2	1.00	36	108	187	258	335	406	485	564	641	721
3	1.05	52	129	205	275	353	425	502	582	660	741
4	0.99	62	139	214	285	362	437	511	593	670	750
5	1.00	74	150	224	298	374	448	524	605	681	763

John Doe wishes to determine the standard time of the operation with an accuracy of 5 percent at a confidence level of 95 percent. The allowances for personal, unavoidable delay, and fatigue are 5 percent, 7 percent, and 6 percent, respectively.

There are two example terminal sessions related to this example. The first session pertains to the initial set of data, and the second session considers the analysis as a result of collecting additional data.

Consider, now, the initial set of data which must be appended to the end of the program before it is executed. For our example, we append the following statements. Note that the line numbers have been chosen such that they follow the source listing of the program.

```
3000 REM-----COMPANY NAME-----
3010 DATA "XYZ COMPANY"
3020 REM-----DEPARTMENT NAME-----
3030 DATA "ASSEMBLY"
3040 REM-----TASK DESCRIPTION-----
3050 DATA "AN ASSEMBLY OPERATION"
3060 REM-----OPERATOR'S NAME-----
3070 DATA "J. DOOER"
3080 REM-----TIME STUDY ANALYST'S NAME-----
3090 DATA "JOHN DOE"
3100 REM-----STUDY DATE-----
3110 DATA "XX/XX/XX"
3120 REM-----PERSONAL ALLOWANCE (PERCENT)-----
3130 DATA 5
3140 REM-----UNAVOIDABLE DELAY (PERCENT)-----
3150 DATA 7
3160 REM-----FATIGUE ALLOWANCE (PERCENT)-----
3170 DATA 6
3180 REM-----NUMBER OF WORK ELEMENTS IN THE TASK-----
3190 DATA 5
3200 REM-----DESCRIPTIONS OF ALL WORK ELEMENTS-----
3205 DATA "FIRST WORK ELEMENT"
3210 DATA "SECOND WORK ELEMENT"
3215 DATA "THIRD WORK ELEMENT"
3220 DATA "FOURTH WORK ELEMENT"
3225 DATA "FIFTH WORK ELEMENT"
3500 REM-----PERFORMANCE RATINGS OF ALL WORK ELEMENTS----
3510 DATA 0.98,1.00,1.05,0.99,1.00
3700 REM-----TYPE OF DATA - "SNAPBACK" OR "CONTINUOUS"---
3710 DATA "CONTINUOUS"
3720 REM
3730 REM    WORK ELEMENT TIME DATA
3740 REM    ----------------------
3750 REM-----FIRST SET OF TIME DATA STARTS HERE-----
3760 REM-----NUMBER OF CYCLES IN THE FIRST DATA SET-----
3770 DATA 10
3780 REM-----OBSERVED TIME DATA IN THE FIRST SET-----
3790 DATA    23,  36,  52,  62,  74
3800 DATA    94, 108, 129, 139, 150
3810 DATA   171, 187, 205, 214, 224
3820 DATA   243, 258, 275, 285, 298
3830 DATA   320, 335, 353, 362, 374
3840 DATA   393, 406, 425, 437, 448
3850 DATA   469, 485, 502, 511, 524
3860 DATA   547, 564, 582, 593, 605
3870 DATA   625, 641, 660, 670, 681
3880 DATA   704, 721, 741, 750, 763
```

The first example terminal session below shows that the program starts by reading the general input data of the first set. The HELP command is entered to display all available commands. The STATUS command shows the default values of the percent accuracy and the confidence level as 5 percent and 95 percent, respectively. The ANALYZE command produces the report which calls for 6 more cycles of observations in order to satisfy the desired accuracy and confidence level.

We assume that John Doe has collected the additional data with the following results.

Element	Performance Rating	1	2	Cycle Number 3	4	5	6
1	0.98	22	94	168	243	318	393
2	1.00	36	110	183	259	333	407
3	1.05	54	127	201	278	352	424
4	0.99	63	137	211	288	362	434
5	1.00	75	147	223	299	373	447

In this case, the performance ratings are assumed to be the same as those of the first set. It may turn out that new ratings are appropriate. Since we will be combining data with the first set, it may be appropriate to take the weighted average of the two sets and use the results as the overall ratings for the two sets of data combined. The corresponding DATA statements in the program (i.e., line number 3510 in our example) should be changed. The weight to be assigned to each set can be based on the number of cycles observed.

Consider the second example terminal session which follows the first one below. The following DATA statements for the second set of time data are appended behind the first set before the program is run:

```
4000 REM-----SECOND SET OF TIME DATA STARTS HERE-----
4010 REM-----NUMBER OF CYCLES IN THE SECOND DATA SET-----
4020 DATA 6
4030 REM-----OBSERVED TIME DATA IN SECOND SET-----
4035 DATA    22,  36,   54,   63,   75
4040 DATA    94, 110,  127,  137,  147
4045 DATA   168, 183,  201,  211,  223
4050 DATA   243, 259,  278,  288,  299
4055 DATA   318, 333,  352,  362,  373
4060 DATA   393, 407,  424,  434,  447
```

After the program reads the first set of data, we issue the ADD command so that the second set of data is combined with the first one before the ANALYZE command is given. The ANALYZE command shows that the degree of accuracy of the time values is within the desired level with 95 percent confidence and no additional data need to be collected. The standard output per hour for this assembly operation is estimated at 66.89 units.

```
TIME.STY EXAMPLE TERMINAL SESSION 1
===================================

T I M E    S T U D Y
--------------------

NEED INTRODUCTION (Y OR N) ... !N

READING GENERAL INPUT DATA ...
READING WORK ELEMENT DESCRIPTIONS ...
READING WORK ELEMENT PERFORMANCE RATINGS ...
READING WORK ELEMENT TIME DATA FOR SET 1
READING 10 CYCLES ...
    CYCLE  1
    CYCLE  2
    CYCLE  3
    CYCLE  4
    CYCLE  5
    CYCLE  6
    CYCLE  7
    CYCLE  8
    CYCLE  9
    CYCLE  10

PLEASE ENTER COMMAND OR TYPE HELP

COMMAND --> !HELP

AVAILABLE COMMANDS
==================
HELP    - PRINT THIS MESSAGE
ACC     - SPECIFY THE DESIRED PERCENT ACCURACY
CL      - SPECIFY THE PERCENT CONFIDENCE LEVEL
ALLOW   - CHANGE THE THREE PERCENT ALLOWANCES
STATUS  - PRINT STATUS OF % ACCURACY, CONFIDENCE LEVEL,
          AND THE PERCENT ALLOWANCES
ANALYZE - PERFORM ANALYSIS ON THE DATA
ADD     - ADD ADDITIONAL DATA TO THE ANALYSIS
EXIT    - EXIT FROM THE PROGRAM

COMMAND --> !STATUS
                TASK : AN ASSEMBLY OPERATION
                NUMBER OF WORK ELEMENTS = 5
                TYPE OF TIME DATA = CONTINUOUS
                NUMBER OF OBSERVED CYCLES = 10

       PERCENT     PERSONAL   UNAVOIDABLE-DELAY    FATIGUE    TOTAL
     ALLOWANCES      5.00           7.00             6.00     18.00

     %ACCURACY =   5.00                  %CONFIDENCE LEVEL = 95.00

COMMAND --> !ANALYZE
```

```
COMPANY : XYZ COMPANY                              DATE : XX/XX/XX
DEPARTMENT : ASSEMBLY
-----------------------------------------------------------------
    OPERATOR : J. DOOER          ANALYST : JOHN DOE
-----------------------------------------------------------------
    TASK : AN ASSEMBLY OPERATION
    NUMBER OF WORK ELEMENTS = 5
    TYPE OF TIME DATA = CONTINUOUS
    NUMBER OF OBSERVED CYCLES = 10

     PERCENT     PERSONAL   UNAVOIDABLE-DELAY   FATIGUE    TOTAL
    ALLOWANCES    5.00            7.00           6.00      18.00

   %ACCURACY =  5.00              %CONFIDENCE LEVEL = 95.00
-----------------------------------------------------------------
ELE-                         AVERAGE  PERFORM  NORMAL   STD.  DESIRED
MENT  WORK ELEMENT DESCRIPTION  TIME   RATING   TIME    TIME   CYCLES
 1    FIRST WORK ELEMENT       .211    .980    .207    .244      9
 2    SECOND WORK ELEMENT      .152   1.000    .152    .179     14
 3    THIRD WORK ELEMENT       .183   1.050    .192    .227     10
 4    FOURTH WORK ELEMENT      .099    .990    .098    .116     16
 5    FIFTH WORK ELEMENT       .118   1.000    .118    .139     12

            T O T A L    F O R    T A S K    .767    .905     16

ADDITIONAL OBSNS FOR DESIRED ACCURACY AND CONF. LEVEL = 6 CYCLES
STANDARD UNITS OF OUTPUT/HOUR = 66.29913373553


COMMAND --> !EXIT
```

TIME.STY EXAMPLE TERMINAL SESSION 2
====================================

T I M E   S T U D Y

NEED INTRODUCTION (Y OR N) ... !N

READING GENERAL INPUT DATA ...
READING WORK ELEMENT DESCRIPTIONS ...
READING WORK ELEMENT PERFORMANCE RATINGS ...
READING WORK ELEMENT TIME DATA FOR SET 1
READING 10 CYCLES ...
   CYCLE   1
   CYCLE   2
   CYCLE   3
   CYCLE   4
   CYCLE   5
   CYCLE   6
   CYCLE   7
   CYCLE   8
   CYCLE   9
   CYCLE 10

PLEASE ENTER COMMAND OR TYPE HELP

COMMAND --> !ADD

READING WORK ELEMENT TIME DATA FOR SET 2
READING 6 CYCLES ...
   CYCLE   1
   CYCLE   2
   CYCLE   3
   CYCLE   4
   CYCLE   5
   CYCLE   6

COMMAND --> !ANALYZE

```
COMPANY : XYZ COMPANY                                    DATE : XX/XX/XX
DEPARTMENT : ASSEMBLY
-----------------------------------------------------------------------
         OPERATOR : J. DOOER          ANALYST : JOHN DOE
-----------------------------------------------------------------------
         TASK : AN ASSEMBLY OPERATION
         NUMBER OF WORK ELEMENTS = 5
         TYPE OF TIME DATA = CONTINUOUS
         NUMBER OF OBSERVED CYCLES = 16

         PERCENT     PERSONAL    UNAVOIDABLE-DELAY    FATIGUE     TOTAL
       ALLOWANCES      5.00           7.00             6.00       18.00

       %ACCURACY =    5.00                %CONFIDENCE LEVEL = 95.00
-----------------------------------------------------------------------
ELE-                           AVERAGE  PERFORM  NORMAL    STD.  DESIRED
MENT   WORK ELEMENT DESCRIPTION   TIME   RATING   TIME     TIME  CYCLES
 1     FIRST WORK ELEMENT         .207    .980    .203     .240      8
 2     SECOND WORK ELEMENT        .151   1.000    .151     .178     11
 3     THIRD WORK ELEMENT         .182   1.050    .191     .225      8
 4     FOURTH WORK ELEMENT        .099    .990    .098     .115     10
 5     FIFTH WORK ELEMENT         .117   1.000    .117     .138     12

              T O T A L    F O R    T A S K    .760     .897     12

ADDITIONAL OBSNS FOR DESIRED ACCURACY AND CONF. LEVEL =  0 CYCLES
STANDARD UNITS OF OUTPUT/HOUR =  66.88639777314


COMMAND --> !EXIT
```

```
10 REM     TIME.STY
20 DIM X(50),Y(50),P(50),D$(50),F$(14),G$(6)
30 M9=50
40 GOSUB 2210
50 PRINT
60 PRINT "T I M E    S T U D Y"
70 PRINT "-------------------"
80 PRINT
90 PRINT "NEED INTRODUCTION (Y OR N) ... ";
100 INPUT Y$
110 IF Y$ <> "Y" THEN 230
120 PRINT
130 PRINT "THIS PROGRAM DETERMINES THE NUMBER OF OBSERVATIONS REQUIRED"
140 PRINT "IN A TIME STUDY FOR A SPECIFIED ACCURACY AND CONFIDENCE"
150 PRINT "LEVEL. IT IS CAPABLE OF ANALYZING THE SNAPBACK TIME DATA"
160 PRINT "AS WELL AS THE CONTINUOUS TIME DATA. IT COMPUTES THE"
170 PRINT "AVERAGE, NORMAL AND STANDARD TIMES FOR EACH WORK ELEMENT"
180 PRINT "AND FOR THE TASK AS A WHOLE."
190 PRINT
200 PRINT "INPUT REQUIREMENTS"
210 PRINT "------------------"
220 PRINT "DETAILS OF INPUT DATA ARE AVAILABLE IN THE DOCUMENT."
230 PRINT
240 PRINT "READING GENERAL INPUT DATA ..."
250 FOR I=1 TO 6
260 READ G$(I)
270 NEXT I
280 REM     PERCENT ALLOWANCES
290 READ F1,F2,F3
300 F0=F1+F2+F3
310 REM     NUMBER OF WORK ELEMENTS IN THE TASK
320 READ M
330 IF M <= M9 THEN 390
340 PRINT
350 PRINT "MAX ALLOWABLE NUMBER OF WORK ELEMENTS = ";M9
360 PRINT "SEE DIMENSIONAL SPECIFICATIONS FOR DETAILS."
370 PRINT "PROGRAM EXECUTION TERMINATED !!!"
380 GOTO 2480
390 REM     DESCRIPTIONS OF ALL WORK ELEMENTS
400 PRINT "READING WORK ELEMENT DESCRIPTIONS ..."
410 FOR I=1 TO M
420 READ D$(I)
430 NEXT I
440 REM     PERFORMANCE RATINGS OF ALL WORK ELEMENTS
450 PRINT "READING WORK ELEMENT PERFORMANCE RATINGS ... "
460 FOR I=1 TO M
470 READ P(I)
480 NEXT I
490 REM     SNAPBACK OR CONTINUOUS TIME DATA
500 READ S$
510 IF S$ = "SNAPBACK" THEN 570
520 IF S$ = "CONTINUOUS" THEN 590
530 PRINT
540 PRINT "*** INPUT DATA ERROR ***"
```

```
550 PRINT "SNAPBACK OR CONTINUOUS TIME DATA ???"
560 GOTO 370
570 S=0
580 GOTO 600
590 S=1
600 GOSUB 1990
610 PRINT
620 PRINT "PLEASE ENTER COMMAND OR TYPE HELP"
630 PRINT
640 PRINT "COMMAND --> ";
650 INPUT C$
660 PRINT
670 IF C$ <> "HELP" THEN 700
680 GOSUB 910
690 GOTO 630
700 IF C$ <> "ACC" THEN 730
710 GOSUB 1040
720 GOTO 630
730 IF C$ <> "CL" THEN 760
740 GOSUB 1090
750 GOTO 630
760 IF C$ <> "ALLOW" THEN 790
770 GOSUB 1260
780 GOTO 630
790 IF C$ <> "STATUS" THEN 820
800 GOSUB 1320
810 GOTO 630
820 IF C$ <> "ANALYZE" THEN 850
830 GOSUB 1450
840 GOTO 630
850 IF C$ <> "ADD" THEN 880
860 GOSUB 1990
870 GOTO 630
880 IF C$ = "EXIT" THEN 2480
890 PRINT "*** ERROR ***   PLEASE TRY AGAIN !!!"
900 GOTO 630
910 REM    *** HELP COMMAND ***
920 PRINT "AVAILABLE COMMANDS"
930 PRINT "=================="
940 PRINT "HELP    - PRINT THIS MESSAGE"
950 PRINT "ACC     - SPECIFY THE DESIRED PERCENT ACCURACY"
960 PRINT "CL      - SPECIFY THE PERCENT CONFIDENCE LEVEL"
970 PRINT "ALLOW   - CHANGE THE THREE PERCENT ALLOWANCES"
980 PRINT "STATUS  - PRINT STATUS OF % ACCURACY, CONFIDENCE LEVEL,"
990 PRINT "          AND THE PERCENT ALLOWANCES"
1000 PRINT "ANALYZE - PERFORM ANALYSIS ON THE DATA"
1010 PRINT "ADD     - ADD ADDITIONAL DATA TO THE ANALYSIS"
1020 PRINT "EXIT    - EXIT FROM THE PROGRAM"
1030 RETURN
1040 REM    *** ACC COMMAND ***
1050 PRINT "ENTER THE NEW DESIRED PERCENT ACCURACY  ";
1060 INPUT A0
1070 A9=A0*A0/10000
1080 RETURN
```

```
1090 REM       *** CL COMMAND ***
1100 PRINT "ENTER THE NEW PERCENT CONFIDENCE LEVEL   ";
1110 INPUT L9
1120 IF L9 <= 0 THEN 1140
1130 IF L9 < 100 THEN 1180
1140 PRINT
1150 PRINT "*** ERROR ***   CL MUST BE > 0 AND < 100"
1160 PRINT
1170 GOTO 1100
1180 C9=0.5-L9/200
1190 A2=LOG(1/C9/C9)
1200 A1=SQR(A2)
1210 Z=2.515517+0.802853*A1+0.010328*A2
1220 Z9=1.0+1.432788*A1+0.189269*A2+0.001308*A1*A2
1230 Z=A1-Z/Z9
1240 Z2=Z*Z
1250 RETURN
1260 REM       *** ALLOW COMMAND ***
1270 PRINT "ENTER PERCENT ALLOWANCES FOR PERSONAL, UNAVOIDABLE DELAY,"
1280 PRINT "AND FATIGUE SEPARATED BY COMMAS --   ";
1290 INPUT F1,F2,F3
1300 F0=F1+F2+F3
1310 RETURN
1320 REM       *** STATUS COMMAND ***
1330 PRINT TAB(15);"TASK : ";G$(3)
1340 PRINT TAB(15);"NUMBER OF WORK ELEMENTS = ";M
1350 PRINT TAB(15);"TYPE OF TIME DATA = ";S$
1360 PRINT TAB(15);"NUMBER OF OBSERVED CYCLES = ";N0
1370 PRINT
1380 PRINT TAB(4);F$(1)
1390 PRINT TAB(4);
1400 PRINT USING F$(2),F1,F2,F3,F0
1410 PRINT
1420 PRINT TAB(4);
1430 PRINT USING F$(3),A0,L9
1440 RETURN
1450 REM       *** ANALYZE COMMAND ***
1460 IF N0 >= 2 THEN 1500
1470 PRINT "NUMBER OF OBSERVED CYCLES MUST BE AT LEAST 2   !!!"
1480 PRINT "PLEASE ADD MORE DATA !!!"
1490 GOTO 1980
1500 PRINT
1510 PRINT
1520 PRINT USING F$(4),G$(1);
1530 PRINT USING F$(5),G$(6)
1540 PRINT USING F$(6),G$(2)
1550 GOSUB 1930
1560 PRINT TAB(8);
1570 PRINT USING F$(7),G$(4),G$(5)
1580 GOSUB 1930
1590 GOSUB 1320
1600 GOSUB 1930
1610 PRINT F$(8);F$(11)
1620 PRINT F$(9);F$(12)
```

```
1630 N9=0
1640 S1=0
1650 T1=0
1660 T3=0
1670 F9=F0/100
1680 FOR I=1 TO M
1690 N=Z2*(Y(I)-X(I)*X(I)/N0)/(N0-1)
1700 N=N/(A9*X(I)*X(I)/N0/N0)
1710 IF N <= N9 THEN 1730
1720 N9=N
1730 W1=P(I)*X(I)/N0
1740 T1=T1+W1
1750 T2=W1*(1+F9)
1760 T3=T3+T2
1770 PRINT USING F$(10),I,D$(I),X(I)/N0;
1780 PRINT USING F$(13),P(I),W1,T2,N
1790 NEXT I
1800 U=60/T3
1810 PRINT
1820 PRINT TAB(18);
1830 PRINT USING F$(14),T1,T3,N9
1840 PRINT
1850 N9=INT(N9-N0+0.5)
1860 IF N9 > 0 THEN 1880
1870 N9=0
1880 PRINT "ADDITIONAL OBSNS FOR DESIRED ACCURACY AND CONF.";
1890 PRINT " LEVEL = ";N9;" CYCLES"
1900 PRINT "STANDARD UNITS OF OUTPUT/HOUR = ";U
1910 PRINT
1920 RETURN
1930 REM    PRINT A LINE ACROSS THE PAGE OF LENGTH L8
1940 FOR J=1 TO L8
1950 PRINT "-";
1960 NEXT J
1970 PRINT
1980 RETURN
1990 REM    *** ADD COMMAND ***
2000 REM    N0 = TOTAL # OF CYCLES,  N5 = ADDITIONAL CYCLES
2010 READ N5
2020 K9=K9+1
2030 PRINT "READING WORK ELEMENT TIME DATA FOR SET ";K9
2040 PRINT "READING ";N5;" CYCLES ..."
2050 W0=0
2060 FOR N=1 TO N5
2070 PRINT TAB(3);"CYCLE   ";N
2080 FOR I=1 TO M
2090 READ W1
2100 IF S=0 THEN 2140
2110 W2=W1-W0
2120 W0=W1
2130 GOTO 2150
2140 W2=W1
2150 X(I)=X(I)+W2/100
2160 Y(I)=Y(I)+W2*W2/10000
```

```
2170 NEXT I
2180 NEXT N
2190 N0=N0+N5
2200 RETURN
2210 REM       *** I N I T I A L I Z A T I O N ***
2220 F$(1)="   PERCENT   PERSONAL   UNAVOIDABLE-DELAY   FATIGUE      TOTAL"
2230 F$(2)="ALLOWANCES      ###.##          ###.##            ###.##     ###.##"
2240 F$(3)="%ACCURACY = ##.##              %CONFIDENCE LEVEL = ##.##"
2250 F$(4)="COMPANY : ##########################################"
2260 F$(5)="DATE : ########"
2270 F$(6)="DEPARTMENT : #######################################"
2280 F$(7)="OPERATOR : ################     ANALYST : ################"
2290 F$(8)="ELE-                                      AVERAGE"
2300 F$(9)="MENT   WORK ELEMENT DESCRIPTION      TIME"
2310 F$(10)="###    #######################   ##.###"
2320 F$(11)=" PERFORM  NORMAL    STD.  DESIRED"
2330 F$(12)="  RATING    TIME    TIME   CYCLES"
2340 F$(13)="   ##.###  ##.###  ##.###     #####"
2350 F$(14)="T O T A L     F O R     T A S K ##.### ##.###    #####"
2360 N0=0
2370 FOR I=1 TO M9
2380 X(I)=0
2390 Y(I)=0
2400 NEXT I
2410 A0=5
2420 A9=A0*A0/10000
2430 L8=68
2440 L9=95
2450 GOSUB 1180
2460 K9=0
2470 RETURN
2480 END
```

II. Work Sampling

The work sampling technique of work measurement was introduced by L.H.C. Tippett in the 1930s. The labor standards for an average operator working under normal conditions are estimated by taking random observations of a single worker to determine the proportion of the time he or she is involved in various activities. It can be applied to nonrepetitive and repetitive existing tasks that are of long duration. The labor standards are usually expressed in minutes per unit of output. Work sampling can also be applied for analyzing group activities. The following procedure is generally employed for setting labor standards for a work sampling study based on observations of a single worker:

1. Select the task and define the standard method.
2. Define working and idle condition. (Working condition refers to when the operator is working on the unit. Idle condition refers to when the operator is deliberately idle.)
3. Select an appropriate operator (operator of good skill and aptitude).
4. Prepare a tour schedule for preliminary observations n (n>=50).
5. Observe and record for each random observation whether the operator is working or idle.
6. Rate the operator performance for each observation when operator is working, with respect to normal pace, PR_i.
7. Calculate the appropriate number of observations (N) for the entire study with the desired accuracy and level of confidence:

$$N = [Z/(Ap/100)]^2 \, p(1-p)$$

where
- Z = Confidence level factor,
- p = Proportion of time spent working = x/n, x is the number of observations when the operator is working, and n is the total number of observations,
- A = Desired accuracy expressed as a percentage of the true proportion (p).

8. Prepare a tour schedule for additional observations equal to (N-n).
9. Observe and record the activity status for additional observations (N-n) if required along with the performance rating of the operator when working.
10. Compute the normal time for the task in minutes:

$$\{NT\} = (T.p.PR)/U$$

where T = Total study time in minutes,
       p = Proportion of time spent working,
       PR = Average performance rating = $\sum_{i=1}^{m} PR_i / m$,
       PR_i = Performance rating for observation i when the operator is working,
       m = Number of observations when operator is working during the entire study,
       U = Number of units produced or completed during the study period.

11. Compute Standard Time in minutes, {ST},

    {ST} = {NT} (1 + Allowance Fraction)

    where Allowance Fraction represents allowance for personal needs of the worker, unavoidable delays, and worker fatigue.

12. Compute standard units of output per hour, {SU}:

    {SU} = 60/{ST} .

The Program WORK.SMPLG

The architecture of this program is very much like that of the program TIME.STY in the preceding section because the input data are entered through the use of DATA statements. There are primarily two usages of WORK.SMPLG. One is to use the program to generate the tour schedule by using the TOUR command. Another usage is to analyze the data collected. The TOUR command will prompt for the start and stop time and the number of observations to be taken during the tour. Clock times must be entered in their natural form using a 24-hour clock - e.g., clock time entered as 16.30 means 4:30 pm. Note that a decimal is used to separate the hour and the minutes instead of a colon.

When no data are entered, the HELP, TOUR, and EXIT commands are the only active ones. Similar to the TIME.STY program, the WORK.SMPLG program contains the ADD, ACC, CL, and ALLOW commands. The ADD command is used to include one data set at a time into the analysis. However, when it is executed the first time, it will also expect to read the general input data first before reading the data set. The ACC, CL, and ALLOW commands are used to modify the percent accuracy, percent confidence level, and three allowance percentages, respectively. The default percent accuracy is equal to 5, and the default confidence level is set at 95 percent.

INPUT

The input data that the user must prepare through the use of DATA statements can be classified as the general input data (1 through 9) and the actual observations. They are
1. The company's name
2. The department's name
3. The task description
4. The operator's name
5. The analyst's name
6. The study date
7. Percent personal allowance
8. Percent unavoidable delay allowance
9. Percent fatigue allowance
10. First set of observed data
11. Second set of observed data
12. Etc.

Each data set contains the following:
a. The study time in minutes.
b. The number of units produced or completed during the study period.
c. The number of observations made in the study period.
d. All observation values in pairs of operator indicator and performance rating separated by commas. Note that when the operator is working, the indicator is set to one. Otherwise it is zero, and its performance rating is also set to zero.

OUTPUT

The TOUR command produces a work sheet that can be used to directly record the actual observations. If the ADD command has not been executed prior to the TOUR command, then the report will not contain the general input information. The ANALYZE command produces a report containing the normal time, standard time, standard units of output per hour, proportion of time the operator is working, the average performance rating, and additional number of observations required in order to satisfy the specified percent accuracy and confidence level.

EXAMPLE

We assume that John Doe of XYZ company has collected 50 observations in the Die Casting department and wishes to use the program WORK.SMPLG to perform the analysis. John has been rating the performance of the operator named Joe Bloe, who performs the grinding operation on metal castings. The observations have been made over a period of 420 minutes, during which the operator has completed grinding 80 castings. The percentages of personal allowance, unavoidable delay, and

fatigue allowance are 8, 8, and 5, respectively. Assume also that John Doe wants to estimate to within 10 percent of the true value of the proportion (p) the operator is working, with 95 percent confidence. The observations are:

IND	PR	IND	PR	IND	PR	IND	PR	IND	PR
1	0.88	1	0.86	1	0.94	0	0.00	1	0.87
1	0.99	1	0.92	1	0.85	1	0.90	0	0.00
1	0.99	1	0.88	1	0.92	1	0.90	1	0.97
1	0.93	1	0.89	1	0.91	1	0.90	0	0.00
0	0.00	1	0.91	1	0.93	1	0.91	1	0.94
0	0.00	1	0.89	1	0.93	1	0.99	1	0.94
1	0.92	1	0.90	1	0.96	1	0.95	1	0.95
1	0.94	0	0.00	1	0.97	1	0.84	1	0.94
1	0.93	1	0.98	1	0.95	1	0.99	0	0.00
1	0.92	1	0.84	1	0.92	1	0.82	1	0.92

where IND = Operator Indicator and PR = Performance Rating.

The corresponding DATA statements that are appended to the end of the WORK.SMPLG program listing are given below. Note that the general input data are combined with the first data set.

```
3500 REM---- COMPANY NAME----
3510 DATA "XYZ COMPANY"
3520 REM---- DEPARTMENT NAME----
3530 DATA "DIE CASTING"
3540 REM---- TASK DISCRIPTION----
3550 DATA "GRINDING"
3560 REM---- OPERATOR'S NAME----
3570 DATA "JOE BLOE"
3580 REM----TIME STUDY ANALYST'S NAME-----
3590 DATA "JOHN DOE"
3600 REM----STUDY DATE----
3610 DATA "XX/XX/XX"
3620 REM----PERSONAL ALLOWANCE (PERCENT)----
3630 DATA 8
3640 REM----UNAVOIDABLE DELAY (PERCENT)----
3650 DATA 8
3660 REM----FATIGUE ALLOWANCE (PERCENT)----
3670 DATA 5
3680 REM **** F I R S T    D A T A    S E T ****
3690 REM----TOTAL STUDY TIME IN MINUTES----
3700 DATA 420
3710 REM----NUMBER OF UNITS PRODUCED IN THAT PERIOD----
3720 DATA 80
3730 REM----NUMBER OF OBSERVATIONS----
3740 DATA 50
3750 REM----OPER. INDICATOR(1/0), PERFORMANCE RATING
3760 DATA   1, 0.88, 1, 0.86, 1, 0.94, 0, 0.00, 1, 0.87
3770 DATA   1, 0.99, 1, 0.92, 1, 0.85, 1, 0.90, 0, 0.00
3780 DATA   1, 0.99, 1, 0.88, 1, 0.92, 1, 0.90, 1, 0.97
```

```
3790 DATA    1, 0.93, 1, 0.89, 1, 0.91, 1, 0.90, 0, 0.00
3800 DATA    0, 0.00, 1, 0.91, 1, 0.93, 1, 0.91, 1, 0.94
3810 DATA    0, 0.00, 1, 0.89, 1, 0.93, 1, 0.99, 1, 0.94
3820 DATA    1, 0.92, 1, 0.90, 1, 0.96, 1, 0.95, 1, 0.95
3830 DATA    1, 0.94, 0, 0.00, 1, 0.97, 1, 0.84, 1, 0.94
3840 DATA    1, 0.93, 1, 0.98, 1, 0.95, 1, 0.99, 0, 0.00
3850 DATA    1, 0.92, 1, 0.84, 1, 0.92, 1, 0.82, 1, 0.92
```

The first example terminal session shows that we issue the HELP command at the first command level prompt. It displays all available commands. Next the ADD command is entered so that data are read into the computer memory. Since the default percent accuracy is 5 and we want a 10 percent accuracy, the ACC command is entered and the desired accuracy is given. The ANALYZE command shows that an additional 13 observations are required. The TOUR command is then entered so that a work sheet of 13 random observation times can be generated between 8:30 am and 10:30 am.

Assume that John Doe has collected the 13 observations with the following results.

IND	PR	IND	PR	IND	PR	IND	PR	IND	PR
1	0.98	0	0.00	1	0.96	1	0.94	1	0.97
1	0.95	1	0.98	1	0.93	1	0.89	1	0.96
1	0.96	1	0.89	1	0.97				

The study time is two hours, which is 120 minutes. During this period the operator has completed grinding 20 castings. The data are translated into BASIC statements as shown below. They are appended to the first set of data before running the program.

```
4000 REM **** S E C O N D    D A T A    S E T ****
4010 REM----STUDY TIME IN MINUTES----
4020 DATA 120
4030 REM----NUMBER OF UNITS PRODUCED IN THIS PERIOD----
4040 DATA 20
4050 REM----NUMBER OF OBSERVATIONS----
4060 DATA 13
4070 REM----OPERATOR INDICATOR, PERFORMANCE RATING----
4080 DATA    1, 0.98, 0, 0.00, 1, 0.96, 1, 0.94, 1, 0.97
4090 DATA    1, 0.95, 1, 0.98, 1, 0.93, 1, 0.89, 1, 0.96
4100 DATA    1, 0.96, 1, 0.89, 1, 0.97
```

The second example terminal session shows that we issue the command ADD twice to bring in the two sets of data for analysis. The ACC command is used to specify the desired accuracy, and the ANALYZE command indicates that no additional observations are required. The standard output per hour for the operation is estimated at about 11.35 castings.

```
WORK.SMPLG EXAMPLE TERMINAL SESSION 1
======================================

W O R K    S A M P L I N G
--------------------------

NEED INTRODUCTION (Y OR N) ... !N

PLEASE ENTER COMMAND OR TYPE HELP

COMMAND --> !HELP

AVAILABLE COMMANDS
==================
HELP    - PRINT THIS MESSAGE
ACC     - SPECIFY THE DESIRED PERCENT ACCURACY
CL      - SPECIFY THE PERCENT CONFIDENCE LEVEL
ALLOW   - CHANGE THE THREE PERCENT ALLOWANCES
ANALYZE - PERFORM ANALYSIS ON THE DATA
ADD     - ADD ADDITIONAL DATA TO THE ANALYSIS
TOUR    - GENERATE TOUR SCHEDULE
EXIT    - EXIT FROM THE PROGRAM

COMMAND --> !ADD

READING GENERAL INPUT DATA ...
READING 50 OBSNS FOR DATA SET  1

COMMAND --> !ACC

ENTER THE NEW DESIRED PERCENT ACCURACY -- !10

COMMAND --> !ANALYZE



W O R K    S A M P L I N G    A N A L Y S I S
---------------------------------------------------------------
COMPANY : XYZ COMPANY                          DATE : XX/XX/XX
DEPARTMENT : DIE CASTING
T A S K  : GRINDING
---------------------------------------------------------------
OPERATOR : JOE BLOE          ANALYST : JOHN DOE
---------------------------------------------------------------
      PERCENT     PERSONAL    UNAVOIDABLE-DELAY   FATIGUE   TOTAL
   ALLOWANCES       8.00              8.00         5.00    21.00
   %ACCURACY =    10.00                       %CONFIDENCE LEVEL = 95.00
---------------------------------------------------------------
             NORMAL TIME (MINUTES) = 4.155899999999
             STANDARD TIME (MINUTES) = 5.028638999999
             STANDARD OUTPUT/HOUR (UNITS) =11.93165785017
             PROPORTION OF TIME OPERATOR WORKING = .86
             AVERAGE PERFORMANCE RATING = .92
             ADDITIONAL OBSNS REQUIRED = 13
---------------------------------------------------------------
```

```
COMMAND --> !TOUR

ENTER START TIME IN A 24 HOUR CLOCK (E.G. 10.30) -- !8.30
ENTER STOP  TIME IN A 24 HOUR CLOCK (E.G. 10.30) -- !10.30
ENTER NUMBER OF OBSERVATIONS TO BE GENERATED -- !13


W O R K    S A M P L I N G    T O U R    S C H E D U L E
-----------------------------------------------------------------
COMPANY : XYZ COMPANY                         DATE : XX/XX/XX
DEPARTMENT : DIE CASTING
T A S K  : GRINDING
-----------------------------------------------------------------
OPERATOR : JOE BLOE          ANALYST : JOHN DOE
-----------------------------------------------------------------
              START TIME :  8.30
              STOP  TIME : 10.30
              NUMBER OF OBSERVATIONS : 13
-----------------------------------------------------------------
         OBSN  OPERATOR  PERFORM         OBSN  OPERATOR  PERFORM
  OBSN   TIME  INDICATOR  RATING   OBSN  TIME  INDICATOR  RATING
-----------------------------------------------------------------
    1    8.32                        2   8.38
-----------------------------------------------------------------
    3    8.40                        4   9.12
-----------------------------------------------------------------
    5    9.18                        6   9.23
-----------------------------------------------------------------
    7    9.28                        8   9.29
-----------------------------------------------------------------
    9    9.37                       10   9.54
-----------------------------------------------------------------
   11   10.09                       12  10.12
-----------------------------------------------------------------
   13   10.16
-----------------------------------------------------------------


COMMAND --> !EXIT
```

```
WORK.SMPLG EXAMPLE TERMINAL SESSION 2
======================================

W O R K    S A M P L I N G
-------------------------

NEED INTRODUCTION (Y OR N) ... !N

PLEASE ENTER COMMAND OR TYPE HELP

COMMAND --> !ADD

READING GENERAL INPUT DATA ...
READING 50 OBSNS FOR DATA SET  1

COMMAND --> !ADD

READING 13 OBSNS FOR DATA SET  2

COMMAND --> !ACC

ENTER THE NEW DESIRED PERCENT ACCURACY -- !10

COMMAND --> !ANALYZE



W O R K    S A M P L I N G    A N A L Y S I S
-----------------------------------------------------------------------
COMPANY : XYZ COMPANY                              DATE : XX/XX/XX
DEPARTMENT : DIE CASTING
T A S K  : GRINDING
-----------------------------------------------------------------------
OPERATOR : JOE BLOE              ANALYST : JOHN DOE
-----------------------------------------------------------------------
     PERCENT     PERSONAL   UNAVOIDABLE-DELAY   FATIGUE    TOTAL
   ALLOWANCES      8.00          8.00             5.00     21.00
   %ACCURACY = 10.00                    %CONFIDENCE LEVEL = 95.00
-----------------------------------------------------------------------
             NORMAL TIME (MINUTES) = 4.367999999999
             STANDARD TIME (MINUTES) = 5.285279999999
             STANDARD OUTPUT/HOUR (UNITS) =11.35228407956
             PROPORTION OF TIME OPERATOR WORKING = .8730158730159
             AVERAGE PERFORMANCE RATING = .926
             ADDITIONAL OBSNS REQUIRED = 0
-----------------------------------------------------------------------


COMMAND --> !EXIT
```

```
10 REM     WORK.SMPLG
20 DIM P(200),F$(12),G$(6)
30 M9=200
40 GOSUB 2750
50 PRINT
60 PRINT "W O R K    S A M P L I N G"
70 PRINT "-------------------------"
80 PRINT
90 PRINT "NEED INTRODUCTION (Y OR N) ... ";
100 INPUT Y$
110 IF Y$ <> "Y" THEN 250
120 PRINT
130 PRINT "THIS PROGRAM CAN BE USED TO PERFORM DUAL FUNCTIONS :"
140 PRINT "1. GENERATE TOUR SCHEDULE SO THAT WORK SAMPLING DATA"
150 PRINT "   CAN BE COLLECTED."
160 PRINT "2. ANALYZE THE WORK SAMPLING DATA WHICH HAVE BEEN"
170 PRINT "   COLLECTED AND DETERMINE THE NUMBER OF ADDITIONAL"
180 PRINT "   OBSERVATIONS REQUIRED TO ESTIMATE THE STANDARD"
190 PRINT "   TIME FOR A DESIRED ACCURACY AND CONFIDENCE LEVEL."
200 PRINT "   IT ALSO COMPUTES THE STANDARD UNITS OF OUTPUT/HR."
210 PRINT
220 PRINT "INPUT REQUIREMENTS"
230 PRINT "------------------"
240 PRINT "DETAILS OF INPUT REQUIREMENTS ARE AVAILABLE IN THE DOCUMENT"
250 PRINT
260 PRINT "PLEASE ENTER COMMAND OR TYPE HELP"
270 PRINT
280 PRINT "COMMAND --> ";
290 INPUT C$
300 PRINT
310 IF C$ <> "HELP" THEN 340
320 GOSUB 560
330 GOTO 270
340 IF C$ <> "ACC" THEN 370
350 GOSUB 670
360 GOTO 270
370 IF C$ <> "CL" THEN 400
380 GOSUB 740
390 GOTO 270
400 IF C$ <> "ALLOW" THEN 430
410 GOSUB 930
420 GOTO 270
430 IF C$ <> "ANALYZE" THEN 460
440 GOSUB 1010
450 GOTO 270
460 IF C$ <> "ADD" THEN 490
470 GOSUB 1330
480 GOTO 270
490 IFC$ <> "TOUR" THEN 520
500 GOSUB 1650
510 GOTO 270
520 IF C$="EXIT" THEN 3010
530 PRINT "*** ERROR *** PLEASE TRY AGAIN !!!"
540 GOTO 270
```

```
550 REM *** HELP COMMAND ***
560 PRINT "AVAILABLE COMMANDS"
570 PRINT "=================="
580 PRINT "HELP   - PRINT THIS MESSAGE"
590 PRINT "ACC    - SPECIFY THE DESIRED PERCENT ACCURACY"
600 PRINT "CL     - SPECIFY THE PERCENT CONFIDENCE LEVEL"
610 PRINT "ALLOW  - CHANGE THE THREE PERCENT ALLOWANCES"
620 PRINT "ANALYZE - PERFORM ANALYSIS ON THE DATA"
630 PRINT "ADD    - ADD ADDITIONAL DATA TO THE ANALYSIS"
640 PRINT "TOUR   - GENERATE TOUR SCHEDULE"
650 PRINT "EXIT   - EXIT FROM THE PROGRAM"
660 RETURN
670 REM *** ACC COMMAND ***
680 GOSUB 2500
690 IF K9 = 0 THEN 730
700 PRINT "ENTER THE NEW DESIRED PERCENT ACCURACY -- ";
710 INPUT A0
720 A9=A0*A0/10000
730 RETURN
740 REM *** CL COMMAND ***
750 GOSUB 2500
760 IF K9 = 0 THEN 920
770 PRINT "ENTER THE NEW PERCENT CONFIDENCE LEVEL -- ";
780 INPUT L9
790 IF L9<=0 THEN 810
800 IF L9<100 THEN 850
810 PRINT
820 PRINT "*** ERROR *** CL MUST BE >0 AND <100"
830 PRINT
840 GOTO 770
850 C9=0.5-L9/200
860 A2=LOG(1/C9/C9)
870 A1=SQR(A2)
880 Z=2.515517+0.802853*A1+0.010328*A2
890 Z9=1.0+1.432788*A1+0.189269*A2+0.001308*A1*A2
900 Z=A1-Z/Z9
910 Z2=Z*Z
920 RETURN
930 REM *** ALLOW COMMAND ***
940 GOSUB 2500
950 IF K9 = 0 THEN 1000
960 PRINT "ENTER PERCENT ALLOWANCES FOR PERSONAL, UNAVOIDABLE DELAY,"
970 PRINT "AND FATIGUE SEPARATED BY COMMAS -- ";
980 INPUT F1,F2,F3
990 F0=F1+F2+F3
1000 RETURN
1010 REM *** ANALYZE COMMAND ***
1020 GOSUB 2500
1030 IF K9 = 0 THEN 1320
1040 C1=S1/N0
1050 P2=S/S1
1060 P1=INT(P2*1000)/1000
1070 N1=INT(Z2*(1-C1)/(A9*C1)+0.5)
1080 N4=(H*C1*P2)/U
```

```
1090 N6=N4*(1+F0/100)
1100 N7=60/N6
1110 GOSUB 2610
1120 PRINT "W O R K    S A M P L I N G    A N A L Y S I S"
1130 GOSUB 2550
1140 GOSUB 2660
1150 PRINT TAB(4);F$(4)
1160 PRINT TAB(4);
1170 PRINT USING F$(5),F1,F2,F3,F0
1180 PRINT TAB(4);
1190 PRINT USING F$(6),A0,L9
1200 GOSUB 2550
1210 PRINT TAB(15);"NORMAL TIME (MINUTES) = ";N4
1220 PRINT TAB(15);"STANDARD TIME (MINUTES) = ";N6
1230 PRINT TAB(15);"STANDARD OUTPUT/HOUR (UNITS) =";N7
1240 PRINT TAB(15);"PROPORTION OF TIME OPERATOR WORKING = ";C1
1250 PRINT TAB(15);"AVERAGE PERFORMANCE RATING = ";P1
1260 N9=0
1270 IF N1-N0 <=0 THEN 1290
1280 N9=N1-N0
1290 PRINT TAB(15);"ADDITIONAL OBSNS REQUIRED = ";N9
1300 GOSUB 2550
1310 GOSUB 2610
1320 RETURN
1330 REM *** ADD COMMAND ***
1340 IF K9 > 0 THEN 1420
1350 PRINT "READING GENERAL INPUT DATA ..."
1360 FOR I = 1 TO 6
1370 READ G$(I)
1380 NEXT I
1390 REM    PERCENT ALLOWANCES
1400 READ F1,F2,F3
1410 F0=F1+F2+F3
1420 REM    H5=STUDY PERIOD, U5=UNITS COMPLETED, N5=# OBSNS
1430 K9=K9+1
1440 READ H5,U5,N5
1450 PRINT "READING ";N5;" OBSNS FOR DATA SET   ";K9
1460 FOR I=1 TO N5
1470 READ C,F
1480 IF C=0 THEN 1530
1490 IF C=1 THEN 1530
1500 K8=1
1510 PRINT "*** ERROR *** OBSN ";I;" OPERATOR INDICATOR = ";C
1520 GOTO 1550
1530 S1=S1+C
1540 S=S+F
1550 NEXT I
1560 H=H+H5
1570 U=U+U5
1580 N0=N0+N5
1590 IF K8=0 THEN 1640
1600 PRINT
1610 PRINT "PLEASE CORRECT INPUT DATA AND RERUN THE PROGRAM !!!"
1620 PRINT "PROGRAM EXECUTION TERMINATED !!!"
```

```
1630 STOP
1640 RETURN
1650 REM     *** TOUR COMMAND ***
1660 Y$="START"
1670 GOSUB 2380
1680 X1=A1
1690 A=A2
1700 Y$="STOP "
1710 GOSUB 2380
1720 X2=A1
1730 B=A2
1740 PRINT "ENTER NUMBER OF OBSERVATIONS TO BE GENERATED -- ";
1750 INPUT T
1760 IF T <= M9 THEN 1800
1770 PRINT "*** ERROR *** NUMBER OF OBSNS MUST BE <= ";M9
1780 PRINT
1790 GOTO 1740
1800 G=B-A
1810 IF G > 0 THEN 1830
1820 G=G+24
1830 FOR I=1 TO T
1840 P(I)=RND(0)
1850 NEXT I
1860 REM    SORT RANDOM NUMBERS FROM LOW TO HIGH
1870 IF T <= 1 THEN 1990
1880 FOR I=2 TO T
1890 H1=P(I-1)
1900 K=I-1
1910 FOR J=I TO T
1920 IF P(J) >= H1 THEN 1950
1930 K=J
1940 H1=P(J)
1950 NEXT J
1960 P(K)=P(I-1)
1970 P(I-1)=H1
1980 NEXT I
1990 GOSUB 2610
2000 PRINT "W O R K   S A M P L I N G   T O U R   S C H E D U L E"
2010 GOSUB 2550
2020 IF K9 <= 0 THEN 2040
2030 GOSUB 2660
2040 PRINT TAB(15);
2050 PRINT USING F$(7),X1
2060 PRINT TAB(15);
2070 PRINT USING F$(8),X2
2080 PRINT TAB(15);"NUMBER OF OBSERVATIONS : ";T
2090 GOSUB 2550
2100 PRINT F$(1);F$(1)
2110 PRINT F$(2);F$(2)
2120 GOSUB 2550
2130 T0=INT(T/2)
2140 IF 2*T0 = T THEN 2160
2150 T0=T0+1
2160 FOR I=1 TO T0
```

```
2170 J=2*I-1
2180 K=2*I
2190 FOR L=J TO K
2200 IF L > T THEN 2330
2210 Y=P(L)*G+A
2220 IF Y < 24 THEN 2240
2230 Y=Y-24
2240 Y1=INT(Y)
2250 Y2=INT(60*(Y-Y1))/100
2260 Y=Y1+Y2
2270 IF L= K THEN 2300
2280 PRINT USING F$(3),L,Y,G$(0);
2290 GOTO 2310
2300 PRINT USING F$(3),L,Y
2310 NEXT L
2320 GOTO 2340
2330 PRINT
2340 GOSUB 2550
2350 NEXT I
2360 GOSUB 2610
2370 RETURN
2380 REM     PROMPT FOR START/STOP TIME
2390 PRINT "ENTER ";Y$;" TIME IN A 24 HOUR CLOCK (E.G. 10.30) -- ";
2400 INPUT A1
2410 IF A1 > 24 THEN 2450
2420 A2=INT(A1)
2430 A3=(A1-A2)/0.6
2440 IF A3 < 1 THEN 2480
2450 PRINT "*** ERROR **   PLEASE TRY AGAIN  !!!"
2460 PRINT
2470 GOTO 2390
2480 A2=A2+A3
2490 RETURN
2500 REM    *** CHECK EXISTENCE OF INPUT DATA ***
2510 IF K9 > 0 THEN 2540
2520 PRINT "COMMAND IS INACTIVE UNLESS OBSERVED DATA ARE AVAILABLE !"
2530 PRINT "PLEASE USE THE ADD COMMAND, FIRST !"
2540 RETURN
2550 REM ***PRINT A LINE OF LENGTH L8 ***
2560 FOR J=1 TO L8
2570 PRINT "-";
2580 NEXT J
2590 PRINT
2600 RETURN
2610 REM    SKIP LINES
2620 FOR L=1 TO 3
2630 PRINT
2640 NEXT L
2650 RETURN
2660 REM    ** PRINT GENERAL INFO. ***
2670 PRINT USING F$(9),G$(1);
2680 PRINT USING F$(10),G$(6)
2690 PRINT USING F$(11),G$(2)
2700 PRINT "T A S K  : ";G$(3)
```

```
2710 GOSUB 2550
2720 PRINT USING F$(12),G$(4),G$(5)
2730 GOSUB 2550
2740 RETURN
2750 REM     I N I T I A L I Z A T I O N
2760 F$(1)="            OBSN  OPERATOR  PERFORM"
2770 F$(2)="    OBSN   TIME INDICATOR   RATING"
2780 F$(3)="    ####   ##.## ##################"
2790 F$(4)="  PERCENT   PERSONAL   UNAVOIDABLE-DELAY   FATIGUE    TOTAL"
2800 F$(5)="ALLOWANCES    ###.##        ###.##                 ###.##   ###.##"
2810 F$(6)="%ACCURACY = ##.##              %CONFIDENCE LEVEL = ##.##"
2820 F$(7)="START TIME : ##.##"
2830 F$(8)="STOP  TIME : ##.##"
2840 F$(9)="COMPANY : ##########################################"
2850 F$(10)="DATE : ########"
2860 F$(11)="DEPARTMENT : #######################################"
2870 F$(12)="OPERATOR : ###############      ANALYST : ###############"
2880 G$(0)=" "
2890 U=0
2900 N0=0
2910 S1=0
2920 S=0
2930 A0=5
2940 A9=A0*A0/10000
2950 L8=68
2960 L9=95
2970 GOSUB 850
2980 K9=0
2990 K8=0
3000 RETURN
3010 END
```

III. Dimensional Specifications

The Program TIME.STY

Let m = maximum number of work elements that the program can handle. Currently, m equals 50. The following statements must be modified if m changes:

Line No.
```
20 DIM X(m),Y(m),P(m),D$(m),F$(14),G$(6)
30 M9=m
```

The Program WORK.SMPLG

Let n be the maximum number of observations to be generated in a tour schedule. Currently, n=200. To increase n, the following statements are affected. Note that the program is not limited by the actual number of observations collected.

Line No.
```
20 DIM P(n),F$(12),G$(6)
30 M9=n
```

IV. References

1. Barnes, R.M., MOTION AND TIME STUDY: DESIGN AND MEASUREMENT OF WORK, 6th ed., John Wiley, New York, 1968.

2. Heiland, R., and W. Richardson, WORK SAMPLING, McGraw-Hill, New York, 1957.

3. 4M DATA, Mod II, Computer-Aided System for Applying MTM-1, MTM Association, 1601 Broadway, Fair Lawn, N.J.

4. Mundell, M.E., MOTION AND TIME STUDY, 5th ed., Prentice-Hall, Englewood Cliffs, N.J., 1978.

5. Neibel, B.W., MOTION AND TIME STUDY, 6th ed., Richard D. Irwin, Homewood, Ill., 1976.

6. WOCOM®, Work Factor Systems®, Science Management Corporation, Bridgewater, N.J.

CHAPTER 7

LAYOUT OF FACILITIES

I. Process Layout Design

   The Program PLAYOUT
   ° Input
   ° Output
   ° Example
   ° PLAYOUT Example Terminal Session
   ° PLAYOUT Program Listing

II. Product Layout Design

   The Program ASSMBLY
   ° Input
   ° Output
   ° Example
   ° ASSMBLY Example Terminal Session
   ° ASSMBLY Program Listing

III. Dimensional Specifications

IV. References

Three classical types of layouts -- process layout, product layout, and fixed-position layout -- are important in today's manufacturing and service industries. Many variations and combinations of the three are commonly employed.

In this chapter, two computer programs, one dealing with process layout, and a special case of product layout called Assembly Line, are presented in their most fundamental forms.

For those readers who wish to compare various process layout commercial packages for large-size problems, a textbook by Francis and White [1] and a publication by Tompkins and Moore [5] may be of interest. For those who are interested in various Assembly Line commercial packages for large-size problems, more information is available in a textbook by Groover [2].

Section I deals with the Process Layout Design and describes the usage of the program "PLAYOUT" with an example. Section II discusses the Assembly Line Design and describes the usage of the program "ASSMBLY" with another example.

I. Process Layout Design

Process layout is a type of layout in which facilities must be shared by multiple products, each with a unique sequence of operations through the system. Classical examples are job shops, batch production, and a hospital or clinic.

The process layout problem involves the determining of the relative location of work centers to achieve a stated decision criterion within certain layout constraints. The decision criterion can be quantitative or qualitative. Process layout problems involving qualitative criteria occur when relationships between departments are specified in qualitative terms.

For those readers who are interested in a qualitative process layout problem, a method of formulation and solution called SLP (Systematic Layout Planning) by Muther [3] may be of interest. The quantitative criteria occur when relationships between departments are specified in quantitative terms, such as minimizing material-handling costs, minimizing distance traveled by customers or employees. Such a criterion can be expressed as follows:

$$K = \sum_{i=1}^{N} \sum_{j=1}^{N} T_{ij} C_{ij} D_{ij}$$

where T_{ij} = Number of trips between work center i and work center j over a time period,

C_{ij} = Cost-per-unit distance per trip from work center i to work center j,

D_{ij} = Distance from work center i to work center j,

K = Total cost in the time period,
N = Number of work centers.

The process layout problem consists of arranging the work center in such a manner that the work centers with large $T_{ij} C_{ij}$ values are relatively close together so that their D_{ij} values are small and total cost is minimized.

The following procedure can be used to establish the nearness priorities. These priorities can be used to establish a desired configuration(s), which can be converted into a block diagram and then to a final layout which satisfies the layout constraints, such as limitations of space, the need to maintain fixed locations for certain work centers, safety regulations, fire regulations, and the aisle requirements.

1. Based on a forecast of demand over the reasonable life of a facility and anticipated product mix, develop the necessary size of each work center and the anticipated average number of trips per period, say per month, from the work center i to j (T_{ij}).

2. Develop cost-per-unit distance, say per hundred feet or per hundred meters, per trip traveled from work center i to j (C_{ij}), direction of movement considered.

3. Compute the per period cost per unit of distance between work centers ($T_{ij} C_{ij}$).

4. Rank $T_{ij} C_{ij}$ in descending order to establish nearness priorities.

5. By trial and error arrange work centers with high nearness priorities adjacent to each other. Several arrangements could be developed that would place work centers with high nearness priorities adjacent to each other.

6. Evaluate the total cost (K) for each arrangement and develop the one with minimum cost into block layout and then into a final layout that satisfies the layout constraints.

The Program PLAYOUT

The program accepts the number of trips per period between each pair of work centers (T_{ij}) and the corresponding cost-per-unit distance per trip traveled (C_{ij}). It determines the nearness priorities by ranking the sum of the product of these two quantities for each pair of i and j. The product can be interpreted as the total cost-per-unit distance per period.

INPUT

1. The problem title
2. Number of work centers
3. Work center names (10 characters maximum)
4. The average number of trips per period between work centers (T_{ij})
5. The cost-per-unit distance per trip between work centers (C_{ij})

Note that T_{ii} and C_{ii} are not prompted by the program.

OUTPUT

After all of the above input are entered, the GO command can be used to generate the nearness priorities in descending order of the cost-per-unit distance per period. Commands are also provided to change and list input cost data or the input trip data between any two work centers.

EXAMPLE

Consider a process layout project called "MASCOR LAYOUT" with 6 work centers. Table 1 shows the average number of trips per month among the work centers. Table 2 shows the cost per hundred feet to move an average load in dollars. We wish to use the program PLAYOUT initially to give the nearness priority ranking.

Table 1. Average Number of Trips per Month

From	To	Work Centers					
		A	B	C	D	E	F
A		-0-	100	200	30	20	90
B		100	-0-	25	100	30	5
C		200	50	-0-	50	5	10
D		5	200	30	-0-	20	35
E		60	35	50	150	-0-	25
F		20	45	40	55	200	-0-

Table 2. Cost-per-Unit Distance per Trip

From	To	Work Centers					
		A	B	C	D	E	F
A		-0-	.10	.10	.20	.10	.20
B		.40	-0-	.20	.10	.10	.10
C		.10	.10	-0-	.10	.10	.20
D		.40	.10	.10	-0-	.10	.20
E		.10	.40	.20	.20	-0-	.10
F		.10	.10	.20	.10	.10	-0-

The example terminal session is illustrated below. After all inputs are entered, the HELP command displays all available commands. The GO command shows that work center A should be closest to B as the top priority. A should also be arranged closer to C as the the second priority, etc. For each alternative of layout, the actual distances can be measured and multiplied by the cost-per-unit distance per period to obtain the total cost per period (K).

```
PLAYOUT EXAMPLE TERMINAL SESSION
================================

PROCESS LAYOUT DESIGN
---------------------

NEED INTRODUCTION (Y OR N)...!N

ENTER PROBLEM TITLE !MASCOR LAYOUT

ENTER NUMBER OF WORK CENTERS (>=2) !6

ENTER WORK CENTER NAMES (10 CHARACTERS MAX)
-------------------------------------------
WORK CENTER   1    !A
WORK CENTER   2    !B
WORK CENTER   3    !C
WORK CENTER   4    !D
WORK CENTER   5    !E
WORK CENTER   6    !F

ENTER AVERAGE NO. OF TRIPS/PERIOD BETWEEN WORK CENTERS
------------------------------------------------------
FROM A   TO  B    !100
FROM A   TO  C    !200
FROM A   TO  D    !30
FROM A   TO  E    !20
FROM A   TO  F    !90
FROM B   TO  A    !100
FROM B   TO  C    !25
FROM B   TO  D    !100
FROM B   TO  E    !30
FROM B   TO  F    !5
FROM C   TO  A    !200
FROM C   TO  B    !50
FROM C   TO  D    !50
FROM C   TO  E    !5
FROM C   TO  F    !10
FROM D   TO  A    !5
FROM D   TO  B    !200
FROM D   TO  C    !30
FROM D   TO  E    !20
FROM D   TO  F    !35
FROM E   TO  A    !60
FROM E   TO  B    !35
FROM E   TO  C    !50
FROM E   TO  D    !150
FROM E   TO  F    !25
FROM F   TO  A    !20
FROM F   TO  B    !45
FROM F   TO  C    !40
FROM F   TO  D    !55
FROM F   TO  E    !200
```

```
ENTER COST PER UNIT DISTANCE PER TRIP TRAVELED
---------------------------------------------------
FROM A    TO    B    !.10
FROM A    TO    C    !.10
FROM A    TO    D    !.20
FROM A    TO    E    !.10
FROM A    TO    F    !.20
FROM B    TO    A    !.40
FROM B    TO    C    !.20
FROM B    TO    D    !.10
FROM B    TO    E    !.10
FROM B    TO    F    !.10
FROM C    TO    A    !.10
FROM C    TO    B    !.10
FROM C    TO    D    !.10
FROM C    TO    E    !.10
FROM C    TO    F    !.20
FROM D    TO    A    !.40
FROM D    TO    B    !.10
FROM D    TO    C    !.10
FROM D    TO    E    !.10
FROM D    TO    F    !.20
FROM E    TO    A    !.10
FROM E    TO    B    !.40
FROM E    TO    C    !.20
FROM E    TO    D    !.20
FROM E    TO    F    !.10
FROM F    TO    A    !.10
FROM F    TO    B    !.10
FROM F    TO    C    !.20
FROM F    TO    D    !.10
FROM F    TO    E    !.10

PLEASE ENTER COMMAND OR TYPE HELP

COMMAND -->!HELP

AVAILABLE COMMANDS
==================
LTRIP    - LIST TRIPS BETWEEN WORK CENTERS
LCOST    - LIST COST/UNIT DISTANCE BETWEEN WORK CENTERS
CTRIP    - CHANGE NO. OF TRIPS BETWEEN TWO WORK CENTERS
CCOST    - CHANGE COST/UNIT DISTANCE BETWEEN TWO
           WORK CENTERS
GO       - COMPUTE THE NEARNESS PRIORITY BETWEEN
           WORK CENTERS AND RANK THEM
RESTART  - START A NEW PROBLEM
EXIT     - EXIT FROM THE PROGRAM
```

```
COMMAND -->!GO

PROCESS LAYOUT DESIGN
=====================
PROBLEM TITLE : MASCOR LAYOUT
NO.OF WORK CENTERS : 6


N E A R N E S S     P R I O R I T Y     R A N K I N G
RANK    FROM         TO         COST/UNIT DISTANCE/PERIOD
  1      A            B                  50.00
  2      A            C                  40.00
  3      D            E                  32.00
  4      B            D                  30.00
  5      E            F                  22.50
  6      A            F                  20.00
  7      B            E                  17.00
  8      D            F                  12.50
  9      C            E                  10.50
 10      B            C                  10.00
 11      C            F                  10.00
 12      C            D                   8.00
 13      A            D                   8.00
 14      A            E                   8.00
 15      B            F                   5.00

COMMAND -->!EXIT
```

```
10 REM     PLAYOUT -  P R O C E S S      L A Y O U T     D E S I G N
20 REM
30 DIM T(10,10),C(10,10),N$(10),F(45),R(45),F$(2)
40 K0=10
50 GOSUB 1960
60 PRINT
70 PRINT "PROCESS LAYOUT DESIGN "
80 PRINT "--------------------"
90 PRINT
100 PRINT "NEED INTRODUCTION (Y OR N)..." ;
110 INPUT Y$
120 IF Y$ <> "Y" THEN 270
130 PRINT
140 PRINT "THIS PROGRAM EVALUATES THE NEARNESS PRIORITY"
150 PRINT "RANKING FOR PROCESS LAYOUT DESIGN. THE CRITERION"
160 PRINT "FOR DETERMINING THE RANKING IS THE COMBINED TRIP"
170 PRINT "TIMES THE COST PER UNIT DISTANCE PER TRIP AMONG"
180 PRINT "THE WORK CENTERS."
190 PRINT
200 PRINT "INPUT REQUIREMENTS"
210 PRINT "-----------------"
220 PRINT "1. PROBLEM TITLE"
230 PRINT "2. NAMES OF ALL WORK CENTERS"
240 PRINT "3. NUMBER OF WORK CENTERS TO BE ANALYZED"
250 PRINT "4. AVERAGE NO. OF TRIPS/PERIOD AMONG WORK CENTERS"
260 PRINT "5. COST PER UNIT DISTANCE PER TRIP AMONG WORK CENTERS."
270 PRINT
280 PRINT "ENTER PROBLEM TITLE ";
290 INPUT P$
300 PRINT
310 PRINT "ENTER NUMBER OF WORK CENTERS (>=2) " ;
320 INPUT K
330 IF K < 2 THEN 300
340 IF K <=K0 THEN 390
350 PRINT
360 PRINT "*** ERROR *** MAXIMUM ALLOWED = "; K0
370 GOTO 300
380 K9 = K*(K-1)/2
390 PRINT
400 PRINT "ENTER WORK CENTER NAMES (10 CHARACTERS MAX)"
410 PRINT "----------------------------------------"
420 FOR I= 1 TO K
430 PRINT "WORK CENTER  ";I;"    ";
440 INPUT N$(I)
450 NEXT I
460 PRINT
470 PRINT "ENTER AVERAGE NO. OF TRIPS/PERIOD BETWEEN WORK CENTERS"
480 PRINT "-----------------------------------------------------"
490 FOR I= 1 TO K
500 FOR J= 1 TO K
510 IF I= J THEN 540
520 PRINT "FROM ";N$(I);"  TO   ";N$(J);"  ";
530 INPUT T(I,J)
540 NEXT J
```

```
550 NEXT I
560 PRINT
570 PRINT "ENTER COST PER UNIT DISTANCE PER TRIP TRAVELED"
580 PRINT "-------------------------------------------"
590 FOR I= 1 TO K
600 FOR J= 1 TO K
610 IF I=J THEN 640
620 PRINT "FROM ";N$(I);"   TO   ";N$(J);"   ";
630 INPUT C(I,J)
640 NEXT J
650 NEXT I
660 PRINT
670 PRINT "PLEASE ENTER COMMAND OR TYPE HELP"
680 PRINT
690 PRINT "COMMAND -->";
700 INPUT C$
710 IF C$<> "HELP" THEN 740
720 GOSUB 940
730 GOTO 680
740 IF C$<> "LTRIP" THEN 770
750 GOSUB 1090
760 GOTO 680
770 IF C$<> "LCOST" THEN 800
780 GOSUB 1180
790 GOTO 680
800 IF C$<> "CTRIP" THEN 830
810 GOSUB 1270
820 GOTO 680
830 IF C$<> "CCOST" THEN 860
840 GOSUB 1270
850 GOTO 680
860 IF C$<> "GO" THEN 890
870 GOSUB 1530
880 GOTO 680
890 IF C$ = "RESTART" THEN 60
900 IF C$ = "EXIT" THEN 2000
910 PRINT
920 PRINT "*** ERROR *** PLEASE TRY AGAIN !"
930 GOTO 680
940 REM *** HELP COMMAND ***
950 PRINT
960 PRINT "AVAILABLE COMMANDS"
970 PRINT "=================="
980 PRINT "LTRIP    - LIST TRIPS BETWEEN WORK CENTERS"
990 PRINT "LCOST    - LIST COST/UNIT DISTANCE BETWEEN WORK CENTERS"
1000 PRINT "CTRIP    - CHANGE NO. OF TRIPS BETWEEN TWO WORK CENTERS"
1010 PRINT "CCOST    - CHANGE COST/UNIT DISTANCE BETWEEN TWO"
1020 PRINT "           WORK CENTERS"
1030 PRINT "GO       - COMPUTE THE NEARNESS PRIORITY BETWEEN "
1040 PRINT "           WORK CENTERS AND RANK THEM"
1050 PRINT "RESTART  - START A NEW PROBLEM"
1060 PRINT "EXIT     - EXIT FROM THE PROGRAM"
1070 RETURN
1080 REM
```

```
1090 REM ***     LTRIP COMMAND    ***
1100 PRINT
1110 FOR I = 1 TO K
1120 FOR J = 1 TO K
1130 IF I=J THEN 1150
1140 PRINT "FROM ";N$(I);"  TO   ";N$(J);"    ";T(I,J)
1150 NEXT J
1160 NEXT I
1170 RETURN
1180 REM ***     LCOST COMMAND    ***
1190 PRINT
1200 FOR I = 1 TO K
1210 FOR J = 1 TO K
1220 IF I = J THEN 1240
1230 PRINT "FROM ";N$(I);"  TO   ";N$(J);"    ";C(I,J)
1240 NEXT J
1250 NEXT I
1260 RETURN
1270 REM    ***    CTRIP AND CCOST COMMANDS   ***
1280 PRINT "FROM WHICH WORK CENTER   ";
1290 INPUT D$
1300 GOSUB 1460
1310 IF L <= 0 THEN 1450
1320 I=L
1330 PRINT "TO WHICH WORK CENTER   ";
1340 INPUT D$
1350 GOSUB 1460
1360 IF L <= 0 THEN 1450
1370 J=L
1380 IF I=J THEN 1450
1390 IF C$="CTRIP" THEN 1430
1400 PRINT "ENTER COST PER UNIT DISTANCE PER TRIP   ";
1410 INPUT C(I,J)
1420 GOTO 1450
1430 PRINT "ENTER AVERAGE NO. OF TRIPS/PERIOD   ";
1440 INPUT T(I,J)
1450 RETURN
1460 L=0
1470 FOR M=1 TO K
1480 IF D$ <> N$(M) THEN 1510
1490 L=M
1500 GOTO 1520
1510 NEXT M
1520 RETURN
1530 REM ***     GO COMMAND    ***
1540 K9 = K*(K-1)/2
1550 K8 = 0
1560 FOR I = 1 TO K-1
1570 FOR J= I+1 TO K
1580 K8 = K8+1
1590 F(K8) =T(I,J)*C(I,J) + T(J,I)*C(J,I)
1600 NEXT J
1610 NEXT I
1620 REM RANK FROM HIGH TO LOW
```

```
1630 FOR I = 1 TO K9
1640 R(I) = I
1650 NEXT I
1660 FOR I = 1 TO K9-1
1670 Z0 = F(R(I))
1680 FOR J = I+1 TO K9
1690 IF Z0 >= F(R(J)) THEN 1740
1700 Z0 = F(R(J))
1710 K1 = R(I)
1720 R(I) = R(J)
1730 R(J) = K1
1740 NEXT J
1750 NEXT I
1760 PRINT
1770 PRINT "PROCESS LAYOUT DESIGN"
1780 PRINT "===================="
1790 PRINT "PROBLEM TITLE : ";P$
1800 PRINT "NO.OF WORK CENTERS : ";K
1810 PRINT
1820 PRINT
1830 PRINT "N E A R N E S S     P R I O R I T Y     R A N K I N G"
1840 PRINT F$(1)
1850 FOR L = 1 TO K9
1860 N = 0
1870 FOR M = K-1 TO 1 STEP -1
1880 N = N+M
1890 IF R(L) <= N THEN 1910
1900 NEXT M
1910 I = K-M
1920 J = K-N+R(L)
1930 PRINT USING F$(2),L,N$(I),N$(J),F(R(L))
1940 NEXT L
1950 RETURN
1960 REM INITIALIZATION
1970 F$(1)="RANK    FROM         TO          COST/UNIT DISTANCE/PERIOD"
1980 F$(2)="####    ##########   ##########    ########.##"
1990 RETURN
2000 END
```

II. Product Layout Design

Product layout is a type of layout in which work activities, determined by steps required to produce the product or service, are arranged in some sort of a line. The classical case of product layout is the assembly line.

The number of work stations and the number of work elements to be performed at each work station on an assembly line depend on the rate of output. The quantitative decision criterion generally used in designing the work station is the "balance delay," which is the idle time on the line caused by uneven allocation of work among the stations.

The assembly line layout problem is to find the minimum number of work stations for a given output and to assign work elements to the work stations in such a manner that the balance delay is minimized without violating any precedence or sequential restrictions.

Assembly lines have been classified as single-model, batch-model, multimodel, and mixed-model. The line balancing methods proposed in the literature range from analytical techniques to heuristic methods. For those readers who are interested in more information on this subject, the textbooks by Prenting and Thomopoulos [4] and Wild [6] are recommended.

In this section, a computer program based on two simple heuristics for the single-model assembly line is discussed.

The following procedure can be used to establish a single-model assembly line:

1. Establish assembly work elements, standard time for each work element, and the total time to assemble one unit,

$$\{AT\} = \sum_{i=1}^{m} t_i$$

where $\{AT\}$ = Total time to assemble one unit in minutes,
m = Number of work elements,
t_i = Standard time associated with work element i.

2. Develop a matrix to represent sequential restrictions on the work elements using assembly drawing.

3. Calculate Theoretical Cycle Time in minutes,

$$\{CT\} = P/D$$

where   P = Production time in minutes per day,
and   D = Number of units to be produced per day.

4. Calculate the theoretical minimum number of work stations:

$$N = \{AT\}/\{CT\}$$

5. Allocate work elements to work stations. This can be done by using one of the following heuristics:

Heuristic 1 (H1) - First allocate those work elements that have the largest number of following work elements.

Heuristic 2 (H2) - First allocate those work elements that have the longest times.

6. Calculate the balance delay:

$$\text{Balance Delay} = 1 - \{AT\}/(n\{SCT\})$$

where n = Actual number of work stations in the line for a given balance, and
{SCT} = Selected Cycle Time which can be equal to {CT}.

The assembly line balance under heuristics 1 and 2 can be compared with respect to the generated balance delays. The line with the minimum balance delay is the one desired.

The Program ASSMBLY

ASSMBLY computes the theoretical cycle time and the theoretical minimum number of work stations required. It allocates work elements to work stations using the two heuristics mentioned above. It automatically numbers the stations sequentially. The balance delays from the heuristics are also given.

INPUT

Because of the large amount of input data required for this program, we have selected to input the data through the use of DATA statements in BASIC. The sequence of inputs is as follows:

1. Problem title
2. Required units of production per day
3. Available production time in minutes per day
4. Number of work elements in the assembly
5. Precedence relationship data

The precedence relationship data consist of

a. The work element number. It is important to note that the work elements must be sequentially numbered starting from 1 up to the number of work elements in the assembly. In addition, in order to maintain the correct precedence logic, the assembly must terminate into a single work element. A dummy work element with zero standard time may be used for this purpose.

b. The standard time in 0.01 minutes for the work element.
c. The number of work elements preceding it.
d. All preceding work element numbers.

OUTPUT

Two output producing commands are H1 and H2 which correspond to heuristics 1 and 2, respectively. The report from these commands shows the detailed allocation of work elements to the required work stations. The LIST command lists the element standard time and its preceding work elements. This serves as a means for verifying the input data easily. Multiple problems can be solved in one computer run if data are appended together. The RESTART command will initiate the reading of input data for the next problem.

EXAMPLE

Consider the precedence network of an assembly operation as shown below. The operation requires 16 active work elements, and one dummy work element to maintain the precedence logic.

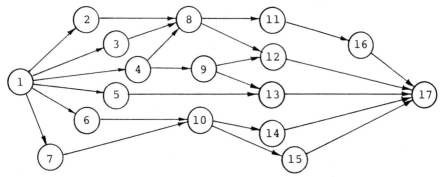

The following table lists the work element number, its standard time in 0.01 minutes, and its precedence relationships. For example, the standard time for work element 10 is 0.12 minutes. There are three work elements that precede work element 10. They are 1, 6, and 7. Note also that work element 17 is a dummy one because its standard time is zero.

This operation needs to be balanced for a required

production of 560 units per day. Available production time per day is given as 420 minutes.

Work Element	Standard Time	Number of Precedence	PRECEDENCE RELATIONSHIP
1	10	0	None
2	11	1	1
3	6	1	1
4	28	1	1
5	4	1	1
6	18	1	1
7	11	1	1
8	11	4	1,2,3,4
9	24	2	1,4
10	12	3	1,6,7
11	8	5	1,2,3,4,8
12	3	6	1,2,3,4,8,9
13	15	4	1,4,5,9
14	3	4	1,6,7,10
15	19	4	1,6,7,10
16	63	6	1,2,3,4,8,11
17	0	16	1,2,3,4,5,6,7,8,9,10,11,12,13,14,15,16

The above data are translated into BASIC and appended to the end of the program ASSMBLY before running. They are shown below.

```
3990 REM -----PROBLEM TITLE-----
3991 DATA "A 17 WORK ELEMENT LINE BALANCING PROBLEM"
3992 REM -----REQUIRED UNITS OF PRODUCTION PER DAY-----
3993 DATA 560
3994 REM -----AVAILABLE PRODUCTION MINUTES PER DAY-----
3995 DATA 420
3996 REM -----NUMBER OF WORK ELEMENTS-----
3997 DATA 17
3998 REM    WORK ELEMENT DATA WHICH CONSIST OF ELEMENT NUMBER,
3999 REM    STANDARD TIME, # OF PRECEDENCE ELEMENTS
4000 REM    PRECEDENCE RELATIONSHIPS FOR THE WORK ELEMENT
4010 REM -----ELEMENT 1 DATA-----
4011 DATA  1,10,0
4020 REM -----ELEMENT 2 DATA-----
4021 DATA  2,11, 1
4022 DATA  1
4030 REM -----ELEMENT 3 DATA-----
4031 DATA  3, 6, 1
4032 DATA  1
4040 REM -----ELEMENT 4 DATA-----
4041 DATA  4,28, 1
4042 DATA  1
```

```
4050 REM -----ELEMENT 5 DATA-----
4051 DATA   5, 4, 1
4052 DATA   1
4060 REM -----ELEMENT 6 DATA-----
4061 DATA   6,18, 1
4062 DATA   1
4070 REM -----ELEMENT 7 DATA-----
4071 DATA   7,11, 1
4072 DATA   1
4080 REM -----ELEMENT 8 DATA-----
4081 DATA   8,11, 4
4082 DATA   1,2,3,4
4090 REM -----ELEMENT 9 DATA-----
4091 DATA   9,24, 2
4092 DATA   1,4
4100 REM -----ELEMENT 10 DATA-----
4101 DATA 10,12, 3
4102 DATA   1, 6, 7
4110 REM -----ELEMENT 11 DATA-----
4111 DATA 11, 8, 5
4112 DATA   1, 2, 3, 4, 8
4120 REM -----ELEMENT 12 DATA-----
4121 DATA 12, 3, 6
4122 DATA   1, 2, 3, 4, 8, 9
4130 REM -----ELEMENT 13 DATA-----
4131 DATA 13,15, 4
4132 DATA   1, 4, 5, 9
4140 REM -----ELEMENT 14 DATA-----
4141 DATA 14, 3, 4
4142 DATA   1, 6, 7,10
4150 REM -----ELEMENT 15 DATA-----
4151 DATA 15,19, 4
4152 DATA   1, 6, 7,10
4160 REM -----ELEMENT 16 DATA-----
4161 DATA 16,63, 6
4162 DATA   1, 2, 3, 4, 8,11
4170 REM -----ELEMENT 17 DATA-----
4171 DATA 17, 0,16
4172 DATA   1, 2, 3, 4, 5, 6, 7, 8, 9,10,11,12,13,14,15,16
```

The example terminal session follows. While the program reads the data, it also performs some input verifications and messages will be printed if invalid data are encountered.

After all of the input are read, we issue the HELP command which lists all available commands. Next the H1 and H2 commands are entered in succession. For the problem, the theoretical number of work stations is 4. Both heuristics generate solutions that require five work stations with identical balance delay of 0.344 minutes. However, the compositions of the work elements within a work station are different between the two solutions. On the basis of the balance delay, we can conclude that both solutions will be equally acceptable.

```
ASSMBLY EXAMPLE TERMINAL SESSION
=================================

A S S E M B L Y    L I N E    B A L A N C E
---------------------------------------------

NEED INTRODUCTION (Y OR N) ... !N

READING GENERAL INPUT DATA ...
NUMBER OF WORK ELEMENTS = 17
READING WORK ELEMENT DATA ...
ELEMENT  1   NUMBER OF PRECEDENCE =  0
ELEMENT  2   NUMBER OF PRECEDENCE =  1
ELEMENT  3   NUMBER OF PRECEDENCE =  1
ELEMENT  4   NUMBER OF PRECEDENCE =  1
ELEMENT  5   NUMBER OF PRECEDENCE =  1
ELEMENT  6   NUMBER OF PRECEDENCE =  1
ELEMENT  7   NUMBER OF PRECEDENCE =  1
ELEMENT  8   NUMBER OF PRECEDENCE =  4
ELEMENT  9   NUMBER OF PRECEDENCE =  2
ELEMENT 10   NUMBER OF PRECEDENCE =  3
ELEMENT 11   NUMBER OF PRECEDENCE =  5
ELEMENT 12   NUMBER OF PRECEDENCE =  6
ELEMENT 13   NUMBER OF PRECEDENCE =  4
ELEMENT 14   NUMBER OF PRECEDENCE =  4
ELEMENT 15   NUMBER OF PRECEDENCE =  4
ELEMENT 16   NUMBER OF PRECEDENCE =  6
ELEMENT 17   NUMBER OF PRECEDENCE = 16

PLEASE ENTER COMMAND OR TYPE HELP

COMMAND --->!HELP

HELP     - PRINT THIS MESSAGE
LIST     - LIST PRECEDENCE RELATIONSHIPS
H1       - LARGEST NUMBER OF FOLLOWING ELEMENT HEURISTIC
H2       - LONGEST WORK ELEMENT TIME HEURISTIC
RESTART  - START A NEW PROBLEM
EXIT     - EXIT FROM THE PROGRAM

COMMAND --->!H1

H1       - LARGEST NUMBER OF FOLLOWING ELEMENT HEURISTIC

PROBLEM TITLE : A 17 WORK ELEMENT LINE BALANCING PROBLEM
REQUIRED UNITS OF PRODUCTION PER DAY = 560
AVAILABLE PRODUCTION TIME PER DAY = 420 MINUTES
NUMBER OF WORK ELEMENTS = 17
PRODUCTION TIME PER UNIT = 2.46 MINUTES
CYCLE TIME PER UNITS = .75 MINUTES
THEORETICAL NUMBER OF WORK STATIONS = 4
```

STATION NUMBER	WORK ELEMENT	STANDARD TIME(0.01 MIN)	REMAINING TIME(0.01 MIN)
1	1	10	65
1	4	28	37
1	2	11	26
1	3	6	20
1	5	4	16
2	6	18	57
2	7	11	46
2	10	12	34
2	8	11	23
3	9	24	51
3	11	8	43
4	16	63	12
4	12	3	9
4	14	3	6
5	13	15	60
5	15	19	41
5	17	0	41

BALANCE DELAY =.344 MINUTES

COMMAND ---> !H2

H2 - LONGEST WORK ELEMENT TIME HEURISTIC

PROBLEM TITLE : A 17 WORK ELEMENT LINE BALANCING PROBLEM
REQUIRED UNITS OF PRODUCTION PER DAY = 560
AVAILABLE PRODUCTION TIME PER DAY = 420 MINUTES
NUMBER OF WORK ELEMENTS = 17
PRODUCTION TIME PER UNIT = 2.46 MINUTES
CYCLE TIME PER UNITS = .75 MINUTES
THEORETICAL NUMBER OF WORK STATIONS = 4

STATION NUMBER	WORK ELEMENT	STANDARD TIME(0.01 MIN)	REMAINING TIME(0.01 MIN)
1	1	10	65
1	4	28	37
1	9	24	13
2	6	18	57
2	2	11	46
2	7	11	35
2	10	12	23
2	15	19	4
3	3	6	69
3	8	11	58
3	11	8	50
4	16	63	12
4	5	4	8
5	13	15	60
5	12	3	57
5	14	3	54
5	17	0	54

BALANCE DELAY =.344 MINUTES

```
10 REM    ASSMBLY
20 DIM F$(9),G$(2)
30 DIM    B(40),C(40),E(40),F(40),G(40,40)
40 N0=40
50 GOSUB 2450
60 PRINT "A S S E M B L Y    L I N E    B A L A N C E"
70 PRINT "------------------------------------------"
80 PRINT
90 PRINT "NEED INTRODUCTION (Y OR N) ... ";
100 INPUT Y$
110 IF Y$ <> "Y" THEN 230
120 PRINT
130 PRINT "THIS PROGRAM PERFORMS SINGLE-MODEL ASSEMBLY LINE"
140 PRINT "BALANCING USING TWO COMMONLY USED HEURISTICS :"
150 PRINT "1. ALLOCATE THE WORK ELEMENT HAVING THE LARGEST"
160 PRINT "   NUMBER OF FOLLOWING WORK ELEMENTS FIRST."
170 PRINT "2. ALLOCATE THE WORK ELEMENT HAVING THE LONGEST"
180 PRINT "   PROCESSING TIME FIRST."
190 PRINT
200 PRINT "INPUT REQUIREMENTS"
210 PRINT "------------------"
220 PRINT "DETAILS OF INPUT FORMAT ARE AVAILABLE THROUGH THE DOCUMENT."
230 PRINT
240 Z1=0
250 PRINT "READING GENERAL INPUT DATA ..."
260 REM    PROBLEM TITLE
270 READ T$
280 REM    REQUIRED UNITS OF PRODUCTION PER DAY
290 READ D
300 IF D > 0 THEN 330
310 PRINT "UNIT OF PRODUCTION MUST BE > 0"
320 Z1=1
330 REM    AVAILABLE PRODUCTION TIME PER DAY IN MINUTES
340 READ P
350 IF P > 0 THEN 380
360 PRINT "AVAILABLE PRODUCTION TIME MUST BE > 0"
370 Z1=1
380 REM    GET THEORETICAL CYCLE TIME
390 C2=P/D
400 C1=C2*100
410 REM    NUMBER OF WORK ELEMENTS
420 READ N
430 PRINT "NUMBER OF WORK ELEMENTS = ";N
440 IF N >= 2 THEN 470
450 PRINT "NUMBER OF ELEMENTS MUST BE >= 2"
460 Z1=1
470 IF N <= N0 THEN 520
480 PRINT "*** ERROR *** MAXIMUM ALLOWED = ";N0
490 PRINT "!!! SEE DIMENSIONAL SPECIFICATIONS FOR MORE DETAILS !!!"
500 PRINT "!!! PROGRAM EXECUTION TERMINATED DUE TO DATA ERROR !!!"
510 GOTO 2630
520 REM    WORK ELEMENT DATA
530 REM    -----------------
540 REM    ELEMENT NUMBER, STANDARD TIME, # OF PRECEDING ELEMENTS.
```

```
550 PRINT "READING WORK ELEMENT DATA ... "
560 S=0
570 FOR I=1 TO N
580 READ K5,T5,N5
590 PRINT "ELEMENT ";K5;"  NUMBER OF PRECEDENCE = ";N5
600 IF K5 > N THEN 670
610 IF K5 <=0 THEN 670
620 B(K5)=T5
630 S=S+T5
640 IF N5 > N THEN 670
650 IF N5 >= 1 THEN 690
660 IF N5 <= 0 THEN 790
670 PRINT "*** ERROR IN ELEMENT NUMBERING ***"
680 Z1=2
690 IF T5 <= C1 THEN 720
700 PRINT "*** ERROR *** ELE. STD.TIME EXCEEDS CYCLE TIME OF ";C1
710 Z1=3
720 FOR J=1 TO N5
730 READ G(K5,J)
740 IF G(K5,J) = K5 THEN 760
750 IF G(K5,J) <= N THEN 780
760 Z1=4
770 PRINT"PRECEDENCE ERROR ";G(K5,J)
780 NEXT J
790 NEXT I
800 IF Z1 <> 0 THEN 500
810 S2=S/100
820 T=INT(S2/C2+1)
830 PRINT
840 PRINT "PLEASE ENTER COMMAND OR TYPE HELP"
850 PRINT
860 PRINT "COMMAND --->";
870 INPUT C$
880 PRINT
890 IF C$ <> "HELP" THEN 920
900 GOSUB 1050
910 GOTO 850
920 IF C$ <> "LIST" THEN 950
930 GOSUB 1200
940 GOTO 850
950 IF C$ <> "H1" THEN 980
960 GOSUB 1410
970 GOTO 850
980 IF C$ <> "H2" THEN 1010
990 GOSUB 1560
1000 GOTO 850
1010 IF C$ = "RESTART" THEN 230
1020 IF C$ ="EXIT" THEN 2630
1030 PRINT "*** INVALID COMMAND *** PLEASE TRY AGAIN !!!"
1040 GOTO 850
1050 REM     *** HELP COMMANDS ***
1060 FOR I=4 TO 9
1070 PRINT F$(I)
1080 NEXT I
```

```
1090 RETURN
1100 REM     PRINT TITLE OF REPORTS
1110 PRINT "PROBLEM TITLE : ";T$
1120 PRINT "REQUIRED UNITS OF PRODUCTION PER DAY = ";D
1130 PRINT "AVAILABLE PRODUCTION TIME PER DAY = ";P;" MINUTES"
1140 PRINT "NUMBER OF WORK ELEMENTS = ";N
1150 PRINT "PRODUCTION TIME PER UNIT = ";S2;" MINUTES"
1160 PRINT "CYCLE TIME PER UNITS = ";C2;" MINUTES"
1170 PRINT "THEORETICAL NUMBER OF WORK STATIONS = ";T
1180 PRINT
1190 RETURN
1200 REM     *** LIST COMMAND ***
1210 GOSUB 1100
1220 PRINT "    WORK   STANDARD"
1230 PRINT "ELEMENT      TIME    P R E C E D E N C E";
1240 PRINT "      R E L A T I O N S H I P"
1250 FOR I = 1 TO N
1260 K7=0
1270 PRINT USING G$(1),I,B(I);
1280 FOR J = 1 TO N
1290 IF G(I,J) = 0 THEN 1360
1300 K7=K7+1
1310 IF K7 <= 12 THEN 1350
1320 K7=1
1330 PRINT
1340 PRINT TAB(18);
1350 PRINT USING G$(2),G(I,J);
1360 NEXT J
1370 PRINT
1380 NEXT I
1390 PRINT
1400 RETURN
1410 REM     *** H1 COMMAND ***
1420 PRINT
1430 PRINT F$(6)
1440 PRINT
1450 GOSUB 1100
1460 FOR I = 1 TO N
1470 FOR J = 1 TO N
1480 H = G(I,J)
1490 IF H = 0 THEN 1520
1500 C(H) = C(H) + 1
1510 NEXT J
1520 NEXT I
1530 GOSUB 1670
1540 GOSUB 2180
1550 GOTO 1630
1560 REM     *** H2 COMMAND ***
1570 PRINT
1580 PRINT F$(7)
1590 PRINT
1600 GOSUB 1100
1610 GOSUB 2040
1620 GOSUB 2180
```

```
1630 M = 1 - S/(N1*C1)
1640 PRINT
1650 PRINT "BALANCE DELAY =";M;" MINUTES"
1660 RETURN
1670 REM    SORT ACCORDING TO HEURISTIC 1
1680 FOR I = 1 TO N
1690 E(I) = I
1700 NEXT I
1710 FOR I = 2 TO N
1720 I1 = I-1
1730 FOR J = 1 TO I1
1740 IF C(J) >= C(I) THEN 1780
1750 T2 = E(I)
1760 E(I) = E(J)
1770 E(J) = T2
1780 NEXT J
1790 NEXT I
1800 I = 1
1810 I2 = I + 1
1820 IF C(E(I)) <> C(E(I2)) THEN 1860
1830 I2 = I2 + 1
1840 IF I2 > N THEN 1900
1850 GOTO 1820
1860 IF (I2 -I) > 1 THEN 1900
1870 I = I + 1
1880 IF I < N THEN 1810
1890 GOTO 2030
1900 K1 = I + 1
1910 K2 = I2 - 1
1920 FOR K = K1 TO K2
1930 K3 = K1 - 1
1940 FOR L  = I TO K3
1950 IF B(E(L)) >= B(E(K)) THEN 1990
1960 T2 = E(K)
1970 E(K) = E(L)
1980 E(L) = T2
1990 NEXT L
2000 NEXT K
2010 I = I2
2020 IF I < N THEN 1810
2030 RETURN
2040 REM    SORT ACCORDING TO HEURISTIC 2
2050 FOR I = 1 TO N
2060 E(I) = I
2070 NEXT I
2080 FOR I = 2 TO N
2090 I1 = I - 1
2100 FOR J = 1 TO I1
2110 IF B(E(J)) >= B(E(I)) THEN 2150
2120 T2 = E(I)
2130 E(I) = E(J)
2140 E(J) = T2
2150 NEXT J
2160 NEXT I
```

```
2170 RETURN
2180 REM     ALLOCATE SUBROUTINE
2190 PRINT F$(1)
2200 PRINT F$(2)
2210 FOR I = 1 TO N
2220 F(I) = 0
2230 NEXT I
2240 N1 = 1
2250 T1 = C1
2260 I = 1
2270 E1 = E(I)
2280 IF F(E1) = 1 THEN 2420
2290 J = 1
2300 IF G(E1,J) <> 0 THEN 2390
2310 IF B(E1) <= T1 THEN 2340
2320 N1 = N1 + 1
2330 T1 = C1
2340 T1 = T1 - B(E1)
2350 PRINT USING F$(3),N1,E1,B(E1),T1
2360 F(E1) =1
2370 I = 1
2380 GOTO 2270
2390 IF F(G(E1,J)) = 0 THEN 2420
2400 J = J +1
2410 GOTO 2300
2420 I = I + 1
2430 IF I <= N THEN 2270
2440 RETURN
2450 REM     I N I T I A L I Z A T I O N
2460 F$(1)="STATION      WORK      STANDARD         REMAINING"
2470 F$(2)=" NUMBER    ELEMENT   TIME(0.01 MIN)   TIME(0.01 MIN)"
2480 F$(3)=" ######    #######     ########         #########"
2490 F$(4)="HELP     - PRINT THIS MESSAGE"
2500 F$(5)="LIST     - LIST PRECEDENCE RELATIONSHIPS"
2510 F$(6)="H1       - LARGEST NUMBER OF FOLLOWING ELEMENT HEURISTIC"
2520 F$(7)="H2       - LONGEST WORK ELEMENT TIME HEURISTIC"
2530 F$(8)="RESTART  - START A NEW PROBLEM"
2540 F$(9)="EXIT     - EXIT FROM THE PROGRAM"
2550 G$(1)="#######   ########"
2560 G$(2)="####"
2570 FOR I=1 TO N0
2580 FOR J=1 TO N0
2590 G(I,J)=0
2600 NEXT J
2610 NEXT I
2620 RETURN
2630 END
```

III. Dimensional Specifications

Program PLAYOUT

Let n be the maximum number of work centers that the program can handle. Also let m=n(n-1)/2. Currently, n is set to 10, which means that m=45. The following statements should be changed if n changes:

Line No.
----
30 DIM T(n,n),C(n,n),N$(n),F(m),R(m),F$(2)
40 K0=n

Program ASSMBLY

Let n be the allowable maximum number of work elements. In this program, n is currently equal to 40. To modify the program to accommodate larger values of n, the following statements are affected:

Line No.
----
30 DIM B(n),C(n),E(n),F(n),G(n,n)
40 N0=n

IV. References
----
1. Francis, R.L., and J.A. White, FACILITY LAYOUT AND LOCATION - AN ANALYTICAL APPROACH, Prentice-Hall, Englewood Cliffs, N.J., 1974.

2. Groover, M.P., AUTOMATION, PRODUCTION SYSTEMS, AND COMPUTER-AIDED MANUFACTURING, Prentice-Hall, Englewood Cliffs, N.J., 1980.

3. Muther, R., SYSTEMATIC LAYOUT PLANNING, Industrial Education Institute, Boston, 1961.

4. Prenting, T.O., and N.T. Thomopoulos, HUMANISM AND TECHNOLOGY IN ASSEMBLY LINE SYSTEMS, Spartan Books, Hayden Book Co., Rochelle Park, N.J., 1974.

5. Tompkins, J.A., and J.M. Moore, COMPUTER AIDED LAYOUT: A USER'S GUIDE, AIIE, Norcross, Ga. 30092.

6. Wild, R., MASS-PRODUCTION MANAGEMENT, John Wiley Ltd., London, 1972.

CHAPTER

INVENTORY ANALYSIS

I. The Program PBREAK
  ○ Input
  ○ Output
  ○ Example
  ○ PBREAK Example Terminal Session
  ○ PBREAK Program Listing

II. The (Q, r) Inventory Policy

III. The Program EXCHNGE
  ○ Input
  ○ Output
  ○ Example
  ○ EXCHNGE Example Terminal Session
  ○ EXCHNGE Program Listing

IV. The Program INVCOST
  ○ Input
  ○ Output
  ○ Example
  ○ INVCOST Example Terminal Session
  ○ INVCOST Program Listing

V. The Program LOTSZ.MRP
  ○ Input
  ○ Output
  ○ Example
  ○ LOTSZ.MRP Example Terminal Session
  ○ LOTSZ.MRP Program Listing

VI. Dimensional Specifications

VII. References

Current literature on inventory control classifies inventory items into two categories as follows.

1. Independent demand items. These are items whose demands do not depend on the demands of other items. Demands can be either deterministic or probabilistic. Two approaches for analyzing these items exist. The first approach analyzes the problem on the basis of cost alone. The objective of the analysis is to determine optimal control parameters which will minimize the sum of the ordering cost, inventory carrying cost, and shortage cost. The second approach finds the optimal control parameters subject to some service level constraints. It can also be shown that under certain conditions the two approaches are equivalent.

2. Dependent demand items. These are items or components of a larger product or item. The forecasts of demand for a larger item may be determined independently, and the forecasts of the smaller items are the result of exploding the requirements down the product structure of the larger item. These items occur more frequently in the context of Material Requirements Planning (MRP), which is currently a popular technique for production and inventory control.

There are basically three questions to be answered in any inventory control system:

1. At what time interval should the inventory of an item be reviewed?
2. When should an order for the item be placed?
3. How large should the order size (quantity) or the production lot size be?

In this chapter, we provide four computer programs that give some answers to the three questions for the two categories of items mentioned above. The first three programs can be used to analyze independent demand items, and the last program deals with lot-sizing techniques in MRP.

The techniques for analyzing inventory items that we have selected have been widely implemented in various inventory control packages. We are, however, interested only in the analysis side of the problem.

The first program is called PBREAK. It can be used to find the optimal order quantity of an item when the vendor offers price breaks or quantity discounts. Demand is assumed to be deterministic and occurs at a constant rate.

The second program is called EXCHNGE. It determines the control parameters (order point, order quantity) for a group of independent items. It is based on the service level concept and contains five decision rules which can be used to determine the trade-offs (exchange) between the investment in safety stocks and various service level measures. The trade-offs can be conveniently expressed in terms of exchange curves. The decision rules contained in this program are those suggested by Brown [2]. Demands are assumed to be probabilistic, but the demand rates are assumed to be stationary.

The third program, INVCOST, can be used to analyze multiple independent items similar to the EXCHNGE program. However, our main emphasis here is on the cost models -i.e., the objective is to determine the optimal control parameters (order point, order quantity) for each item such that the sum of the expected ordering cost per year, the inventory carrying cost per year, and the expected shortage cost per year is minimized.

The last program is called LOTSZ.MRP. It contains a number of commonly used lot-sizing techniques in MRP. In addition to being able to determine the production lot sizes of an item for a fixed planning horizon of T periods, the program also contains a simulation feature that allows the user to evaluate the performance of these lot-sizing techniques.

For consistency, the following notations will be used throughout this chapter:

n = Number of items
D = Demand rate (units/year)
I = Inventory carrying rate (100I percent per year)
V = Unit cost or buying price or value of an item ($)
H = IV = Inventory carrying or holding cost per unit per year ($/unit/year)
P = Penalty cost per unit short ($/unit)
A = Setup cost or cost of placing an order ($/order)
Q = Order quantity or order size or lot size (units)
r = Order point or reorder point (units)
k = Safety factor
M = Mean demand during the procurement lead time (units)
S = Standard deviation of demand during the lead time

The distribution of demand during the lead time for all items (when applicable) is assumed to be normally distributed. Since the above notations apply to a single item of concern, each of the above variables may have a subscript attached to it when we deal with multiple items in the programs EXCHNGE and INVCOST.

I. The Program PBREAK

PBREAK determines the optimal order quantity of an item for which the vendor offers m price breaks or quantity discounts -i.e., the unit price of the item (V) decreases as the order quantity increases. Demand for the item is assumed to be deterministic. No shortage is allowed. Replenishment of stock is assumed to occur instantaneously and completely whenever an order is placed. The relevant cost components are the procurement cost plus the inventory carrying cost.

There are basically two types of quantity discounts:
1. All Units Discount. This means that there is one unit cost that will be applicable for all purchased units whenever the order quantity falls into an applicable range.
2. Incremental Quantity Discount. In this case, if the order quantity exceeds a price break quantity, there will be more than one applicable unit cost. Each unit cost will be applicable only to the amount that lies between the range of the quantity schedule.

Figure 1 below illustrates the variable purchasing cost, K(Q), as a function of the order quantity (Q) when there are three price breaks (m=3) for the two types of discounts mentioned. Note that when there are m price breaks, there will be m+1 values of unit costs.

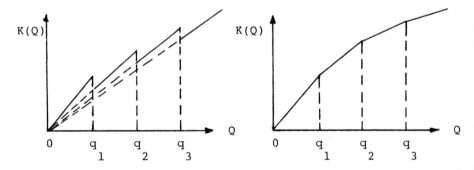

ALL UNITS DISCOUNT INCREMENTAL QUANTITY DISCOUNT

Figure 1. Total variable purchasing cost functions for All Units Discount and Incremental Quantity Discount

The total relevant cost per year to be minimized is C(Q), which is the sum of the setup cost, the variable purchasing cost, and the inventory carrying cost:

(1) $C(Q) = AD/Q + K(Q) D/Q + [IK(Q)/Q]Q/2$,

for $q_{i-1} <= Q < q_i$; i=1,2,...,m. Let V_i be the corresponding unit cost. Then, for the all units discount case, we have

(2) $\quad K(Q) = V_i Q, \quad q_{i-1} <= Q < q_i.$

For the incremental quantity discount case, we have

(3) $\quad K(Q) = \sum_{j=1}^{i-1} V_j(q_j - q_{j-1}) + V_i(Q - q_{i-1})$

$\qquad\qquad = K(q_{i-1}) + V_j(Q - q_{i-1}), \quad q_{i-1} <= Q < q_i.$

It should be noted that in the all units discount case the optimal order quantity can occur at one of the price breaks, whereas in the incremental quantity discount case it cannot. This is illustrated in Figure 2, which displays the total relevant cost per year for the two types of discount.

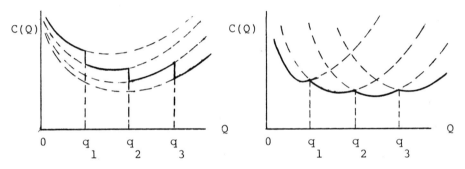

ALL UNITS DISCOUNT   INCREMENTAL QUANTITY DISCOUNT

Figure 2. Total relevant cost per year, C(Q), for All Units Discount and Incremental Quantity Discount

The computation algorithms for determining the optimal order quantity may be found in books such as Hadley and Whitin [7] and also in [8]. The program PBREAK prompts for all relevant data, and the optimal solutions can be determined via the use of the ALL (All Units Discount) or the INC (Incremental Quantity Discount) commands.

INPUT
-----
1. Item description
2. Annual demand in units (D)
3. Setup cost per order (A)
4. Inventory carrying rate in percentage per year (100I)
5. Number of price breaks (m)
6. m values of quantity schedule at price breaks (q_i)
7. m+1 values of pricing schedule (unit costs)

OUTPUT

Output from PBREAK consists of the optimal order quantity, the time between placing an order in years, the average procurement cost per year which is the setup cost plus the variable cost due to the cost of the item, the average inventory carrying cost per year, and the average total cost per year. Since the input data for price breaks can be interpreted as being of either the all units discount type or the incremental quantity discount type, their solutions are both available through the use of appropriate commands.

EXAMPLE

Consider a metal casing (part number A11232FR) whose demand rate is 2000 units per year. The setup cost per order is $50. The company uses an annual inventory carrying rate of 30 percent. The vendor of this part offers a quantity discount schedule as follows.

Quantity Schedule	Unit Price ($)
0 < Q < 1000	18.00
1000 <= Q < 3000	17.00
3000 <= Q < 5000	16.00
5000 <= Q	15.00

We wish to find the optimal solutions assuming that the schedule applies to both discount types.

The example terminal session for this problem follows. We note that there are three price breaks occurring at quantities of 1000, 3000, and 5000 units, respectively. After all of the data are entered, the HELP command is issued. Next we enter the ALL command to determine the optimal solution of the all units discount case. The optimal order quantity occurs at the first price break and equals 1000 units. The average total cost per year is $36,650, which also includes the cost of buying 2000 units at a cost of $17 per unit.

The command INC is entered next. Here the quantity discount schedule is being interpreted as that of the incremental quantity discount type. We note that the optimal solution in this case is to order more often at less amount -i.e., the optimal Q is 192.45 units and the order cycle is 0.962 years, or approximately every 1.15 months. The cost per unit is $18. The average total cost per year is $37,039.23, which confirms our intuition that under identical conditions as in our problem the incremental quantity discount will always cost us more.

```
PBREAK EXAMPLE TERMINAL SESSION
================================

QUANTITY DISCOUNTS - INVENTORY ANALYSIS
---------------------------------------

NEED INTRODUCTION (Y OR N) ... !N

ENTER ITEM DESCRIPTION    !METAL CASING - PART NO. A11232FR

ENTER ANNUAL DEMANDS IN UNITS    !2000

ENTER COST IN SETTING UP AN ORDER    !50

ENTER INVENTORY CARRYING RATE IN PERCENTAGE   !30

ENTER NUMBER OF PRICE BREAKS (M>=1)    !3

ENTER   3  VALUES OF QUANTITY SCHEDULE
PRICE BREAK   1    QUANTITY = !1000
PRICE BREAK   2    QUANTITY = !3000
PRICE BREAK   3    QUANTITY = !5000

ENTER ALL UNIT PRICES
FOR QUANTITY BETWEEN   0  AND  1000   UNIT PRICE =!18
FOR QUANTITY BETWEEN  1000 AND  3000  UNIT PRICE =!17
FOR QUANTITY BETWEEN  3000 AND  5000  UNIT PRICE =!16
FOR QUANTITY GREATER THAN  5000  UNIT PRICE = !15

PLEASE ENTER COMMAND OR TYPE HELP

COMMAND -->!HELP

AVAILABLE COMMANDS
==================
HELP      - PRINT THIS MESSAGE
ALL       - FIND ALL-UNITS-DISCOUNTS SOLUTION
INC       - FIND INCREMENTAL-DISCOUNTS SOLUTION
RESTART   - START A NEW PROBLEM
EXIT      - EXIT FROM THE PROGRAM

COMMAND -->!ALL
```

```
A L L    U N I T S    D I S C O U N T
ITEM DESCRIPTION : METAL CASING - PART NO. Al1232FR
ANNUAL DEMAND IN UNITS : 2000
SETUP COST PER ORDER : 50
INVENTORY CARRYING RATE (%) : 30
     QUANTITY-DISCOUNT  SCHEDULE
     FROM         TO        UNIT-PRICE
        0       1000          18.00
     1000       3000          17.00
     3000       5000          16.00
     5000    INFINITY         15.00
* OPTIMAL ORDER QUANTITY = 1000
* TIME BETWEEN ORDER = .5  YEARS
* AVERAGE PROCUREMENT COST/YEAR = 34100
* AVERAGE INVENTORY CARRYING COST/YEAR = 2550
* AVERAGE TOTAL COST/YEAR = 36650


COMMAND -->!INC

I N C R E M E N T A L    D I S C O U N T
ITEM DESCRIPTION : METAL CASING - PART NO. Al1232FR
ANNUAL DEMAND IN UNITS : 2000
SETUP COST PER ORDER : 50
INVENTORY CARRYING RATE (%) : 30
     QUANTITY-DISCOUNT  SCHEDULE
     FROM         TO        UNIT-PRICE
        0       1000          18.00
     1000       3000          17.00
     3000       5000          16.00
     5000    INFINITY         15.00
* OPTIMAL ORDER QUANTITY = 192.4500897299
* TIME BETWEEN ORDER = .09622504486494  YEARS
* AVERAGE PROCUREMENT COST/YEAR = 36519.61524227
* AVERAGE INVENTORY CARRYING COST/YEAR = 519.6152422706
* AVERAGE TOTAL COST/YEAR = 37039.23048454


COMMAND -->!EXIT
```

```
10 REM     QUANTITY DISCOUNTS (PRICE BREAKS)
20 REM
30 DIM P(10),Q(10),T(10),S(10),F$(5)
40 M0=10
50 GOSUB 1810
60 PRINT
70 PRINT "QUANTITY DISCOUNTS - INVENTORY ANALYSIS"
80 PRINT "-------------------------------------"
90 PRINT
100 PRINT "NEED INTRODUCTION (Y OR N) ... ";
110 INPUT Y$
120 IF Y$ <> "Y" THEN 320
130 PRINT
140 PRINT "THIS PROGRAM DETERMINES THE OPTIMAL ORDER QUANTITY"
150 PRINT "OF AN INVENTORY ITEM WHEN THE VENDOR OFFERS PRICE"
160 PRINT "BREAKS FOR LARGER ORDER SIZE.  TWO TYPES OF QUANTITY"
170 PRINT "DISCOUNTS CAN BE ANALYZED:"
180 PRINT "   1. ALL UNITS DISCOUNT"
190 PRINT "   2. INCREMENTAL QUANTITY DISCOUNT"
200 PRINT
210 PRINT "INPUT REQUIREMENTS"
220 PRINT "------------------"
230 PRINT "1. ITEM DESCRIPTION"
240 PRINT "2. ANNUAL DEMANDS IN UNITS"
250 PRINT "3. COST OF SETTING UP AN ORDER INDEPENDENT OF QUANTITY"
260 PRINT "4. INVENTORY CARRYING RATE IN PERCENTAGE"
270 PRINT "5. NUMBER OF PRICE BREAKS"
280 PRINT "6. QUANTITY SCHEDULE IN INCREASING ORDER"
290 PRINT "7. CORRESPONDING PRICING SCHEDULE IN DECREASING ORDER."
300 PRINT
310 PRINT "NOTE : NUMBER OF PRICE BREAKS IS LIMITED TO ";M0
320 PRINT
330 PRINT "ENTER ITEM DESCRIPTION    ";
340 INPUT D$
350 PRINT
360 PRINT "ENTER ANNUAL DEMANDS IN UNITS   ";
370 INPUT R
380 PRINT
390 PRINT "ENTER COST IN SETTING UP AN ORDER   ";
400 INPUT A
410 PRINT
420 PRINT "ENTER INVENTORY CARRYING RATE IN PERCENTAGE   ";
430 INPUT P0
440 PRINT
450 PRINT "ENTER NUMBER OF PRICE BREAKS (M>=1)   ";
460 INPUT M
470 M=INT(M)
480 IF M <= 0 THEN 440
490 IF M <= M0 THEN 530
500 PRINT
510 PRINT "NUMBER OF PRICE BREAKS IS LIMITED TO ";M0
520 GOTO 440
530 PRINT
540 PRINT "ENTER ";M;" VALUES OF QUANTITY SCHEDULE"
```

```
550 Q(0)=0
560 FOR I=1 TO M
570 PRINT "PRICE BREAK   ";I;"    QUANTITY = ";
580 INPUT Q(I)
590 IF Q(I) > Q(I-1) THEN 620
600 GOSUB 1980
610 GOTO 570
620 NEXT I
630 PRINT
640 PRINT "ENTER ALL UNIT PRICES"
650 FOR I=1 TO M
660 PRINT "FOR QUANTITY BETWEEN   ";Q(I-1);
670 PRINT "   AND   ";Q(I);"  UNIT PRICE =";
680 INPUT P(I-1)
690 IF I=1 THEN 730
700 IF P(I-1) < P(I-2) THEN 730
710 GOSUB 1980
720 GOTO 660
730 NEXT I
740 PRINT "FOR QUANTITY GREATER THAN   ";Q(M);"  UNIT PRICE = ";
750 INPUT P(M)
760 IF P(M) < P(M-1) THEN 790
770 GOSUB 1980
780 GOTO 740
790 PRINT
800 PRINT "PLEASE ENTER COMMAND OR TYPE HELP"
810 PRINT
820 PRINT "COMMAND -->";
830 INPUT C$
840 IF C$ <> "HELP" THEN 870
850 GOSUB 1880
860 GOTO 810
870 IF C$ <> "ALL" THEN 900
880 GOSUB 970
890 GOTO 810
900 IF C$ <> "INC" THEN 930
910 GOSUB 1290
920 GOTO 810
930 IF C$ = "EXIT" THEN 2020
940 IF C$ = "RESTART" THEN 320
950 GOSUB 1980
960 GOTO 800
970 REM    *** ALL COMMAND    ***
980 K=M
990 Q9=SQR(2*A*R/(P0*P(K)/100))
1000 GOSUB 1240
1010 IF Q9 > Q(M) THEN 1210
1020 Q9=Q(M)
1030 GOSUB 1240
1040 FOR I=M-1 TO 0 STEP -1
1050 Z8=Z9
1060 Q9=SQR(2*A*R/(P0*P(I)/100))
1070 IF Q9 >= Q(I+1) THEN 1120
1080 IF Q9 < Q(I) THEN 1160
```

```
1090 K=I
1100 GOSUB 1240
1110 IF Z9 < Z8 THEN 1210
1120 Q9=Q(I+1)
1130 K=I+1
1140 GOSUB 1240
1150 GOTO 1210
1160 K=I
1170 Q9=Q(I)
1180 GOSUB 1240
1190 IF Z9 >= Z8 THEN 1120
1200 NEXT I
1210 L=1
1220 GOSUB 1600
1230 RETURN
1240 REM     ALL-UNITS-DISCOUNTS COST FUNCTION EVALUATION
1250 P9=R*P(K) + R*A/Q9
1260 H9=P0*P(K)*Q9/200
1270 Z9=P9+H9
1280 RETURN
1290 REM     ***   INC COMMAND   ***
1300 S(0)=0
1310 FOR I=1 TO M
1320 S(I)=S(I-1)+P(I-1)*(Q(I)-Q(I-1))
1330 NEXT I
1340 FOR I=0 TO M
1350 T(I)=SQR((200*R*(A+S(I)-P(I)*Q(I)))/(P0*P(I)))
1360 NEXT I
1370 Z8=1.E30
1380 Q8=Z8
1390 FOR I=M TO 0 STEP -1
1400 Q9=Q8
1410 Q8=Q(I)
1420 IF T(I) > Q9 THEN 1480
1430 IF T(I) < Q8 THEN 1480
1440 GOSUB 1550
1450 IF Z9 >= Z8 THEN 1480
1460 Z8=Z9
1470 K=I
1480 NEXT I
1490 I=K
1500 Q9=T(I)
1510 GOSUB 1550
1520 L=2
1530 GOSUB 1600
1540 RETURN
1550 REM     INCREMENTAL-QUANTITY-DISCOUNTS COST FUNCTION EVAL.
1560 P9=R*(P(I)+(A+S(I)-P(I)*Q(I))/Q9)
1570 H9=P0*(S(I)+P(I)*(Q9-Q(I)))/200
1580 Z9=P9+H9
1590 RETURN
1600 REM     REPORTING OPTIMAL SOLUTION
1610 PRINT
1620 PRINT
```

```
1630 PRINT F$(L)
1640 PRINT "ITEM DESCRIPTION : ";D$
1650 PRINT "ANNUAL DEMAND IN UNITS : ";R
1660 PRINT "SETUP COST PER ORDER : ";A
1670 PRINT "INVENTORY CARRYING RATE (%) : ";P0
1680 PRINT F$(3)
1690 PRINT F$(4)
1700 FOR I=1 TO M
1710 PRINT USING F$(5),Q(I-1),Q(I),P(I-1)
1720 NEXT I
1730 PRINT USING F$(5),Q(M)," INFINITY",P(M)
1740 PRINT "* OPTIMAL ORDER QUANTITY = ";Q9
1750 PRINT "* TIME BETWEEN ORDER = ";Q9/R;"   YEARS"
1760 PRINT "* AVERAGE PROCUREMENT COST/YEAR = ";P9
1770 PRINT "* AVERAGE INVENTORY CARRYING COST/YEAR = ";H9
1780 PRINT "* AVERAGE TOTAL COST/YEAR = ";Z9
1790 PRINT
1800 RETURN
1810 REM    INITIALIZATION
1820 F$(1)="A L L      U N I T S      D I S C O U N T"
1830 F$(2)="I N C R E M E N T A L      D I S C O U N T"
1840 F$(3)="      QUANTITY-DISCOUNT SCHEDULE"
1850 F$(4)="      FROM      TO    UNIT-PRICE"
1860 F$(5)="########## ########## #########.##"
1870 RETURN
1880 REM    ***   HELP COMMAND   ***
1890 PRINT
1900 PRINT "AVAILABLE COMMANDS"
1910 PRINT "=================="
1920 PRINT "HELP    - PRINT THIS MESSAGE"
1930 PRINT "ALL     - FIND ALL-UNITS-DISCOUNTS SOLUTION"
1940 PRINT "INC     - FIND INCREMENTAL-DISCOUNTS SOLUTION"
1950 PRINT "RESTART - START A NEW PROBLEM"
1960 PRINT "EXIT    - EXIT FROM THE PROGRAM "
1970 RETURN
1980 PRINT
1990 PRINT "*** ERROR *** PLEASE TRY AGAIN !!!"
2000 PRINT
2010 RETURN
2020 END
```

II. The (Q,r) Inventory Policy

One of the most common inventory models being used for independent demand items today is the lot size (Q), reorder point (r) model, or the (Q,r) model. The characteristics of the model are:

1. The system operates a continuous review inventory system. This means that the inventory position of the item is updated whenever there is a transaction. The transaction may be either a demand of the item or an order placed with the vendor. The inventory position is defined as the amount on hand plus the amount on order minus any amount backordered. Under practical conditions, the system essentially assumes that the review period is very short; perhaps daily updates of the inventory status may be adequate.

2. Demands are assumed to be probabilistic but stationary. Under practical conditions, this implies that the demand pattern is horizontal. The most popular forecasting model for this type of demand pattern is the single exponential smoothing model. If $F(t)$ is the forecast for any month t, the annual demand is $12F(t)$ units. If the replenishment lead time is L months, the mean demand during the lead time (M) and the standard deviation of the demand during the lead time (S) can be estimated using equations (4) and (5):

(4) $$M = LF(t)$$

and

(5) $$S = \sqrt{L}\, s$$

where s is the standard error of the forecast and is approximately equal to 1.25MAD. MAD is the mean absolute deviation (error). Periodic updates of the model parameters (Q and r) can be made as appropriate when new forecasts are available.

The inventory policy operates as follows: Whenever the inventory position of the item reaches the reorder point (r), an order quantity (Q) is placed with the vendor and the replenishment is received in stock at some later date. The usual assumptions are that unsatisfied demands are either completely backlogged or completely lost. The exact development under the Poisson demand process can be found in Hadley and Whitin [7]. It is also well known that under continuous review process and when demand occurs one unit at a time, the (Q,r) policy is indeed the optimal policy to use.

When unsatisfied demands are completely backlogged, there are three well-known approximate models. One is due to Brown [1], which is based on the service level concept. The program

EXCHNGE in the next section is based on a similar idea, but the emphasis is on setting the appropriate levels of safety stocks [2, 4]. Another model is due to Hadley and Whitin [7]. It is based on the cost model concept and is the basis for the program INVCOST. Another cost model is due to Wagner [13]. Nahmias [9] has shown that the three models are equivalent when service constraints are imposed on the cost models. He also provides computationally efficient algorithms to determine the solutions. An excellent example of the actual implementation of the (Q,r) model can be found in Gross et al. [5].

III. The Program EXCHNGE

In this section, we will follow Brown's work very closely. In particular, the computer program is based on the material contained in Chapter 10 of reference 2, which deals with invesment strategies in safety stocks. Safety stocks are necessary in providing adequate service to customers during the replenishment lead time. The goal of studying various strategies is for management to determine the minimum level of investment and yet maintain the desired service levels.

Two measures of service level are commonly used:

1. The fraction of demands that can be directly filled from available stock. This is equivalent to specifying the probability that any unit demanded can be filled without having to wait. It is also equivalent to specifying the average time of a stockout.

2. The probability of a shortage occurring during the replenishment lead time. This is equivalent to specifying the average number of shortage occurrences per year or the average number of order cycles having stockouts.

Since the idea is to determine the optimal trade-off point between service levels and inventory investment, the use of exchange curves can be very helpful. For example, an exchange curve may be that of plotting the dollars invested in safety stocks versus the fraction of demands that can be filled from stocks.

Consider an inventory item. It is well known that the total cost function (inventory carrying plus ordering costs) is relatively flat around the Economic Order Quantity (EOQ) when the demand is deterministic where

(6) $\quad\quad EOQ = Q = \sqrt{2DA/(IV)}$.

The same behavior is observed in the case of probabilistic demands. In other words, the total cost function is relatively insensitive to Q as long as it is close to the EOQ.

The program EXCHNGE uses the EOQ to obtain order quantities.

The total inventory investment consists of the cycle stock plus the safety stock for all items:

(7) $$\text{Total Investment} = Y = \sum_{i=1}^{n} (Q_i V_i /2 + k_i S_i V_i)$$

The first term on the right-hand side of equation (7) is the average dollar value of the cycle stock. The second term is the safety stock investment. In Brown's approach, the order quantity can be determined independently, and the safety stock is adjusted taking into consideration the order quantity selected.

There are five decision rules contained in EXCHNGE. Each decision rule has associated with it a management policy variable. We denote X as the management policy variable. The trick in each decision rule lies in the determination of the safety factor (k) for each item in equation (7). Each rule produces a different combination of the safety factors depending on the value of X. Once the safety factor is determined for the item, the safety stock can easily be obtained. The reorder point is simply the sum of the mean demand during the lead time and the safety stock:

(8) $$r_i = M_i + k_i S_i \quad .$$

The rules are as follows.

1. RULE1 - Specify the fraction of demands to be filled from stock. The management policy variable (X) is the service level. For example, if X=0.95, then on the average for each item, 95 percent of the demands can be directly filled from the shelf. The safety factor for each item can be determined from

(9) $$E(k) = Q(1-X)/S$$

where

(10) $$E(k) = \int_{k}^{\infty} (z-k) f(z) dz \quad .$$

Brown refers to E(k) as the "partial expectation." Here, f(z) is the density function of the normal distribution. Equation (9) is a nonlinear function of k. The EXCHNGE program uses Newton's method to solve for it. The use of this rule is equivalent to specifying the first service level measure mentioned above.

2. RULE2 - Specify the number of shortages per item per

year. This implies that the average time between shortage occurrences is equal for every item. Under this rule, the total value of backorders per year will be minimized for a given safety stock investment [2, 4]. The management policy variable, X, is the number of shortages per item per year. The item safety factor is determined by solving equation (11):

(11) $$F(k_i) = X/N_i$$

where $F(k_i)$ is the probability of shortage and $N_i = D_i/Q_i$, which is the average number of order cycles per year. Brown [2] has mentioned that this rule has a tendency to give small safety stocks to slow-moving items. The use of this rule is equivalent to specifying the second measure of service level mentioned earlier.

3. RULE3 - Use one safety factor for all items. Here the management policy variable, X, is the safety factor. Under this rule, the probability of a shortage is the same for every item.

4. RULE4 - Specify days of supply of safety stocks as the management policy variable. We assume a 360-day year. Since the average demand rate is known, the safety stock for an item is XD/360. But the safety stock is equal to kS. Hence the safety factor is

(12) $$k = XD/360S .$$

5. RULE5 - Specify Lagrange Multiplier to minimize the number of replenishment orders that requires expediting. The Lagrange Multiplier, which is the management policy variable, arises from the problem of minimizing the total expected number of shortage cycles subject to the constraint of fixed inventory investment of equation (7). The equivalent unconstrained problem to be minimized is

(13) $$K = \sum_{i=1}^{n} D_i F(k_i)/Q_i - X [Y - \sum_{i=1}^{n} (Q_i V_i/2 + k_i S_i V_i)]$$

where X is the Lagrange Multiplier and Y is the total inventory investment. If Q_i is fixed, equation (13) can be minimized by differentiating K with respect to each k_i and setting the result to zero. This yields

(14) $$f(k_i) = XV_i S_i Q_i/D_i .$$

$f(k_i)$ is the standard normal density function. The k_i can be easily obtained by solving equation (14) if X is specified. There is really no good way of guessing at the appropriate values of X. We simply have to try several values and narrow down the range of the search. Since the dimensionality of the Lagrange Multiplier is the inverse of the imputed marginal shortage cost [2] it is quite likely that, for practical problems, X will be much less than one.

Depending on the characteristics of the inventory items and also on the service level objectives of the management, it is difficult to recommend one best rule. Brown [2] has given some general guidelines for applying these rules. Interested readers should consult his book [2].

Since there may be several thousands of items within an inventory system, it is probably not practical to perform an analysis of all items simultaneously. A recommended procedure is to randomly select a small percentage of items, perform the analysis on the sample, and extrapolate the results to the whole inventory system. The extrapolation factor may simply be N/n where N is the number of items in the entire inventory and n is the size of the sample.

INPUT
-----
1. Number of items (n)
2. Annual inventory carrying rate in percentages
3. Data for all items entered one item at a time separated by commas, that is, A,V,D,M,S, where A = setup cost, V = unit cost of the item, D = annual demand, M = mean lead time demand, S = standard deviation of lead time demand

OUTPUT

Two sets of output are given by EXCHNGE. The first set produces detailed results by items, such as order quantity, reorder point, safety factor, and $ backorder, for a particular value of the management policy variable. The second set provides summary results of the analysis so that exchange curves can be developed. If the user desires to produce exchange curves, the command XON will activate the exchange curve output option. If individual item output is desired, the command XOFF should be entered before selecting a rule. This is the default output option when the program is first executed. The commands RULE1, RULE2,..., RULE5 correspond to the description given earlier.

EXAMPLE

Consider analyzing the following five items assuming that the annual inventory carrying rate is 20 percent.

Item	A	V	D	M	S
1	12	3.00	7000	3500	705
2	12	4.50	2500	1250	275
3	12	2.75	700	350	82
4	15	5.80	250	125	30
5	15	6.20	70	35	8

The example terminal session is shown below. The HELP command is entered after the above data are given. Since the default output option is the detailed output by items, we issue the command RULE1 and specify the fraction of the demand to be filled from stock as 0.95. Under this rule the investment in safety stock is $4869.61. The average worth of backorders per year is $1802.95.

Next the RULE2 command is entered and we specify the desired number of shortage occurrence equal to 0.75. Note that the last column in the table yields the required management policy variable. RULE3 is entered with the safety factor specified as 1.0. When RULE4 is entered, the number of days of supply of safety stock is set at 90. A Lagrange Multiplier value of 0.001 is entered for RULE5.

Next we turn the exchange curve option on by entering the command XON. RULE1 is entered. A range of management policy variable of 0.85 to 0.99 with an increment of 0.02 is entered. The program generates a table of the safety stock investment, the value of the backorders, the shortage occurrences for the five items per year, and the average number of days during a stockout period. RULE2, RULE3, RULE4, and RULE5 are executed in that sequence with the given range and increment of the corresponding management policy variable. Please see the example terminal session for details. The resulting table from each rule allows the user to plot the desired exchange curve easily for each of the five rules.

```
EXCHNGE EXAMPLE TERMINAL SESSION
================================

E X C H A N G E    C U R V E S
-------------------------------

NEED INTRODUCTION (Y OR N) ... !N

ENTER NUMBER OF ITEMS   !5

ENTER ANNUAL INVENTORY CARRYING RATE IN %   !20

ENTER A,V,D,M,S FOR ALL ITEMS
-----------------------------
ITEM  1   !12,3.00,7000,3500,705
ITEM  2   !12,4.50,2500,1250,275
ITEM  3   !12,2.75,700,350,82
ITEM  4   !15,5.80,250,125,30
ITEM  5   !15,6.20,70,35,8

PLEASE ENTER COMMAND OR TYPE HELP

COMMAND --> !HELP

AVAILABLE COMMANDS
==================
HELP    - PRINT THIS MESSAGE
XON     - SET EXCHANGE CURVE OPTION ON
XOFF    - GENERATE INDIVIDUAL ITEM RESULTS
RULE1   - SPECIFY FRACTION DEMANDS TO BE FILLED FROM STOCK
RULE2   - SPECIFY NUMBER OF SHORTAGES/ITEM/YEAR
RULE3   - USE ONE SAFETY FACTOR FOR ALL ITEMS
RULE4   - SPECIFY DAYS OF SUPPLY OF SAFETY STOCKS
RULE5   - SPECIFY LAGRANGE MULTIPLIER TO MINIMIZE
          NUMBER OF ORDERS WHICH REQUIRES EXPEDITING
RESTART - START A NEW PROBLEM
EXIT    - EXIT FROM THE PROGRAM

COMMAND --> !RULE1

ENTER FRACTION OF DEMAND TO BE FILLED FROM STOCK   !.95

MANAGEMENT POLICY VARIABLE = .95
```

ITEM	ORDERS PER YR	ORDER SIZE	REORDER POINT	SAFETY FACTOR	$SAFETY STOCK	$ BACK ORDERS	PROB.OF SHORT.	SHORTAGE OCCUR/YR
1	13.23	529	4480	1.389	2938.68	1050.00	.0823	1.089
2	9.68	258	1604	1.286	1590.86	562.50	.0993	.961
3	4.01	175	421	.867	195.60	96.25	.1929	.772
4	3.11	80	147	.738	128.33	72.50	.2304	.716
5	1.70	41	38	.325	16.14	21.70	.3725	.634

```
TOTAL $ COST OF SALES PER YEAR = 36059
```

```
TOTAL $ AVERAGE STOCK  = 6845.353945939
TOTAL $ CYCLE STOCK    = 1975.743176613
TOTAL $ SAFETY STOCKS  = 4869.610769326
TOTAL $ BACKORDERS     = 1802.95
TOTAL NUMBER OF SHORTAGE OCCURRENCES/YR = 4.173152094782
TOTAL NUMBER OF ORDERS/YR = 31.72652632839
TOTAL AVERAGE ORDERING COST/YR = 395.1486353227
AVERAGE NUMBER OF DAYS SHORT/REPLENISHMENT = 4.056046918559


COMMAND --> !RULE2

ENTER NUMBER OF SHORTAGES PER ITEM PER YEAR !.75

MANAGEMENT POLICY VARIABLE = .75

ITEM  ORDERS  ORDER  REORDER  SAFETY  $SAFETY   $ BACK   PROB.OF  SHORTAGE
      PER YR  SIZE   POINT    FACTOR  STOCK     ORDERS   SHORT.   OCCUR/YR
 1    13.23   529    4616     1.583   3349.05    674.36  .0567    .750
 2     9.68   258    1641     1.423   1760.50    417.28  .0775    .750
 3     4.01   175     423      .888    200.23     92.76  .1873    .750
 4     3.11    80     146      .702    122.16     77.06  .2412    .750
 5     1.70    41      36      .148      7.36     27.77  .4409    .750

TOTAL $ COST OF SALES PER YEAR = 36059
TOTAL $ AVERAGE STOCK  = 7415.044152735
TOTAL $ CYCLE STOCK    = 1975.743176613
TOTAL $ SAFETY STOCKS  = 5439.300976122
TOTAL $ BACKORDERS     = 1289.218903616
TOTAL NUMBER OF SHORTAGE OCCURRENCES/YR = 3.75
TOTAL NUMBER OF ORDERS/YR = 31.72652632839
TOTAL AVERAGE ORDERING COST/YR = 395.1486353227
AVERAGE NUMBER OF DAYS SHORT/REPLENISHMENT = 4.509330174802


COMMAND --> !RULE3

ENTER ONE SAFETY FACTOR FOR ALL ITEMS !1.0

MANAGEMENT POLICY VARIABLE = 1

ITEM  ORDERS  ORDER  REORDER  SAFETY  $SAFETY   $ BACK   PROB.OF  SHORTAGE
      PER YR  SIZE   POINT    FACTOR  STOCK     ORDERS   SHORT.   OCCUR/YR
 1    13.23   529    4205     1.000   2115.00   2331.07  .1587    2.099
 2     9.68   258    1525     1.000   1237.50    998.29  .1587    1.536
 3     4.01   175     432     1.000    225.50     75.25  .1587     .635
 4     3.11    80     155     1.000    174.00     45.07  .1587     .493
 5     1.70    41      43     1.000     49.60      7.03  .1587     .270

TOTAL $ COST OF SALES PER YEAR = 36059
TOTAL $ AVERAGE STOCK  = 5777.343176613
TOTAL $ CYCLE STOCK    = 1975.743176613
TOTAL $ SAFETY STOCKS  = 3801.6
TOTAL $ BACKORDERS     = 3456.709850419
```

```
TOTAL NUMBER OF SHORTAGE OCCURRENCES/YR = 5.033579262293
TOTAL NUMBER OF ORDERS/YR = 31.72652632839
TOTAL AVERAGE ORDERING COST/YR = 395.1486353227
AVERAGE NUMBER OF DAYS SHORT/REPLENISHMENT = 2.710069777063


COMMAND --> !RULE4

ENTER DAYS OF SUPPLY OF SAFETY STOCKS  !90

MANAGEMENT POLICY VARIABLE = 90
```

ITEM	ORDERS PER YR	ORDER SIZE	REORDER POINT	SAFETY FACTOR	$SAFETY STOCK	$ BACK ORDERS	PROB.OF SHORT.	SHORTAGE OCCUR/YR
1	13.23	529	5250	2.482	5250.00	59.23	.0065	.086
2	9.68	258	1875	2.273	2812.50	47.51	.0115	.112
3	4.01	175	525	2.134	481.25	5.31	.0164	.066
4	3.11	80	187	2.083	362.50	3.66	.0186	.058
5	1.70	41	53	2.188	108.50	.43	.0144	.024

```
TOTAL $ COST OF SALES PER YEAR = 36059
TOTAL $ AVERAGE STOCK  = 10990.49317661
TOTAL $ CYCLE STOCK    = 1975.743176613
TOTAL $ SAFETY STOCKS  = 9014.75
TOTAL $ BACKORDERS     = 116.1438527439
TOTAL NUMBER OF SHORTAGE OCCURRENCES/YR = .3459269756229
TOTAL NUMBER OF ORDERS/YR = 31.72652632839
TOTAL AVERAGE ORDERING COST/YR = 395.1486353227
AVERAGE NUMBER OF DAYS SHORT/REPLENISHMENT = .1309287606429


COMMAND --> !RULE5

ENTER LAGRANGE MULTIPLIER (YEAR/$)  !.001

MANAGEMENT POLICY VARIABLE = .001
```

ITEM	ORDERS PER YR	ORDER SIZE	REORDER POINT	SAFETY FACTOR	$SAFETY STOCK	$ BACK ORDERS	PROB.OF SHORT.	SHORTAGE OCCUR/YR
1	13.23	529	4453	1.352	2860.18	1138.50	.0881	1.166
2	9.68	258	1665	1.509	1867.18	344.15	.0657	.636
3	4.01	175	512	1.979	446.25	8.11	.0239	.096
4	3.11	80	184	1.982	344.86	4.82	.0237	.074
5	1.70	41	53	2.287	113.45	.32	.0111	.019

```
TOTAL $ COST OF SALES PER YEAR = 36059
TOTAL $ AVERAGE STOCK  = 7607.66670696
TOTAL $ CYCLE STOCK    = 1975.743176613
TOTAL $ SAFETY STOCKS  = 5631.923530347
TOTAL $ BACKORDERS     = 1495.903278814
TOTAL NUMBER OF SHORTAGE OCCURRENCES/YR = 1.99021543766
TOTAL NUMBER OF ORDERS/YR = 31.72652632839
TOTAL AVERAGE ORDERING COST/YR = 395.1486353227
AVERAGE NUMBER OF DAYS SHORT/REPLENISHMENT = .5246941861956
```

```
COMMAND --> !XON

COMMAND --> !RULE1

POLICY VARIABLE = FRACTION OF DEMAND TO BE FILLED FROM STOCK
ENTER MANAGEMENT POLICY AS LOW,HIGH,INCREMENT -- !.85,.99,.02
```

POLICY VARIABLE	$ SAFETY STOCK	$ BACK ORDERS	SHORTAGE OCCUR/YR	DAYS/ SHORT
.85000	2685	5376	9.12	10.20
.87000	2996	4665	8.30	9.19
.89000	3347	3952	7.43	8.16
.91000	3753	3240	6.49	7.15
.93000	4242	2524	5.43	5.84
.95000	4870	1803	4.17	4.06
.97000	5756	1082	2.76	2.33
.99000	7453	361	1.09	.71

```
COMMAND --> !RULE2

POLICY VARIABLE = NUMBER OF SHORTAGES PER ITEM PER YEAR
ENTER MANAGEMENT POLICY AS LOW,HIGH,INCREMENT -- !.80,1.2,0.1
```

POLICY VARIABLE	$ SAFETY STOCK	$ BACK ORDERS	SHORTAGE OCCUR/YR	DAYS/ SHORT
.80000	5305	1393	4.00	4.98
.90000	5057	1604	4.45	5.68
1.00000	4832	1817	4.85	6.11
1.10000	4623	2037	5.25	6.58
1.20000	4426	2264	5.65	7.07

```
COMMAND --> !RULE3

POLICY VARIABLE = ONE SAFETY FACTOR FOR ALL ITEMS
ENTER MANAGEMENT POLICY AS LOW,HIGH,INCREMENT -- !0.8,2.4,0.2
```

POLICY VARIABLE	$ SAFETY STOCK	$ BACK ORDERS	SHORTAGE OCCUR/YR	DAYS/ SHORT
.80000	3041	4987	6.72	4.03
1.00000	3802	3457	5.03	2.71
1.20000	4562	2328	3.65	1.77
1.40000	5322	1521	2.56	1.12
1.60000	6083	964	1.74	.69
1.80000	6843	592	1.14	.41
2.00000	7603	352	.72	.24
2.20000	8364	203	.44	.13
2.40000	9124	113	.26	.07

```
COMMAND --> !RULE4

POLICY VARIABLE = DAYS OF SUPPLY OF SAFETY STOCKS
ENTER MANAGEMENT POLICY AS LOW,HIGH,INCREMENT -- !15,90,15

  POLICY  $ SAFETY   $ BACK  SHORTAGE   DAYS/
  VARIABLE  STOCK    ORDERS  OCCUR/YR   SHORT
  15.00000   1502     9543    11.09      8.57
  30.00000   3005     4971     6.98      4.51
  45.00000   4507     2320     3.93      2.16
  60.00000   6010      964     1.97       .94
  75.00000   7512      356      .87       .37
  90.00000   9015      116      .35       .13


COMMAND --> !RULE5
POLICY VARIABLE = LAGRANGE MULTIPLIER (YEAR/$)
ENTER MANAGEMENT POLICY AS LOW,HIGH,INCREMENT -- !.0001,.0012,.0001

  POLICY  $ SAFETY   $ BACK  SHORTAGE   DAYS/
  VARIABLE  STOCK    ORDERS  OCCUR/YR   SHORT
   .00010    9933      67      .13       .02
   .00020    8864     160      .29       .05
   .00030    8174     271      .46       .09
   .00040    7646     397      .64       .13
   .00050    7209     540      .84       .18
   .00060    6830     697     1.04       .24
   .00070    6493     871     1.26       .30
   .00080    6185    1062     1.49       .37
   .00090    5900    1269     1.74       .44
   .00100    5632    1496     1.99       .52
   .00110    5377    1743     2.26       .61
   .00120    5133    2011     2.54       .71


COMMAND --> !EXIT
```

```
10 REM    E X C H A N G E     C U R V E S
20 REM
30 DIM F$(16),P(3),Y(5)
40 DIM A(20),V(20),D(20),M(20),S(20),Q(20),E(20),F(20),W(20),K(20)
50 N0=20
60 K0=0
70 G9=0.3989423
80 GOSUB 3210
90 PRINT
100 PRINT "E X C H A N G E    C U R V E S"
110 PRINT "----------------------------"
120 PRINT
130 PRINT "NEED INTRODUCTION (Y OR N) ... ";
140 INPUT Y$
150 PRINT
160 IF Y$ <> "Y" THEN 360
170 PRINT "THIS PROGRAM CAN BE USED TO DETERMINE THE ECONOMIC"
180 PRINT "ORDER QUANTITY AND REORDER POINT BASED ON THE"
190 PRINT "SERVICE LEVEL CONCEPT FOR A NUMBER OF INVENTORY ITEMS."
200 PRINT "IT ALSO CONTAINS FIVE DECISION RULES FOR THE PURPOSE"
210 PRINT "OF INVESTIGATING THE SAFETY STOCK INVESTMENT STRATEGIES."
220 PRINT "DEMANDS DURING THE LEADTIME FOR ALL ITEMS ARE ASSUMED"
230 PRINT "TO BE NORMALLY DISTRIBUTED."
240 PRINT
250 PRINT "INPUT REQUIREMENTS"
260 PRINT "------------------"
270 PRINT "1. NUMBER OF ITEMS"
280 PRINT "2. INVENTORY CARRYING RATE IN % FOR THE COMPANY"
290 PRINT "3. FOR EACH ITEM, THE FOLLOWING DATA ARE REQUIRED :"
300 PRINT "    - COST OF PLACING AN ORDER (A)"
310 PRINT "    - UNIT COST OF THE ITEM (V)"
320 PRINT "    - DEMAND RATE IN # OF UNITS/YEAR (D)"
330 PRINT "    - MEAN DEMAND DURING THE ORDER LEADTIME (M)"
340 PRINT "    - STANDARD DEVIATION OF LEADTIME DEMAND (S)"
350 PRINT
360 PRINT "ENTER NUMBER OF ITEMS   ";
370 INPUT N
380 IF N <= N0 THEN 410
390 PRINT "*** ERROR *** ALLOWABLE NUMBER OF ITEMS = ";N0
400 GOTO 350
410 PRINT
420 PRINT "ENTER ANNUAL INVENTORY CARRYING RATE IN %   ";
430 INPUT I1
440 PRINT
450 PRINT "ENTER A,V,D,M,S FOR ALL ITEMS"
460 PRINT "-----------------------------"
470 FOR I=1 TO N
480 PRINT "ITEM    ";I;"   ";
490 INPUT A(I),V(I),D(I),M(I),S(I)
500 NEXT I
510 GOSUB 860
520 PRINT
530 PRINT "PLEASE ENTER COMMAND OR TYPE HELP"
540 PRINT
```

```
550 PRINT "COMMAND --> ";
560 INPUT C$
570 PRINT
580 IF C$ <> "HELP" THEN 610
590 GOSUB 1000
600 GOTO 540
610 IF C$ <> "XON" THEN 640
620 K0=1
630 GOTO 550
640 IF C$ <> "XOFF" THEN 670
650 K0=0
660 GOTO 550
670 IF C$ <> "RULE1" THEN 700
680 GOSUB 1160
690 GOTO 540
700 IF C$ <> "RULE2" THEN 730
710 GOSUB 1560
720 GOTO 540
730 IF C$ <> "RULE3" THEN 760
740 GOSUB 1890
750 GOTO 540
760 IF C$ <> "RULE4" THEN 790
770 GOSUB 2150
780 GOTO 540
790 IF C$ <> "RULE5" THEN 820
800 GOSUB 2390
810 GOTO 540
820 IF C$ = "RESTART" THEN 90
830 IF C$ = "EXIT" THEN 3650
840 GOSUB 3510
850 GOTO 550
860 REM    CALCULATE EOQ FOR EACH ITEM
870 S0=0
880 S1=0
890 S2=0
900 S3=0
910 FOR I=1 TO N
920 Q(I)=SQR(200*D(I)*A(I)/(I1*V(I)))
930 W(I)=D(I)/Q(I)
940 S0=S0+D(I)*V(I)
950 S1=S1+V(I)*Q(I)/2
960 S2=S2+W(I)
970 S3=S3+A(I)*W(I)
980 NEXT I
990 RETURN
1000 REM    *** HELP COMMAND ***
1010 PRINT
1020 PRINT "AVAILABLE COMMANDS"
1030 PRINT "=================="
1040 PRINT "HELP  - PRINT THIS MESSAGE"
1050 PRINT "XON   - SET EXCHANGE CURVE OPTION ON"
1060 PRINT "XOFF  - GENERATE INDIVIDUAL ITEM RESULTS"
1070 PRINT "RULE1 - SPECIFY FRACTION DEMANDS TO BE FILLED FROM STOCK'
1080 PRINT "RULE2 - SPECIFY NUMBER OF SHORTAGES/ITEM/YEAR"
```

```
1090 PRINT "RULE3 - USE ONE SAFETY FACTOR FOR ALL ITEMS"
1100 PRINT "RULE4 - SPECIFY DAYS OF SUPPLY OF SAFETY STOCKS"
1110 PRINT "RULE5 - SPECIFY LAGRANGE MULTIPLIER TO MINIMIZE"
1120 PRINT "        NUMBER OF ORDERS WHICH REQUIRES EXPEDITING"
1130 PRINT "RESTART - START A NEW PROBLEM"
1140 PRINT "EXIT    - EXIT FROM THE PROGRAM"
1150 RETURN
1160 REM    *** RULE1 COMMAND ***
1170 J=1
1180 IF K0 = 1 THEN 1240
1190 GOSUB 2700
1200 GOSUB 1520
1210 GOSUB 1340
1220 GOSUB 2790
1230 GOTO 1330
1240 GOSUB 3390
1250 GOSUB 3150
1260 FOR F1 = P(1) TO P(2) STEP P(3)
1270 GOSUB 1520
1280 GOSUB 1340
1290 GOSUB 2850
1300 PRINT USING F$(9),F1,Y(1),Y(2),Y(3),Y(4)
1310 NEXT F1
1320 PRINT
1330 RETURN
1340 REM FIND SAFETY FACTOR USING NEWTON'S METHOD
1350 FOR I=1 TO N
1360 IF E(I) < G9 THEN 1410
1370 E(I)=G9
1380 K(I)=0
1390 F(I)=0.5
1400 GOTO 1500
1410 Z9=0
1420 Z=Z9
1430 GOSUB 3560
1440 E0=(H9-Z*C9-E(I))/C9
1450 IF ABS(E0) < .0001 THEN 1480
1460 Z9=Z9+E0
1470 GOTO 1420
1480 K(I)=Z9-E0
1490 F(I)=C9
1500 NEXT I
1510 RETURN
1520 FOR I=1 TO N
1530 E(I)=(1-F1)*Q(I)/S(I)
1540 NEXT I
1550 RETURN
1560 REM    *** RULE2 COMMAND ***
1570 J=2
1580 IF K0 = 1 THEN 1630
1590 GOSUB 2700
1600 GOSUB 1720
1610 GOSUB 2790
1620 GOTO 1710
```

```
1630 GOSUB 3390
1640 GOSUB 3150
1650 FOR Fl = P(1) TO P(2) STEP P(3)
1660 GOSUB 1720
1670 GOSUB 2850
1680 PRINT USING F$(9),Fl,Y(1),Y(2),Y(3),Y(4)
1690 NEXT Fl
1700 PRINT
1710 RETURN
1720 FOR I=1 TO N
1730 C9=Fl/W(I)
1740 IF C9 < 0.5 THEN 1790
1750 K(I)=0
1760 F(I)=0.5
1770 E(I)=G9
1780 GOTO 1870
1790 A2=LOG(1/C9/C9)
1800 A1=SQR(A2)
1810 Z=2.515517 + 0.802853*A1 + 0.010328*A2
1820 Z9=1.0 + 1.432788*A1 + 0.189269*A2 + 0.001308*A1*A2
1830 Z=A1-Z/Z9
1840 K(I)=Z
1850 F(I)=C9
1860 E(I)=G9*EXP(-Z*Z/2) - Z*F(I)
1870 NEXT I
1880 RETURN
1890 REM    *** RULE3 COMMAND ***
1900 J=3
1910 IF K0 = 1 THEN 1980
1920 GOSUB 2700
1930 IF Fl <= 4 THEN 1950
1940 Fl=4
1950 GOSUB 2070
1960 GOSUB 2790
1970 GOTO 2060
1980 GOSUB 3390
1990 GOSUB 3150
2000 FOR Fl = P(1) TO P(2) STEP P(3)
2010 GOSUB 2070
2020 GOSUB 2850
2030 PRINT USING F$(9),Fl,Y(1),Y(2),Y(3),Y(4)
2040 NEXT Fl
2050 PRINT
2060 RETURN
2070 Z=Fl
2080 GOSUB 3560
2090 FOR I=1 TO N
2100 K(I)=Fl
2110 E(I)=H9-Fl*C9
2120 F(I)=C9
2130 NEXT I
2140 RETURN
2150 REM    *** RULE4 COMMAND ***
2160 J=4
```

```
2170 IF K0 = 1 THEN 2220
2180 GOSUB 2700
2190 GOSUB 2310
2200 GOSUB 2790
2210 GOTO 2300
2220 GOSUB 3390
2230 GOSUB 3150
2240 FOR F1 = P(1) TO P(2) STEP P(3)
2250 GOSUB 2310
2260 GOSUB 2850
2270 PRINT USING F$(9),F1,Y(1),Y(2),Y(3),Y(4)
2280 NEXT F1
2290 PRINT
2300 RETURN
2310 FOR I=1 TO N
2320 K(I)=F1*D(I)/(360*S(I))
2330 Z=K(I)
2340 GOSUB 3560
2350 F(I)=C9
2360 E(I)=H9-K(I)*C9
2370 NEXT I
2380 RETURN
2390 REM     *** RULE5 COMMAND ***
2400 J=5
2410 IF K0 = 1 THEN 2460
2420 GOSUB 2700
2430 GOSUB 2550
2440 GOSUB 2790
2450 GOTO 2540
2460 GOSUB 3390
2470 GOSUB 3150
2480 FOR F1 = P(1) TO P(2) STEP P(3)
2490 GOSUB 2550
2500 GOSUB 2850
2510 PRINT USING F$(9),F1,Y(1),Y(2),Y(3),Y(4)
2520 NEXT F1
2530 PRINT
2540 RETURN
2550 FOR I=1 TO N
2560 A2=F1*V(I)*S(I)*2.5066283/W(I)
2570 IF A2 >= 1 THEN 2650
2580 A2=-2*LOG(A2)
2590 K(I)=SQR(A2)
2600 Z=K(I)
2610 GOSUB 3560
2620 F(I)=C9
2630 E(I)=H9-K(I)*C9
2640 GOTO 2680
2650 K(I)=0
2660 F(I)=0.5
2670 E(I)=G9
2680 NEXT I
2690 RETURN
2700 REM     *** PROMPT WHEN XOFF OPTION IS ACTIVE ***
```

```
2710 PRINT F$(16);F$(10+J);
2720 INPUT Fl
2730 IF Fl > 0 THEN 2760
2740 GOSUB 3510
2750 GOTO 2710
2760 IF J <> 1 THEN 2780
2770 IF Fl >= 1 THEN 2740
2780 RETURN
2790 REM      *** REPORT BY ITEM ***
2800 PRINT
2810 PRINT "MANAGEMENT POLICY VARIABLE = ";Fl
2820 PRINT
2830 PRINT F$(1);F$(4)
2840 PRINT F$(2);F$(5)
2850 FOR I=1 TO 4
2860 Y(I)=0
2870 NEXT I
2880 FOR I=1 TO N
2890 A4=M(I)+K(I)*S(I)
2900 A6=V(I)*K(I)*S(I)
2910 A7=V(I)*W(I)*S(I)*E(I)
2920 A9=W(I)*F(I)
2930 Y(1)=Y(1)+A6
2940 Y(2)=Y(2)+A7
2950 Y(3)=Y(3)+A9
2960 Y(4)=Y(4)+360*S(I)*E(I)*M(I)/(D(I)*(M(I)+S(I)*(K(I)+E(I))))
2970 IF K0 =1 THEN 3000
2980 PRINT USING F$(3),I,W(I),Q(I),A4,K(I);
2990 PRINT USING F$(6),A6,A7,F(I),A9
3000 NEXT I
3010 Y(4)=Y(4)/N
3020 IF K0 = 1 THEN 3140
3030 PRINT
3040 PRINT "TOTAL $ COST OF SALES PER YEAR = ";S0
3050 PRINT "TOTAL $ AVERAGE STOCK = ";S1+Y(1)
3060 PRINT "TOTAL $ CYCLE STOCK    = ";S1
3070 PRINT "TOTAL $ SAFETY STOCKS = ";Y(1)
3080 PRINT "TOTAL $ BACKORDERS     = ";Y(2)
3090 PRINT "TOTAL NUMBER OF SHORTAGE OCCURRENCES/YR = ";Y(3)
3100 PRINT "TOTAL NUMBER OF ORDERS/YR = ";S2
3110 PRINT "TOTAL AVERAGE ORDERING COST/YR = ";S3
3120 PRINT "AVERAGE NUMBER OF DAYS SHORT/REPLENISHMENT = ";Y(4)
3130 PRINT
3140 RETURN
3150 REM      *** PRINT EXCHANGE CURVE HEADING ***
3160 PRINT
3170 PRINT
3180 PRINT F$(7)
3190 PRINT F$(8)
3200 RETURN
3210 REM      *** I N I T I A L I Z A T I O N ***
3220 F$(1)="ITEM   ORDERS   ORDER REORDER SAFETY"
3230 F$(2)="       PER YR    SIZE   POINT FACTOR"
3240 F$(3)="### #####.## ###### ####### ##.###"
```

```
3250 F$(4)=" $SAFETY   $ BACK   PROB.OF SHORTAGE"
3260 F$(5)="   STOCK    ORDERS    SHORT.  OCCUR/YR"
3270 F$(6)="#####.## ####.## ###.#### ####.###"
3280 F$(7)="    POLICY  $ SAFETY    $ BACK SHORTAGE    DAYS/"
3290 F$(8)=" VARIABLE    STOCK    ORDERS OCCUR/YR    SHORT"
3300 F$(9)="###.##### ######## ######### #####.## ####.##"
3310 F$(10)="POLICY VARIABLE = "
3320 F$(11)="FRACTION OF DEMAND TO BE FILLED FROM STOCK   "
3330 F$(12)="NUMBER OF SHORTAGES PER ITEM PER YEAR   "
3340 F$(13)="ONE SAFETY FACTOR FOR ALL ITEMS   "
3350 F$(14)="DAYS OF SUPPLY OF SAFETY STOCKS   "
3360 F$(15)="LAGRANGE MULTIPLIER (YEAR/$)   "
3370 F$(16)="ENTER "
3380 RETURN
3390 REM    * PROMPT FOR POLICY RANGE ***
3400 PRINT F$(10);F$(10+J)
3410 PRINT "ENTER MANAGEMENT POLICY AS LOW,HIGH,INCREMENT -- ";
3420 INPUT P(1),P(2),P(3)
3430 IF P(1) >= P(2) THEN 3480
3440 IF P(3) <= 0 THEN 3480
3450 IF P(1) <= 0 THEN 3480
3460 IF J <> 1 THEN 3500
3470 IF P(2) < 1 THEN 3500
3480 GOSUB 3510
3490 GOTO 3410
3500 RETURN
3510 REM    *** PRINT ERROR MESSAGE ***
3520 PRINT
3530 PRINT "*** ERROR *** PLEASE TRY AGAIN !!!"
3540 PRINT
3550 RETURN
3560 REM    *** GIVEN Z, CALCULATE DENSITY H9 AND DECUM. C9
3570 Z0=ABS(Z)
3580 H9=G9*EXP(-Z*Z/2)
3590 T=1.0/(1.0+.2316419*Z0)
3600 C9=((1.330274*T-1.821256)*T+1.781478)*T
3610 C9=1-H9*T*((C9-0.3565638)*T+0.3193815)
3620 IF Z <=0 THEN 3640
3630 C9=1-C9
3640 RETURN
3650 END
```

IV. The Program INVCOST

In this section we consider two approximate cost models of Hadley and Whitin [7], the backorder case and the lost sales case for the (Q,r) policy. These models assume that the time during which the system will be out of stock will be very short. In addition, for the backorder model, when a replenishment order is received, it is large enough to bring the inventory position up to at least the reorder point. The objective is to determine the optimal control parameters (Q and r) which will minimize the sum of the ordering cost, inventory carrying cost, and penalty cost for shortage. Let $C(Q,r)$ be the objective function for the backorder case:

(15) $\qquad C(Q,r) = DA/Q + IV(Q/2+r-M) + N(r)PD/Q$

where $N(r)$ is the expected number of units short per order cycle and is equal to

(16) $\qquad N(r) = \int_r^\infty (x-r)g(x)dx$.

$g(x)$ is the distribution of the demand during the lead time. Taking partial derivatives of $C(Q,r)$ with respect to Q and r, setting them to zero, and solving, we get

(17) $\qquad Q = \sqrt{2D(A + N(r)P)/(IV)}$

and

(18) $\qquad G(r) = QIV/(PD)$.

$G(r)$ is the probability of shortage during an order cycle. The optimal Q and r can be solved using an iterative procedure using the EOQ as the initial estimate for Q. The procedure is to alternately solve for r using (18) and Q using (17) until convergence is reached. Note that the right-hand side of (18) must be less than one for the solution to make sense.

It is obvious that the solution of Q in equation (17) will always be greater than the EOQ because $N(r)$ is greater than zero. In fact, because the algorithm starts with the EOQ, the value of Q will increase monotonically as the iterations continue. This implies that $G(r)$ will also increase monotonically. Thus $G(r)$ can approach and possibly exceed 1 during the iterative procedure. In order to ensure convergence of the algorithm, as soon as the value of $G(r)$ reaches 1, $G(r)$ is set to 1 and the optimal reorder point (r) is set to zero [7]. The optimal Q is determined by replacing $N(r)$ in (17) with the mean demand during the lead time (M).

For the lost sales case, the cost function is

(19) $\qquad C(Q,r) = DA/Q + IV(Q/2 + r - M + N(r)) + N(r)PD/Q$.

The solution to (19) is to calculate Q using equation (17), but the reorder point is determined from

(20) $\quad G(r) = (QIV)/(PD+QIV)$.

The same iterative procedure is required to solve for Q and r.

The total average stock of an item can be defined as the sum of the cycle stock and safety stock as usual. The cycle stock for both the backorder and the lost sales case is equal to Q/2. The calculation of the safety stock for the two cases differs. For the backorder case, the safety stock (SS) is

(21) $\quad SS = r - M = kS$

which is obvious from (15) and (8). For the lost sales case, by examining (19) and using (8), we have

(22) $\quad SS = r - M + N(r) = kS + N(r)$.

The program INVCOST refers to the value of k in (21) and (22) as the safety factor. Hence its value must be interpreted correspondingly.

It is interesting to compare the solution obtained for the BACKORDER model using this program and the one using the EXCHNGE program. The trick is to determine the imputed unit backorder (shortage) cost (P) for each item. Suppose we are interested in determining the appropriate value of P that corresponds to specifying the average number of shortage occurrences per year -i.e., the second measure of service level as defined in RULE2 of the EXCHNGE program. This can easily be accomplished by solving the value of P from equation (18). We get

(23) $\quad P = IV/(G(r)D/Q)$.

Now, $G(r)D/Q$ is simply the average number of shortage occurrences per year for the item and will be specified. Hence P would correspond to the imputed unit backorder cost to produce the desired service level.

In fact, if we solve the same set of items with the program EXCHNGE using different rules, it will be possible to determine the corresponding imputed unit backorder costs of all items for every rule. This implies that for the backorder case, the program INVCOST can be used to develop exchange curves that correspond to every rule contained in the EXCHNGE program if we are willing to accept this measure of service level. The program INVCOST has been designed such that a comparable output to the EXCHNGE program is available.

INPUT
-----
1. Number of items
2. Annual inventory carrying rate in percentage
3. Item data seperated by commas -i.e.,
   setup cost (A), unit penalty cost for shortage (P),
   unit cost of the item (V), annual demand rate (D),
   mean lead time demand (M), standard deviation of
   lead time demand (S)

OUTPUT

Two output options are available. The first option gives the optimal solutions by items and the breakdowns of the cost components in the objective function -i.e., average ordering cost per year, average carrying cost per year, and average penalty cost per year for shortage. The second option produces a table similar to the output obtained from the EXCHNGE program when the exchange curve option is turned off (XOFF). The commands BACK1 and BACK2 generate solutions for the backorder model with output options 1 and 2, respectively. Similarly the commands LOST1, LOST2 produce results for the lost sales case.

EXAMPLE

We will use the same data as in the EXCHNGE program except that the imputed unit shortage cost or penalty cost for each item is determined from equation (23). Consider the EXCHNGE example terminal session output for RULE2 in which the number of shortages per item per year is specified as 0.75. The corresponding unit shortage costs calculated using (23) are 0.800, 1.200, 0.733, 1.546, and 1.653, respectively, for the five items. For RULE5 with Lagrange Multiplier equals 0.001, the corresponding unit shortage costs are estimated as being 0.515, 1.416, 5.730, 15.676, and 65.264, respectively.

The example terminal session shown below starts with the equivalent data of RULE2 in the EXCHNGE program. The HELP command is issued after the data are entered. The BACK1 and BACK2 commands are entered in succession to show the two types of output. It is easily verified that the solution from BACK2 command produces the average shortage occurrence per year for each item of 0.75 as the desired service level (i.e., the last column in the table). Next the command RESTART is issued so that we may enter the second set of data based on RULE5 above. Again, the BACK2 command produces the result comparable to that of the EXCHNGE program in terms of the average shortage occurrences per year of each item. The LOST1 and LOST2 commands are entered next in order to display the solutions for the lost sales case. Note that their solutions are independent of the backorder case.

```
INVCOST EXAMPLE TERMINAL SESSION
=================================

COST MODELS - INVENTORY ANALYSIS
--------------------------------

NEED INTRODUCTION (Y OR N) ... !N

ENTER NUMBER OF ITEMS  !5

ENTER ANNUAL INVENTORY CARRYING RATE IN %  !20

ENTER A,P,V,D,M,S FOR ALL ITEMS
-------------------------------
ITEM  1   !12,0.800,3.00,7000,3500,705
ITEM  2   !12,1.20,4.50,2500,1250,275
ITEM  3   !12,0.733,2.75,700,350,82
ITEM  4   !15,1.546,5.80,250,125,30
ITEM  5   !15,1.653,6.20,70,35,8

PLEASE ENTER COMMAND OR TYPE HELP

COMMAND --> !HELP


AVAILABLE COMMANDS
==================
HELP     - PRINT THIS MESSAGE
BACK1    - BACKORDER SOLUTION WITH OUTPUT OPTION 1
BACK2    - BACKORDER SOLUTION WITH OUTPUT OPTION 2
LOST1    - LOSTSALES SOLUTION WITH OUTPUT OPTION 1
LOST2    - LOSTSALES SOLUTION WITH OUTPUT OPTION 2
RESTART  - START A NEW PROBLEM
EXIT     - EXIT FROM THE PROGRAM

COMMAND --> !BACK1



B A C K O R D E R    S Y S T E M    - OUTPUT OPTION 1

ITEM  ORDERS  ORDER  REORDER  INVENTORY  PENALTY  ORDERING    TOTAL
      PER YR  SIZE   POINT    $/YEAR     $/YEAR   $/YEAR     $/YEAR
  1    7.28   962    4391      823.4      27.5     87.3       938.4
  2    5.81   430    1561      473.5      21.3     69.8       564.5
  3    3.04   230     406       94.2       8.8     36.5       139.5
  4    2.45   102     140       76.7       9.1     36.8       122.6
  5    1.45    48      35       29.5       5.5     21.8        56.8

TOTAL INVENTORY COST ($/YEAR) = 1497.285009808
TOTAL PENALTY   COST ($/YEAR) =   72.36698607128
TOTAL ORDERING  COST ($/YEAR) =  252.2096134625
T O T A L    C O S T ($/YEAR) = 1821.861609342
```

```
COMMAND --> !BACK2


B A C K O R D E R    S Y S T E M    - OUTPUT OPTION 2
ITEM  ORDERS   ORDER  REORDER  SAFETY   $SAFETY  $ SHORT  PROB.OF  SHORTAGE
      PER YR   SIZE   POINT    FACTOR   STOCK    PER YR   SHORT.   OCCUR/YR
 1    7.28     962    4391     1.264    2674.24  754.80   .1031    .750
 2    5.81     430    1561     1.131    1400.00  463.86   .1290    .750
 3    3.04     230     406      .684     154.34  100.62   .2468    .750
 4    2.45     102     140      .508      88.35   83.50   .3057    .750
 5    1.45      48      35  -   .040  -    2.00   30.24   .5161    .750

TOTAL $ COST OF SALES PER YEAR = 36059
TOTAL $ VALUE AVERAGE STOCK = 7486.425049039
TOTAL $ VALUE CYCLE STOCK = 3171.499634233
TOTAL $ VALUE SAFETY STOCK = 4314.925414806
TOTAL $ VALUE OF SHORTAGES/YEAR = 1433.017683777
TOTAL NUMBER OF SHORTAGE OCCURRENCES/YR = 3.750815719554
TOTAL NUMBER OF ORDERS/YEAR = 20.04043834784
AVERAGE NUMBER OF DAYS SHORT/REPLENISHMENT = 6.342580018598


COMMAND --> !RESTART

ENTER NUMBER OF ITEMS  !5

ENTER ANNUAL INVENTORY CARRYING RATE IN %  !20

ENTER A,P,V,D,M,S FOR ALL ITEMS
-------------------------------
ITEM  1   !12,.515,3.00,7000,3500,705
ITEM  2   !12,1.416,4.50,2500,1250,275
ITEM  3   !12,5.730,2.75,700,350,82
ITEM  4   !15,15.676,5.80,250,125,30
ITEM  5   !15,65.264,6.20,70,35,8

PLEASE ENTER COMMAND OR TYPE HELP

COMMAND --> !BACK2


B A C K O R D E R    S Y S T E M    - OUTPUT OPTION 2
ITEM  ORDERS   ORDER  REORDER  SAFETY   $SAFETY  $ SHORT  PROB.OF  SHORTAGE
      PER YR   SIZE   POINT    FACTOR   STOCK    PER YR   SHORT.   OCCUR/YR
 1    6.81     1027   4170      .950    2009.62  1319.18  .1710    1.165
 2    5.93      422   1591     1.241    1536.16   377.44  .1073     .636
 3    3.35      209    506     1.901     428.78     8.28  .0286     .096
 4    2.70       93    183     1.921     334.20     4.90  .0274     .074
```

```
     5       1.59      44        53   2.260    112.07      .32    .0119     .019

TOTAL $ COST OF SALES PER YEAR = 36059
TOTAL $ VALUE AVERAGE STOCK = 7603.433654177
TOTAL $ VALUE CYCLE STOCK = 3182.611601796
TOTAL $ VALUE SAFETY STOCK = 4420.822052381
TOTAL $ VALUE OF SHORTAGES/YEAR = 1710.111705875
TOTAL NUMBER OF SHORTAGE OCCURRENCES/YR = 1.989626026262
TOTAL NUMBER OF ORDERS/YEAR = 20.37995657073
AVERAGE NUMBER OF DAYS SHORT/REPLENISHMENT = 1.013634189673


COMMAND --> !LOST1



L O S T     S A L E S     S Y S T E M    - OUTPUT OPTION 1

ITEM  ORDERS   ORDER  REORDER  INVENTORY  PENALTY  ORDERING     TOTAL
      PER YR   SIZE   POINT    $/YEAR     $/YEAR   $/YEAR       $/YEAR
  1    7.61    920    4285     775.2      24.2     91.3         890.8
  2    6.26    399    1615     518.9      16.7     75.2         610.7
  3    3.37    208     507     144.0       5.0     40.4         189.4
  4    2.71     92     183     121.1       4.7     40.7         166.5
  5    1.59     44      53      49.7       2.1     23.9          75.7

TOTAL INVENTORY COST ($/YEAR) = 1609.004601411
TOTAL PENALTY COST   ($/YEAR) = 52.72291103675
TOTAL ORDERING COST  ($/YEAR) = 271.4830349453
T O T A L    C O S T ($/YEAR) = 1933.210547393


COMMAND --> !LOST2



L O S T     S A L E S     S Y S T E M    - OUTPUT OPTION 2

ITEM  ORDERS   ORDER  REORDER  SAFETY   $SAFETY   $ SHORT  PROB.OF  SHORTAGE
      PER YR   SIZE   POINT    FACTOR    STOCK    PER YR   SHORT.   OCCUR/YR
  1    7.61    920    4285     1.114    2496.45  1075.13   .1327    1.010
  2    6.26    399    1615     1.328    1696.30   332.07   .0921     .577
  3    3.37    208     507     1.916     434.48     8.01   .0277     .093
  4    2.71     92     183     1.934     338 30     4.74   .0266     .072
  5    1.59     44      53     2.264     112.52      .32   .0118     .019

TOTAL $ COST OF SALES PER YEAR = 36059
TOTAL $ VALUE AVERAGE STOCK = 8045.023007052
TOTAL $ VALUE CYCLE STOCK = 2966.978588683
TOTAL $ VALUE SAFETY STOCK = 5078.04441837
TOTAL $ VALUE OF SHORTAGES/YEAR = 1420.274398878
TOTAL NUMBER OF SHORTAGE OCCURRENCES/YR = 1.771559822954
TOTAL NUMBER OF ORDERS/YEAR = 21.54760083671
AVERAGE NUMBER OF DAYS SHORT/REPLENISHMENT = .7943719056086
```

```
10 REM     INVENTORY COST MODELS - INVCOST
20 REM
30 DIM F$(12),Y(11)
40 DIM A(20),P(20),V(20),D(20),M(20),S(20)
50 DIM Q(20),R(20),K(20),E(20),F(20)
60 N0=20
70 G9=0.3989423
80 GOSUB 1880
90 PRINT
100 PRINT "COST MODELS - INVENTORY ANALYSIS"
110 PRINT "-------------------------------"
120 PRINT
130 PRINT "NEED INTRODUCTION (Y OR N) ... ";
140 INPUT Y$
150 PRINT
160 IF Y$ <> "Y" THEN 380
170 PRINT "THIS PROGRAM DETERMINES THE OPTIMAL ORDER QUANTITY"
180 PRINT "AND REORDER POINT BASED ON THE COST MODEL CONCEPT"
190 PRINT "FOR A NUMBER OF INVENTORY ITEMS."
200 PRINT "ASSUMPTIONS : CONTINUOUS REVIEW INVENTORY SYSTEM,"
210 PRINT "DEMANDS DURING THE LEAD TIME FOR ALL ITEMS ARE"
220 PRINT "NORMALLY DISTRIBUTED.  UNSATISFIED DEMANDS ARE"
230 PRINT "EITHER COMPLETELY BACKLOGGED (BACKORDERED) OR"
240 PRINT "COMPLETELY LOST."
250 PRINT
260 PRINT "INPUT REQUIREMENTS"
270 PRINT "------------------"
280 PRINT "1. NUMBER OF ITEMS"
290 PRINT "2. INVENTORY CARRYING RATE IN % FOR THE COMPANY"
300 PRINT "3. FOR EACH ITEM, THE FOLLOWING DATA ARE REQUIRED :"
310 PRINT "    - COST OF PLACING AN ORDER (A)"
320 PRINT "    - PENALTY COST FOR SHORTAGE/UNIT (P)"
330 PRINT "    - UNIT COST OF THE ITEM (V)"
340 PRINT "    - DEMAND RATE IN NUMBER OF UNITS/YEAR (D)"
350 PRINT "    - MEAN DEMAND DURING THE ORDER LEAD TIME (M)"
360 PRINT "    - STANDARD DEVIATION OF LEAD TIME DEMAND (S)"
370 PRINT
380 PRINT "ENTER NUMBER OF ITEMS   ";
390 INPUT N
400 IF N <=N0 THEN 430
410 PRINT "*** ERROR *** ALLOWABLE NUMBER OF ITEMS = ";N0
420 GOTO 370
430 PRINT
440 PRINT "ENTER ANNUAL INVENTORY CARRYING RATE IN %   ";
450 INPUT I1
460 I1=I1/100
470 PRINT
480 PRINT "ENTER A,P,V,D,M,S FOR ALL ITEMS"
490 PRINT "-------------------------------"
500 FOR I=1 TO N
510 PRINT "ITEM   ";I;"   ";
520 INPUT A(I),P(I),V(I),D(I),M(I),S(I)
530 NEXT I
540 PRINT
```

```
550 PRINT "PLEASE ENTER COMMAND OR TYPE HELP"
560 PRINT
570 PRINT "COMMAND --> ";
580 INPUT C$
590 PRINT
600 IF C$ <> "HELP" THEN 630
610 GOSUB 880
620 GOTO 560
630 IF C$ <> "BACK1" THEN 660
640 B0=1
650 GOTO 780
660 IF C$ <> "BACK2" THEN 690
670 B0=1
680 GOTO 740
690 IF C$ <> "LOST1" THEN 720
700 B0=2
710 GOTO 780
720 IF C$ <> "LOST2" THEN 830
730 B0=2
740 L1=4
750 L2=5
760 L9=2
770 GOTO 810
780 L1=7
790 L2=8
800 L9=1
810 GOSUB 1000
820 GOTO 560
830 IF C$ = "RESTART" THEN 370
840 IF C$ = "EXIT" THEN 2160
850 PRINT
860 PRINT "*** ERROR *** PLEASE TRY AGAIN !!!"
870 GOTO 370
880 REM    *** HELP COMMAND ***
890 PRINT
900 PRINT "AVAILABLE COMMANDS"
910 PRINT "=================="
920 PRINT "HELP    - PRINT THIS MESSAGE"
930 PRINT "BACK1   - BACKORDER SOLUTION WITH OUTPUT OPTION 1"
940 PRINT "BACK2   - BACKORDER SOLUTION WITH OUTPUT OPTION 2"
950 PRINT "LOST1   - LOSTSALES SOLUTION WITH OUTPUT OPTION 1"
960 PRINT "LOST2   - LOSTSALES SOLUTION WITH OUTPUT OPTION 2"
970 PRINT "RESTART - START A NEW PROBLEM"
980 PRINT "EXIT    - EXIT FROM THE PROGRAM"
990 RETURN
1000 REM    *** BACKORDER/LOSTSALE COMMANDS ***
1010 FOR I=1 TO N
1020 R1=2*D(I)
1030 R2=I1*V(I)
1040 R3=D(I)*P(I)
1050 Q(I)=SQR(2*D(I)*A(I)/R2)
1060 Q0=Q(I)
1070 IF B0 <> 1 THEN 1150
1080 F0=Q0*R2/R3
```

```
1090 IF F0 < 1 THEN 1170
1100 F(I)=1
1110 K(I)=-M(I)/S(I)
1120 Q(I)=SQR(R1*(A(I)+P(I)*M(I))/R2)
1130 E(I)=M(I)
1140 GOTO 1260
1150 F0=Q0*R2
1160 F0=F0/(R3+F0)
1170 C0=1-F0
1180 GOSUB 2040
1190 E(I)=S(I)*(G9*EXP(-Z*Z/2)-Z*F0)
1200 Q0=SQR(R1*(A(I)+P(I)*E(I))/R2)
1210 IF ABS(Q0-Q(I)) < .01 THEN 1240
1220 Q(I)=Q0
1230 GOTO 1070
1240 K(I)=Z
1250 F(I)=F0
1260 NEXT I
1270 REM     *** REPORTING ***
1280 PRINT
1290 PRINT
1300 PRINT F$(9+B0);F$(12);L9
1310 PRINT
1320 PRINT F$(1);F$(L1)
1330 PRINT F$(2);F$(L2)
1340 FOR I=1 TO 11
1350 Y(I)=0
1360 NEXT I
1370 FOR I=1 TO N
1380 A0=M(I)+K(I)*S(I)
1390 A1=V(I)*Q(I)/2
1400 A2=V(I)*E(I)
1410 A3=V(I)*K(I)*S(I)
1420 IF B0=1 THEN 1440
1430 A3=A3+A2
1440 A4=D(I)/Q(I)
1450 A5=A4*F(I)
1460 IF L1 <> 4 THEN 1570
1470 A6=360*E(I)*M(I)/(D(I)*(M(I)+K(I)*S(I)+E(I)))
1480 B8=A2*A4
1490 Y(1)=Y(1)+V(I)*D(I)
1500 Y(3)=Y(3)+A1
1510 Y(4)=Y(4)+A3
1520 Y(5)=Y(5)+B8
1530 Y(6)=Y(6)+A5
1540 Y(7)=Y(7)+A4
1550 Y(8)=Y(8)+A6
1560 GOTO 1630
1570 A8=P(I)*E(I)
1580 A9=A(I)*A4
1590 B9=I1*(A1+A3)
1600 Y(9)=Y(9)+B9
1610 Y(10)=Y(10)+A8
1620 Y(11)=Y(11)+A9
```

```
1630 PRINT USING F$(3),I,A4,Q(I),A0;
1640 IF L1 <> 4 THEN 1670
1650 PRINT USING F$(6),K(I),A3,B8,F(I),A5
1660 GOTO 1680
1670 PRINT USING F$(9),B9,A8,A9,B9+A8+A9
1680 NEXT I
1690 Y(8)=Y(8)/N
1700 Y(2)=Y(3)+Y(4)
1710 PRINT
1720 IF L1 <> 4 THEN 1820
1730 PRINT "TOTAL $ COST OF SALES PER YEAR = ";Y(1)
1740 PRINT "TOTAL $ VALUE AVERAGE STOCK = ";Y(2)
1750 PRINT "TOTAL $ VALUE CYCLE STOCK = ";Y(3)
1760 PRINT "TOTAL $ VALUE SAFETY STOCK = ";Y(4)
1770 PRINT "TOTAL $ VALUE OF SHORTAGES/YEAR = ";Y(5)
1780 PRINT "TOTAL NUMBER OF SHORTAGE OCCURRENCES/YR = ";Y(6)
1790 PRINT "TOTAL NUMBER OF ORDERS/YEAR = ";Y(7)
1800 PRINT "AVERAGE NUMBER OF DAYS SHORT/REPLENISHMENT = ";Y(8)
1810 GOTO 1860
1820 PRINT "TOTAL INVENTORY COST ($/YEAR) = ";Y(9)
1830 PRINT "TOTAL PENALTY COST   ($/YEAR) = ";Y(10)
1840 PRINT "TOTAL ORDERING COST  ($/YEAR) = ";Y(11)
1850 PRINT "T O T A L    C O S T ($/YEAR) = ";Y(9)+Y(10)+Y(11)
1860 PRINT
1870 RETURN
1880 REM    *** I N I T I A L I Z A T I O N ***
1890 F$(1)=" ITEM   ORDERS    ORDER  REORDER"
1900 F$(2)="        PER YR    SIZE   POINT"
1910 F$(3)="### #####.## ###### #######"
1920 F$(4)="  SAFETY   $SAFETY $ SHORT PROB.OF SHORTAGE"
1930 F$(5)="  FACTOR    STOCK   PER YR  SHORT. OCCUR/YR"
1940 F$(6)=" -#.### -####.## ####.## ##.#### ####.###"
1950 F$(7)="  INVENTORY PENALTY ORDERING     TOTAL"
1960 F$(8)="   $/YEAR   $/YEAR  $/YEAR     $/YEAR"
1970 F$(9)=" #######.# #####.# ######.# #######.#"
1980 F$(10)="B A C K O R D E R    S Y S T E M"
1990 F$(11)="L O S T    S A L E S    S Y S T E M"
2000 F$(12)="     - OUTPUT OPTION "
2010 L1=7
2020 L2=8
2030 RETURN
2040 REM    *** CALCULATE Z ***
2050 C9=C0
2060 IF C9 <= 0.5 THEN 2080
2070 C9=1.0-C9
2080 A2=LOG(1/C9/C9)
2090 A1=SQR(A2)
2100 Z=2.515517+0.802853*A1+0.010328*A2
2110 Z9=1.0+1.432788*A1+0.189269*A2+0.001308*A1*A2
2120 Z=A1-Z/Z9
2130 IF C0 > 0.5 THEN 2150
2140 Z=-Z
2150 RETURN
2160 END
```

V. The Program LOTSZ.MRP

The problem considered in this section is the determination of lot sizes for deterministic demands over a fixed planning horizon. The planning horizon is divided into a number of periods, and the demand forecasts for all periods are available. This is a common problem in Material Requirements Planning (MRP). Orlicky [10] describes most of the well-known techniques for lot sizing. The program LOTSZ.MRP, on the other hand, contains the following techniques:

1. Period Order Quantity (POQ)
2. Least Unit Cost (LUC)
3. Least Total Cost (LTC)
4. Part-Period Algorithm (PPA)
5. Silver-Meal Heuristic (S+M)
6. Wagner and Whitin Algorithm (W+W)

The cost structure of the problem is such that there is a fixed setup cost for every production lot size. In addition, the inventory carrying cost incurred is based on the amount of ending inventory in a period. The prime objective of these lot-sizing techniques is to determine how much to produce in a period and when to produce.

The program LOTSZ.MRP is capable of solving for the solutions of the above techniques for an individual set of demands and cost structure. In addition, it has a simulation capability in which the program will generate random demands from a uniform distribution. The range of the demands is specified by the user as well as the ratio of the setup cost to the inventory carrying cost per unit per period (h). In the simulation, h is assumed to be one. The user can obtain the performance of each algorithm with respect to the Wagner and Whitin algorithm. The performance is given in terms of the frequency of percentage of cost above the Wagner and Whitin algorithm because this algorithm always produces the optimal solution for a fixed planning horizon, whereas the other algorithms may not.

Detailed explanations of the POQ, LUC, and LTC algorithms can be found in Orlicky [10]. In the POQ procedure, our program differs slightly from that described in Orlicky in that the number of periods that each lot size covers is equal to the largest integer smaller than the ratio of EOQ to the average demand per period. However, the number of periods must be at least one. The LUC and the LTC algorithms follow Orlicky precisely.

The Part-Period Algorithm (PPA) contains the look-ahead and look-back features as originally described by DeMatteis [3]. The Silver-Meal Heuristic (S+M) is described in Peterson and Silver [11]. The Wagner and Whitin algorithm is described in many sources [7, 8, 12, 13].

INPUT

When the individual problem solution is desired through the use of the SOLVE command, the following input data are required:

1. Item desciption
2. Length of planning horizon in periods
3. Unit holding (carrying) cost per period
4. Forecasts of demands for all periods

When the simulation option of the program is activated using the SIMULATE command or the RESIM command, the program requires

1. The number of times to simulate
2. The length of the planning horizon
3. The ratio of the setup cost to the unit holding cost per period
4. The lower and upper limits for the demand range

OUTPUT

When the SOLVE command is used, output consists of the production lot sizes from all six techniques and the corresponding total costs of the production schedules.

When the SIMULATE or the RESIM commands are entered, the output is the performance summary in terms of the frequency distribution of the percentage of cost above the Wagner and Whitin algorithm. If the RESIM command is entered, the program assumes that a new simulation problem is desired and all previous statistics are destroyed. If the user wants to accumulate the statistics for the same problem, the command SIMULATE should be used and the program will print the latest statistics when the simulation is completed.

EXAMPLE

Given the monthly requirements for the 16 x 16 aluminum grill for a year as 160, 60, 50, 0, 100, 130, 78, 45, 90, 89, 0, and 105, repectively, we wish to use the program LOTSZ.MRP to determine the production lot sizes via the six techniques mentioned. The setup cost is $200, and the unit holding cost per period is $1.

The example terminal session is shown below. The first command entered is the HELP command, which displays all available commands. We proceed to issue the SOLVE command.

The program then prompts for all necessary data, and the results of production lot sizes are listed. We see that the Wagner and Whitin algorithm generates a solution with the total cost of $1417, which is the optimal solution. The production lot sizes are 270, 100, 253, 179, and 105 to be produced in periods 1, 5, 6, 9, and 12, respectively.

The SIMULATE command is entered next, and we simulate for 100 times the problems in which the planning horizon is 12 periods, the ratio of the setup cost to the unit holding cost per period is 200, and the demands are uniformly distributed between 0 and 250 units. The result of the simulation shows that 99 percent of the solutions from the Silver-Meal heuristic come within 10 percent of the W+W algorithm. Also 96 percent of the PPA solutions come within 10 percent of the W+W algorithm, etc.

The users should be cautioned that when the simulation feature is used, it is quite possible to observe some small percentages of the solutions of some algorithms to be greater than 100 percent of the Wagner and Whitin algorithm. This is due to the fixed planning horizon effect and the decision rules which we have implemented when the heuristic algorithmic procedures do not apply toward the last few periods of the planning horizon.

For example, consider a particular problem in which we are using the LUC algorithm. Suppose that the ratio of the setup cost to the inventory carrying cost per period is 30. Also assume that the algorithm has just finished determining a lot size and there are two periods remaining in the planning horizon. If we assume that the two periods have demands of 3 and 3000, respectively, then the calculated unit costs for the two periods will be 30/3=10 and (30+3000)/3003=1.009. Hence the algorithm will terminate with a lot size of 3003 units in the next to last period. This implies that an additional $3000 of inventory carrying cost is incurred due to the last period demand. Obviously, if the requirements of each period are produced individually, the additional inventory carrying cost will be zero according to the lot-sizing technique assumptions. The additional $3000 can certainly cause the total cost from the LUC to be many times over the W+W algorithm.

The cause of this phenomenon is twofold. One is due to the assumption of the fixed planning horizon employed in this program. The other is due to the range of the random demands that the user specifies when simulating. If the range of the demands is very large when compared with the setup cost as in the example we have mentioned, the user will experience the above phenomenon in algorithms such as the LUC and the PPA.

```
LOTSZ.MRP EXAMPLE TERMINAL SESSION
==================================

M R P    LOT-SIZING TECHNIQUES
-----------------------------

NEED INTRODUCTION (Y OR N)...!N

PLEASE ENTER COMMAND OR TYPE HELP

COMMAND -->!HELP

AVAILABLE COMMANDS
==================
HELP     - PRINT THIS MESSAGE
SOLVE    - SOLVE AN INDIVIDUAL PROBLEM
SIMULATE - CONTINUE SIMULATION
RESIM    - START A NEW SIMULATION PROBLEM
EXIT     - EXIT FROM THE PROGRAM

COMMAND -->!SOLVE

ENTER ITEM DESCRIPTION : !16X16 ALUMINUM GRILL

ENTER PLANNING HORIZON, N(>=2) !12

ENTER THE SETUP COST   !200

ENTER THE UNIT HOLDING COST PER PERIOD  !1

ENTER DEMANDS FOR ALL PERIODS
PERIOD  1   DEMAND  !160
PERIOD  2   DEMAND  !60
PERIOD  3   DEMAND  !50
PERIOD  4   DEMAND  !0
PERIOD  5   DEMAND  !100
PERIOD  6   DEMAND  !130
PERIOD  7   DEMAND  !78
PERIOD  8   DEMAND  !45
PERIOD  9   DEMAND  !90
PERIOD 10   DEMAND  !89
PERIOD 11   DEMAND  !0
PERIOD 12   DEMAND  !105

M R P    L O T - S I Z I N G
----------------------------
ITEM DESCRIPTION : 16X16 ALUMINUM GRILL
PLANNING HORIZON = 12
COST PER SETUP = 200
UNIT HOLDING COST PER PERIOD = 1

  PERIOD DEMAND    POQ     LUC     LTC     PPA     S+M     W+W
       1    160    220     220     270     270     270     270
       2     60      0       0       0       0       0       0
```

3	50	50	150	0	0	0	0
4	0	0	0	0	0	0	0
5	100	230	0	230	230	353	100
6	130	0	208	0	0	0	253
7	78	123	0	213	123	0	0
8	45	0	224	0	0	0	0
9	90	179	0	0	179	179	179
10	89	0	0	194	0	0	0
11	0	0	0	0	0	0	0
12	105	105	105	0	105	105	105

TOTAL COST   1524.0 1606.0 1525.0 1424.0 1470.0 1417.0

COMMAND -->!SIMULATE

ENTER NUMBER OF TIMES TO SIMULATE   !100

ENTER THE PLANNING HORIZON, N(>=2)   !12

ENTER RATIO OF SETUP COST/UNIT HOLDING COST   !200

ENTER RANGE OF DEMANDS (LOW,HIGH)   !0,250

S I M U L A T I N G !!! PLEASE STAND BY ...

FREQUENCY OF PERCENTAGE OF COST ABOVE THE W+W

NUMBER OF TIMES SIMULATED = 100
PLANNING HORIZON = 12 PERIODS
RATIO OF SETUP COST/UNIT HOLDING COST = 200
DEMANDS RANGE FROM 0 TO 250

FROM(%)	TO(%)	POQ	LUC	LTC	PPA	S+M
0	1	1	1	22	49	50
1	5	1	4	16	33	36
5	10	3	10	17	12	13
10	20	7	31	29	2	1
20	30	16	29	11	3	0
30	40	29	16	4	1	0
40	50	33	4	1	0	0
50	75	10	5	0	0	0
75	100	0	0	0	0	0
100	ABOVE	0	0	0	0	0

COMMAND -->!EXIT

```
10 REM     M R P    LOT-SIZING TECHNIQUES
20 REM
30 DIM D(12),K(13),C(12),F(13),Q(12,6)
40 DIM A(10,5),B(9),G(6),F$(7)
50 DATA 1,5,10,20,30,40,50,75,100
60 N0=12
70 GOSUB 4870
80 PRINT
90 PRINT "M R P    LOT-SIZING TECHNIQUES"
100 PRINT "---------------------------"
110 PRINT
120 PRINT "NEED INTRODUCTION (Y OR N)...";
130 INPUT Y$
140 IF Y$ <> "Y" THEN 370
150 PRINT
160 PRINT "THIS PROGRAM DETERMINES THE PRODUCTION LOT-SIZES"
170 PRINT "OF AN ITEM FOR A PLANNING HORIZON OF N PERIODS"
180 PRINT "USING THE FOLLOWING TECHNIQUES :"
190 PRINT "1. PERIOD ORDER QUANTITY (POQ)"
200 PRINT "2. LEAST UNIT COST (LUC)"
210 PRINT "3. LEAST TOTAL COST (LTC)"
220 PRINT "4. PART-PERIOD ALGORITHM (PPA)"
230 PRINT "5. THE SILVER-MEAL HEURISTIC (S+M)"
240 PRINT "6. THE WAGNER AND WHITIN ALGORITHM (W+W)"
250 PRINT
260 PRINT "IN ADDITION THE PROGRAM CAN BE USED TO SIMULATE"
270 PRINT "UNIFORM DISTRIBUTION TYPE DEMANDS AND EVALUATE"
280 PRINT "THE PERFORMANCE OF EACH TECHNIQUE RELATIVE TO THE"
290 PRINT "WAGNER AND WHITIN ALGORITHM."
300 PRINT
310 PRINT "INPUT REQUIREMENTS"
320 PRINT "------------------"
330 PRINT "1. ITEM DESCRIPTION"
340 PRINT "2. LENGTH OF THE PLANNING HORIZON = N PERIODS"
350 PRINT "3. UNIT HOLDING COST PER PERIOD"
360 PRINT "4. FORECASTS OF DEMAND FOR PERIOD T, T=1,2,...,N"
370 PRINT
380 PRINT "PLEASE ENTER COMMAND OR TYPE HELP"
390 PRINT
400 PRINT "COMMAND -->";
410 INPUT C$
420 IF C$ <> "HELP" THEN 450
430 GOSUB 4770
440 GOTO 390
450 IF C$ <> "SOLVE" THEN 480
460 GOSUB 590
470 GOTO 390
480 IF C$ <> "SIMULATE" THEN 510
490 GOSUB 860
500 GOTO 390
510 IF C$ <> "RESIM" THEN 550
520 M9=0
530 GOSUB 860
540 GOTO 390
```

```
550 IF C$ ="EXIT" THEN 5010
560 PRINT
570 PRINT "*** ERROR ***   PLEASE TRY AGAIN !!!"
580 GOTO 390
590 REM    *** SOLVE    C O M M A N D ***
600 PRINT
610 PRINT "ENTER ITEM DESCRIPTION : ";
620 INPUT N$
630 PRINT
640 PRINT "ENTER PLANNING HORIZON, N(>=2) ";
650 INPUT N
660 IF N <= N0 THEN 690
670 PRINT "*** ERROR ***   HORIZON LIMITED TO ";N0
680 GOTO 630
690 PRINT
700 PRINT "ENTER THE SETUP COST   ";
710 INPUT S
720 PRINT
730 PRINT "ENTER THE UNIT HOLDING COST PER PERIOD   ";
740 INPUT H
750 PRINT
760 PRINT "ENTER DEMANDS FOR ALL PERIODS"
770 FOR T=1 TO N
780 PRINT "PERIOD   ";T;"  DEMAND   ";
790 INPUT D(T)
800 NEXT T
810 P=S/H
820 GOSUB 1670
830 GOSUB 1810
840 GOTO 390
850 REM
860 REM    SIMULATE AND RESIM    C O M M A N D S
870 GOSUB 970
880 PRINT
890 PRINT "S I M U L A T I N G  !!!  PLEASE STAND BY ... ";
900 FOR J8=1 TO M8
910 GOSUB 1250
920 GOSUB 1670
930 GOSUB 1300
940 NEXT J8
950 GOSUB 1400
960 RETURN
970 REM    *** INITIALIZATION OF SIMULATION ***
980 PRINT
990 PRINT "ENTER NUMBER OF TIMES TO SIMULATE    ";
1000 INPUT M8
1010 IF M9 > 0 THEN 1190
1020 PRINT
1030 PRINT "ENTER THE PLANNING HORIZON, N(>=2)    ";
1040 INPUT N9
1050 IF N9 <= N0 THEN 1080
1060 PRINT "*** ERROR *** HORIZON LIMITED TO ";N0
1070 GOTO 1020
1080 PRINT
```

```
1090 PRINT "ENTER RATIO OF SETUP COST/UNIT HOLDING COST  ";
1100 INPUT R9
1110 PRINT
1120 PRINT "ENTER RANGE OF DEMANDS (LOW,HIGH  ";
1130 INPUT D8,D9
1140 FOR I=1 TO 10
1150 FOR J=1 TO 5
1160 A(I,J)=0
1170 NEXT J
1180 NEXT I
1190 M9=M9+M8
1200 S=R9
1210 H=1
1220 P=R9
1230 N=N9
1240 RETURN
1250 REM    *** UNIFORMLY DISTRIBUTED DEMANDS ***
1260 FOR T=1 TO N
1270 D(T)=D8+(D9-D8)*RND(W9)
1280 NEXT T
1290 RETURN
1300 REM    *** COUNT FREQUENCY ***
1310 FOR J=1 TO 5
1320 R5=(G(J)-G(6))/G(6)*100
1330 FOR I=1 TO 9
1340 IF R5 < B(I) THEN 1370
1350 NEXT I
1360 I=10
1370 A(I,J)=A(I,J)+1
1380 NEXT J
1390 RETURN
1400 REM    *** SIMULATION OUTPUT ***
1410 PRINT
1420 PRINT
1430 PRINT "FREQUENCY OF PERCENTAGE OF COST ABOVE THE W+W"
1440 PRINT "----------------------------------------------"
1450 PRINT "NUMBER OF TIMES SIMULATED = ";M9
1460 PRINT "PLANNING HORIZON = ";N9;" PERIODS"
1470 PRINT "RATIO OF SETUP COST/UNIT HOLDING COST = ";R9
1480 PRINT "DEMANDS RANGE FROM ";D8;"   TO   ";D9
1490 PRINT
1500 PRINT F$(6);F$(2)
1510 FOR I=1 TO 9
1520 PRINT USING F$(4),B(I-1);
1530 PRINT USING F$(4),B(I);
1540 FOR J=1 TO 5
1550 PRINT USING F$(4),A(I,J);
1560 NEXT J
1570 PRINT
1580 NEXT I
1590 PRINT USING F$(4),B(9);
1600 PRINT USING F$(4),"   ABOVE";
1610 FOR J=1 TO 5
1620 PRINT USING F$(4),A(10,J);
```

```
1630 NEXT J
1640 PRINT
1650 PRINT
1660 RETURN
1670 REM     *** DETERMINE LOT-SIZES FOR ALL TECHNIQUES ***
1680 J=1
1690 GOSUB 2080
1700 J=2
1710 GOSUB 2340
1720 J=3
1730 GOSUB 2610
1740 J=4
1750 GOSUB 2960
1760 J=5
1770 GOSUB 3550
1780 J=6
1790 GOSUB 3830
1800 RETURN
1810 REM *** OUTPUT SOLVE COMMAND ***
1820 PRINT
1830 PRINT "M R P    L O T - S I Z I N G"
1840 PRINT "--------------------------"
1850 PRINT "ITEM DESCRIPTION : ";N$
1860 PRINT "PLANNING HORIZON = ";N
1870 PRINT "COST PER SETUP = ";S
1880 PRINT "UNIT HOLDING COST PER PERIOD = ";H
1890 PRINT
1900 PRINT F$(1);F$(2);F$(3)
1910 FOR T=1 TO N
1920 PRINT USING F$(4),T;
1930 PRINT USING F$(4),D(T);
1940 FOR J=1 TO 6
1950 PRINT USING F$(4),Q(T,J);
1960 NEXT J
1970 PRINT
1980 NEXT T
1990 PRINT
2000 PRINT F$(7);
2010 FOR J=1 TO 6
2020 PRINT USING F$(5),G(J);
2030 NEXT J
2040 PRINT
2050 PRINT
2060 RETURN
2070 REM
2080 REM     P E R I O D    O R D E R    Q U A N T I T Y
2090 GOSUB 4540
2100 Q1=SQR(2*D0*P)
2110 T1=INT(Q1/D0)
2120 IF T1 >= 1 THEN 2140
2130 T1=1
2140 T=0
2150 T=T+1
2160 IF T>N THEN 2310
```

```
2170 IF D(T)>0 THEN 2200
2180 Q(T,J)=0
2190 GOTO 2150
2200 T2=T+T1-1
2210 IF T2<=N THEN 2230
2220 T2=N
2230 Q0=0
2240 FOR L=T TO T2
2250 Q0=Q0+D(L)
2260 Q(L,J)=0
2270 NEXT L
2280 Q(T,J)=Q0
2290 T=T2
2300 GOTO 2150
2310 GOSUB 4630
2320 RETURN
2330 REM
2340 REM     L E A S T     U N I T     C O S T
2350 Q0=0
2360 T=0
2370 T=T+1
2380 IF T>N THEN 2570
2390 IF D(T)<=0 THEN 2550
2400 IF Q0>0 THEN 2460
2410 T0=T
2420 U=S
2430 Q0=D(T)
2440 U0=U/Q0
2450 GOTO 2370
2460 Q0=Q0+D(T)
2470 Q(T,J)=0
2480 U=U+H*(T-T0)*D(T)
2490 U1=U/Q0
2500 IF U1 > U0 THEN 2530
2510 U0=U1
2520 GOTO 2370
2530 Q(T0,J)=Q0-D(T)
2540 GOTO 2410
2550 Q(T,J)=0
2560 GOTO 2370
2570 Q(T0,J)=Q0
2580 GOSUB 4630
2590 RETURN
2600 REM
2610 REM     L E A S T     T O T A L     C O S T
2620 Q0=0
2630 T=0
2640 T=T+1
2650 IF T>N THEN 2910
2660 IF D(T)<=0 THEN 2890
2670 IF Q0>0 THEN 2730
2680 T0=T
2690 Q0=D(T)
2700 P0=0
```

```
2710 U0=-P
2720 GOTO 2640
2730 P0=P0+(T-T0)*D(T)
2740 Q(T,J)=0
2750 Q0=Q0+D(T)
2760 U1=P0-P
2770 IF U1=0 THEN 2830
2780 IF U1>0 THEN 2810
2790 U0=U1
2800 GOTO 2640
2810 U0=ABS(U0)
2820 IF U0<U1 THEN 2870
2830 Q(T0,J)=Q0
2840 Q0=0
2850 Q(T,J)=0
2860 GOTO 2640
2870 Q(T0,J)=Q0-D(T)
2880 GOTO 2680
2890 Q(T,J)=0
2900 GOTO 2640
2910 IF Q0<=0 THEN 2930
2920 Q(T0,J)=Q0
2930 GOSUB 4630
2940 RETURN
2950 REM
2960 REM    P A R T - P E R I O D    A L G O R I T H M
2970 Q0=0
2980 T=0
2990 T=T+1
3000 IF T > N THEN 3530
3010 Q(T,J)=0
3020 IF D(T) > 0 THEN 3050
3030 IF T=N THEN 3520
3040 GOTO 2990
3050 IF Q0 > 0 THEN 3110
3060 T0=T
3070 Q0=D(T)
3080 P0=0
3090 IF T=N THEN 3510
3100 GOTO 2990
3110 P0=P0+(T-T0)*D(T)
3120 IF P0 > P THEN 3160
3130 Q0=Q0+D(T)
3140 IF T=N THEN 3510
3150 GOTO 2990
3160 REM    LOOK-AHEAD FEATURE
3170 IF T-1 <> T0 THEN 3200
3180 Q(T0,J)=Q0
3190 GOTO 3060
3200 IF T=N THEN 3490
3210 P2=(T-T0)*D(T)
3220 IF D(T+1) < P2 THEN 3250
3230 IF T+1=N THEN 3410
3240 GOTO 3270
```

```
3250 IF T+1 = N THEN 3410
3260 IF D(T+2) < P2 THEN 3410
3270 FOR M=2 TO 4
3280 IF T+M > N THEN 3350
3290 IF D(T+M) >= (T+M-1-T0)*D(T+M-1) THEN 3320
3300 T=T+M-2
3310 GOTO 3360
3320 Q0=Q0+D(T+M-2)
3330 Q(T+M-2,J)=0
3340 NEXT M
3350 T=T+M-1
3360 Q(T,J)=0
3370 Q0=Q0+D(T)
3380 Q(T0,J)=Q0
3390 Q0=0
3400 GOTO 2990
3410 REM    LOOK-BACK FEATURE
3420 IF 2*D(T) <= D(T-1) THEN 3450
3430 T=T-1
3440 GOTO 3380
3450 Q(T0,J)=Q0-D(T-1)
3460 Q0=0
3470 T=T-2
3480 GOTO 2990
3490 IF (N-T0)*D(N) < P THEN 3370
3500 Q(N,J)=D(N)
3510 Q(T0,J)=Q0
3520 GOSUB 4630
3530 RETURN
3540 REM
3550 REM     S I L V E R - M E A L    H E U R I S T I C
3560 Q0=0
3570 T=0
3580 T=T+1
3590 IF T>N THEN 3790
3600 IF D(T)<=0 THEN 3750
3610 IF Q0>0 THEN 3660
3620 T0=T
3630 Q0=D(T)
3640 F(T)=P
3650 GOTO 3580
3660 U=T-T0
3670 U2=U*U
3680 IF U2*D(T) > F(T-1) THEN 3730
3690 Q0=Q0+D(T)
3700 F(T)=F(T-1)+U*D(T)
3710 Q(T,J)=0
3720 GOTO 3580
3730 Q(T0,J)=Q0
3740 GOTO 3620
3750 Q(T,J)=0
3760 IF T<=1 THEN 3580
3770 F(T)=F(T-1)
3780 GOTO 3580
```

```
3790 Q(T0,J)=Q0
3800 GOSUB 4630
3810 RETURN
3820 REM
3830 REM     W A G N E R + W H I T I N     A L G O R I T H M
3840 M=N+1
3850 FOR I=1 TO N
3860 C(I)=0
3870 NEXT I
3880 FOR I=1 TO N
3890 F(I)=0
3900 K(I)=I-1
3910 IF D(I) > 0 THEN 3940
3920 D(I)=0
3930 NEXT I
3940 I0=I
3950 K(I0+1)=I0
3960 F(I0+1)=C(I0)
3970 IF I0 = N THEN 4470
3980 C(I0)=S
3990 F(I0+1)=C(I0)
4000 I0=I0+2
4010 IF I0 > M THEN 4290
4020 FOR I = I0 TO M
4030 J1=I-1
4040 IF D(J1) > 0 THEN 4090
4050 K(I)=K(J1)
4060 F(I)=F(J1)
4070 C(J1)=1.E30
4080 GOTO 4270
4090 L=K(J1)
4100 A0=1.E30
4110 FOR T=L TO J1
4120 IF T=J1 THEN 4150
4130 C(T)=C(T)+(J1-T)*H*D(J1)
4140 GOTO 4180
4150 C(T)=C(T)+S
4160 REM
4170 REM     C A L C U L A T E     P O L I C Y     C O S T
4180 A=F(T)+C(T)
4190 IF A>=A0 THEN 4220
4200 A0=A
4210 T0=T
4220 NEXT T
4230 REM
4240 REM     MINIMUM POLICY COST FOUND FOR HORIZON = J1
4250 K(I)=T0
4260 F(I)=A0
4270 NEXT I
4280 REM
4290 REM     TRACE OPTIMAL SOLUTION AND STORE IN Q(.,J)
4300 M0=K(M)
4310 M1=N
4320 FOR T=2 TO M
```

```
4330 L=M-T+1
4340 IF M0<L THEN 4430
4350 Z=0
4360 FOR I=M0 TO M1
4370 Z=Z+D(I)
4380 NEXT I
4390 Q(L,J)=Z
4400 M1=M0-1
4410 M0=K(L)
4420 GOTO 4440
4430 Q(L,J)=0
4440 NEXT T
4450 G(J)=F(M)
4460 GOTO 4520
4470 FOR T=1 TO N-1
4480 Q(T,J)=0
4490 NEXT T
4500 Q(N,J)=D(N)
4510 G(J)=S
4520 RETURN
4530 REM
4540 REM    CALCULATE AVERAGE DEMAND PER PERIOD = D0
4550 REM
4560 D0=0
4570 FOR T=1 TO N
4580 D0=D0+D(T)
4590 NEXT T
4600 D0=D0/N
4610 RETURN
4620 REM
4630 REM    CALCULATE SOLUTION COST
4640 G(J)=0
4650 I0=0
4660 T=0
4670 T=T+1
4680 IF T>N THEN 4760
4690 IF Q(T,J)<=0 THEN 4730
4700 I0=I0+Q(T,J)-D(T)
4710 G(J)=G(J)+S+H*I0
4720 GOTO 4670
4730 I0=I0-D(T)
4740 G(J)=G(J)+H*I0
4750 GOTO 4670
4760 RETURN
4770 REM    ***   HELP   C O M M A N D   ***
4780 PRINT
4790 PRINT "AVAILABLE COMMANDS"
4800 PRINT "=================="
4810 PRINT "HELP      - PRINT THIS MESSAGE"
4820 PRINT "SOLVE     - SOLVE AN INDIVIDUAL PROBLEM"
4830 PRINT "SIMULATE  - CONTINUE SIMULATION"
4840 PRINT "RESIM     - START A NEW SIMULATION PROBLEM"
4850 PRINT "EXIT      - EXIT FROM THE PROGRAM"
4860 RETURN
```

```
4870 REM     INITIALIZATION
4880 F$(1)=" PERIOD DEMAND"
4890 F$(2)="     POQ     LUC      LTC     PPA     S+M"
4900 F$(3)="     W+W"
4910 F$(4)="#######"
4920 F$(5)="#####.#"
4930 F$(6)="FROM(%)   TO(%)"
4940 F$(7)="  TOTAL COST   "
4950 M9=0
4960 B(0)=0
4970 FOR I=1 TO 9
4980 READ B(I)
4990 NEXT I
5000 RETURN
5010 END
```

VI. Dimensional Specifications

Program PBREAK

Let m be the maximum number of price breaks. Currently, m is 10. The following lines must be modified to change it:

Line No.

```
30 DIM P(m),Q(m),T(m),S(m),F$(5)
40 M0=m
```

Program EXCHNGE

Let n be the maximum number of items that EXCHNGE can handle. Currently, n equals 20 items. The following statements can be modified to accommodate other values of n:

Line No.

```
40 DIM A(n),V(n),D(n),M(n),S(n),Q(n),E(n),F(n),W(n),K(n)
50 N0=n
```

Program INVCOST

Similar to the EXCHNGE program, n is set at 20. To modify n, the following statements are affected.

Line No.

```
40 DIM A(n),P(n),V(n),D(n),M(n),S(n)
50 DIM Q(n),R(n),K(n),E(n),F(n)
60 N0=n
```

Program LOTSZ.MRP

Let p be the maximum length of the planning horizon that this program can accommodate. Also let q=p+1. The following statements should be modified if p exceeds the current setting of 12 periods:

Line No.

```
30 DIM D(p),K(q),C(p),F(q),Q(p,6)
60 N0=p
```

Also note that line number 50 of LOTSZ.MRP contains 9 class limits for the frequency distribution in the simulation output.

VII. REFERENCES

1. Brown, R.G., DECISION RULES FOR INVENTORY MANAGEMENT, Holt, Rinehart & Winston, New York, 1967.

2. Brown, R.G., MATERIALS MANAGEMENT SYSTEMS, John Wiley, New York, 1977.

3. DeMatteis, J.J., "An Economic Lot Sizing Technique," IBM SYSTEMS JOURNAL, 7, No.1 (1968), 30-46.

4. Gerson, G., and R.G. Brown, "Decision Rules for Equal Shorttage Policies," NAVAL RESEARCH LOGISTICS QUARTERLY, 17, No.3 (1970), 351-58.

5. Gross, D., C. Harris, and P. Robers, "Bridging the Gap between Mathematical Inventory Theory and the Construction of a Workable Model," INTERNATIONAL JOURNAL OF PRODUCTION RESEARCH, 10 (1972), 201-14.

6. Gross, D., and R. Ince, "A Comparison and Evaluation of Approximate Continuous Review Inventory Models," INTERNATIONAL JOURNAL OF PRODUCTION RESEARCH, 13 (1975), 9-23.

7. Hadley, G., and T.M. Whitin, ANALYSIS OF INVENTORY SYSTEMS, Prentice-Hall, Englewood Cliffs, N.J., 1963.

8. Johnson, L.A., and D.C. Montgomery, OPERATIONS RESEARCH IN PRODUCTION PLANNING, SCHEDULING AND INVENTORY CONTROL, John Wiley, New York, 1974.

9. Nahmias, S., "On the Equivalence of Three Approximate Continuous Review Inventory Models," NAVAL RESEARCH LOGISTICS QUARTERLY, 23, No.1 (1976), 31-36.

10. Orlicky, J., MATERIAL REQUIREMENTS PLANNING, McGraw-Hill, New York, 1975.

11. Peterson, R., and E.A. Silver, DECISION SYSTEMS FOR INVENTORY MANAGEMENT AND PRODUCTION PLANNING, John Wiley, New York, 1979.

12. Wagner, H.M., and T.M. Whitin, "Dynamic Version of the Economic Lot Size Model", MANAGEMENT SCIENCE, 5 (1958), 89-96.

13. Wagner, H.M., PRINCIPLES OF OPERATIONS RESEARCH, 2nd ed., Prentice-Hall, Englewood Cliffs, N.J., 1975.

CHAPTER 9

STATISTICAL QUALITY CONTROL

I. Process Control

   The Program XBARR
   ° Input
   ° Output
   ° Example
   ° XBARR Example Terminal Session
   ° XBARR Program Listing

   The Program PUC
   ° Input
   ° Output
   ° Examples
   ° PUC Example Terminal Session
   ° PUC Program Listing

II. Product Control

   Lot-by-Lot Single-Sampling Acceptance Plans by Attributes

   The Program LBL.AOQL
   ° Input
   ° Output
   ° Example
   ° LBL.AOQL Example Terminal Session
   ° LBL.AOQL Program Listing

   The Program AQL.LTFD
   ° Input
   ° Output
   ° Example

- ○ AQL.LTFD Example Terminal Session
- ○ AQL.LTFD Program Listing

Lot-by-Lot Single-Sampling Acceptance Plans by Variables

The Program LBL.SDV
- ○ Input
- ○ Output
- ○ Example
- ○ LBL.SDV Example Terminal Session
- ○ LBL.SDV Program Listing

III. Published Sampling Plans

IV. Dimensional Specifications

V. References

Statistical quality control combines inspection and testing with the principles of probability and statistics to decide the acceptability of a process or a product. The need for quality control arises due to variability in manufacturing. Quality control techniques enable us to recognize some causes of this variability. Steps can then be taken to control contributing factors so as to maintain quality within acceptable limits. Complete elimination of the variability is usually infeasible or uneconomical. Instead the producer's philosophy is based on tolerable, statistically predictable levels of imperfection.

The acceptability of a process implies that it conforms to the limits imposed on it. For a product to be acceptable it must meet the specifications, standards, and workmanship criteria imposed in the original design. Control charts are used for process control, and acceptance sampling is used for product control.

The first section of this chapter contains computer programs that are used in the design of statistical control charts for process control. The second section contains computer programs that are used in designing acceptance sampling plans for product (material) control.

I. Process Control

The basis of process control is the realization that variation in manufactured items exists. We must understand and distinguish between variability due to chance and variability due to assignable causes. When only chance causes are present, the process is considered "in control" or "under control." A process "in control" implies a random pattern of variation. If assignable causes are present, then the variation will be excessive and the process is said to be "out of control."

The computer programs for statistical control charts are based on the procedure shown in Table 1.* These programs can be used to analyze the manufacturing process for out-of-control conditions and for establishing 3-sigma limits for future process control (3-sigma limits are the 99.87 percent confidence intervals for the model selected). The control chart used will depend upon the product and the quality characteristics. In all cases, preparation for the charting includes some management decision.

In this section, two computer programs are provided: The first program, XBARR, can be used to develop control charts by

*Page 318-19

variables. The second program, PUC, is used for control charts by attributes. These programs can handle the cases when the standard values of the quality characteristics are known or unknown.

If the standard values are known, the program calculates the control limits that can immediately be used for process control. If the standard values are unknown, the relevant past data are used to calculate trial limits. If all past data points are within the trial control limits, the chart will be called a REVISED control chart and can be used for future process control. If at least one of these points is outside the trial control limits, the chart will be called a TRIAL control chart.

Using the TRIAL control chart the process may be furthur analyzed for out-of-control condition. If out-of-control points are to be discarded, they must have assignable causes. Under the convention of our programs, if all out-of-control points are discarded and the newly calculated control chart exhibits no out-of-control points, the new chart will be called a REVISED control chart; otherwise it will still be called a TRIAL control chart. In the latter instance, the control chart can still be used for future process control if deemed appropriate.

Both the programs XBARR and PUC contain commands that allow the user to edit the data. Commands are also provided for discarding and reinstating a discarded point in the control chart. If a point is discarded, the programs will prompt for a short description of the assignable cause. The assignable causes will be documented with the subsequent control chart. In this respect, the programs allow the user to simulate as closely as possible the manual procedure used in practice.

PROCEDURE	$\bar{X}$-CHART	R-CHART	p-CHART
1. Quality Characteristic	$\bar{X}$ - Sample Mean	R - Sample Range	p - Fraction Defective
2A. Rational Subgroup	n = 4 or more, 5 is quite common	n = 4 or more, 5 is quite common	The size is a function of the fraction defective
2B. Number of Subgroups	m = 25 or more	m = 25 or more	m = 25 or more
STANDARD VALUES (given)	$\bar{X}'$ - Standard Value of $\bar{X}$ (given) σ_x' - Standard Deviation Value of X (given)	σ_x' - Standard Deviation of x (given) $E[R] = d_2 \sigma_x'$ $\sigma_R' = d_3 \sigma_x'$	p' - Standard Value of p (given) $\sigma_p' = \sqrt{\dfrac{p'(1-p')}{n}}$
3A. Control Limits			
Central Value	$\bar{X}'$	$d_2 \sigma_x'$	p'
Control Limits	$\bar{X}' \pm 3 \sigma_{\bar{X}}'$ $= \bar{X}' \pm \dfrac{3 \sigma_x'}{\sqrt{n}}$ $= \bar{X}' \pm A \sigma_x'$	$d_2 \sigma_x' + 3 d_3 \sigma_x' = D_2 \sigma_x'$ $d_2 \sigma_x' - 3 d_3 \sigma_x' = D_1 \sigma_x'$	$p' \pm 3 \sqrt{\dfrac{p'(1-p')}{n}}$
STANDARD VALUES (not given)			
3B. Collect Past Data	$\bar{X}_j = \dfrac{\sum_{i=1}^{n} X_i}{n}$	R_j = Range of Subgroup j	p_j = Fraction Defective in Subgroup j
4. Compute Sample Statistics	$\bar{\bar{X}} = \dfrac{\sum_{j=1}^{m} \bar{X}_j}{m}$ $\bar{R} = \dfrac{\sum_{j=1}^{m} R_j}{m}$	$\bar{R} = \dfrac{\sum_{j=1}^{m} R_j}{m}$ $\hat{\sigma}_R = \dfrac{d_3 \bar{R}}{d_2}$	$\bar{p} = \dfrac{\sum_{j=1}^{m} p_j}{m}$ $\hat{\sigma}_p = \sqrt{\dfrac{\bar{p}(1-\bar{p})}{n}}$
5. Trial Control Limits			
Central Value	$\bar{\bar{X}}$	$\bar{R}$	$\bar{p}$
Control Limits	$\bar{\bar{X}} \pm 3 \hat{\sigma}_{\bar{X}}$ $= \bar{\bar{X}} \pm \dfrac{3 \bar{R}}{d_2 \sqrt{n}}$ $= \bar{\bar{X}} \pm A_2 \bar{R}$	$\bar{R} + 3 \hat{\sigma}_R = D_4 \bar{R}$ and $\bar{R} - 3 \hat{\sigma}_R = D_3 \bar{R}$	$\bar{p} \pm 3 \sqrt{\dfrac{\bar{p}(1-\bar{p})}{n}}$
6. Post the Data	$\bar{X}_j$, j = 1, 2, ... m	R_j, j = 1, 2, ... m	p_j, j = 1, 2, ... m
7A. Process State (in control). All points fall within control limits. Control limits for future use.	Use central value and control limits as computed in Step No. 5.	⇒ = = = = = = = = ⇒ = = = = = =	
7B. Process State (out-of-control) Some points falling outside the control limits or unnatural pattern of variation exists. Compute revised sample statistics.	Discard $\bar{X}_j$ and R_j out-of-control values. $\bar{\bar{X}}_{Revised} = \dfrac{\sum_{j=1}^{m} \bar{X}_j - \bar{X}_d}{m - m_d}$ $\bar{R}_{Revised} = \dfrac{\sum_{j=1}^{m} R_j - R_d}{m - m_d}$	$\bar{R}_{Revised} = \dfrac{\sum_{j=1}^{m} R_j - R_d}{m - m_d}$	Discard p_j out-of-control values $\bar{p}_{Revised} = \dfrac{\sum_{j=1}^{m} p_j - p_d}{m - m_d}$
CONTROL LIMITS FOR FUTURE USE			
Central Value	$\bar{\bar{X}}_{Revised}$	$\bar{R}_{Revised}$	$\bar{p}_{Revised}$
Control Limits	$\bar{\bar{X}}_{Revised} \pm \dfrac{3 \bar{R}_{Revised}}{d_2 \sqrt{n}}$ $\bar{\bar{X}}_{Revised} \pm A_2 \bar{R}_{Revised}$	$\bar{R}_{Revised} \pm 3 \hat{\sigma}_R$ $= D_4 \bar{R}_{Revised}$ and $D_3 \bar{R}_{Revised}$	$\bar{p}_{Revised} \pm 3 \sqrt{\dfrac{\bar{p}_{Revised}(1-\bar{p}_{Revised})}{n}}$

PROCEDURE	c-CHART	u-CHART
1. Quality Characteristic	c - Number of Defects per Single Unit	u - Number of Defects per Unit = $\frac{\sum c_i}{n}$
		c_i - Number of Defects in Unit i
		Subgroup Size = n > 1, n is constant
2A. Rational Subgroup	Single Unit, n = 1	
2B. Number of Subgroups	m = 25 or more Number of Subgroups (m) = Number of Single Units	m = 25 or more
STANDARD VALUES (given)	c' - Standard Value of c (given) $\sigma_c' = \sqrt{c'}$	u' - Standard Value of u (given) $\sigma_u' = \sqrt{\frac{u'}{n}}$
3A. Control Limits		
Central Value	c'	u'
Control Limits	$c' \pm 3\sqrt{c'}$	$u' \pm 3\sqrt{\frac{u'}{n}}$
STANDARD VALUES (not given)		
3B. Collect Past Data	c_j = Number of Defects in Unit j	$u_j = \frac{\sum_{i=1}^{n} c_i}{n}$
4. Compute Sample Statistics	$\bar{c} = \frac{\sum_{j=1}^{m} c_j}{m}$	$\bar{u} = \frac{\sum_{j=1}^{m} u_j}{m}$
	$\hat{\sigma}_c = \sqrt{\bar{c}}$	$\hat{\sigma}_u = \sqrt{\frac{\bar{u}}{n}}$
5. Trial Control Limits		
Central Value	$\bar{c}$	$\bar{u}$
Control Limits	$\bar{c} \pm 3\sqrt{\bar{c}}$	$\bar{u} \pm 3\sqrt{\frac{\bar{u}}{n}}$
6. Post the Data	c_j, j = 1, 2, ... m	u_j, j = 1, 2, ... m
7A. Process State (in control). All points fall within control limits. Control limits for future use.	====▷========================▷	
7B. Process State (out-of-control) Some points falling outside the control limits or unnatural pattern of variation exists. Compute revised sample statistics.	Discard c_j out-of-control values $\bar{c}_{Revised} = \frac{\sum_{j=1}^{m} c_j - c_d}{m - m_d}$	Discard u_j out-of-control values $\bar{u}_{Revised} = \frac{\sum_{j=1}^{m} u_j - u_d}{m - m_d}$
CONTROL LIMITS FOR FUTURE USE		
Central Value	$\bar{c}_{Revised} \pm 3\sqrt{\bar{c}_{Revised}}$	$\bar{u}_{Revised} \pm 3\sqrt{\frac{\bar{u}_{Revised}}{n}}$
Control Limits		

SYMBOLS USED IN TABLE 1

Symbol		Definition
A	$=$	$\dfrac{3}{\sqrt{n}}$
A_2	$=$	$\dfrac{3}{d_2\sqrt{n}}$
c	$=$	Number of defects per single unit
$\bar{c}$	$=$	Average number of defects per unit
c'	$=$	Process (population) value of c
c_d	$=$	Number of defects in discarded units
c_k	$=$	Number of defects in unit k
D_1	$=$	$d_2 - 3d_3$
D_2	$=$	$d_2 + 3d_3$
D_3	$=$	$1 - \dfrac{3d_3}{d_2}$
D_4	$=$	$1 + \dfrac{3d_3}{d_2}$
d_2	$=$	Mean value of relative range $\omega = \dfrac{\bar{R}}{\sigma_x'}$
d_3	$=$	$\dfrac{d_2\hat{\sigma}_R}{\bar{R}}$
E	$=$	Expected population value symbol
m	$=$	Number of subgroups
m_d	$=$	Number of discarded subgroups
n	$=$	Sample size
p	$=$	Fraction defective
p_j	$=$	Fraction defective in subgroup j
$\bar{p}$	$=$	Average fraction defective in m subgroups $= \dfrac{\sum_{j=1}^{m} p_j}{m}$
p'	$=$	Process (population) fraction defective
p_d	$=$	Fraction defective in discarded groups
R	$=$	Sample range
R_d	$=$	Ranges of discarded groups
R_j	$=$	Range of subgroup j
$\bar{R}$	$=$	Average range of m subgroups $= \dfrac{\sum_{j=1}^{m} R_j}{m}$
u	$=$	Number of defects per unit when subgroups size is more than 1,

u'	=	Process (population) value of u
u_d	=	Number of defects per unit in discarded groups
u_j	=	Number of defects per unit in subgroup j
$\bar{u}$	=	Average number of defects per unit of m subgroups $= \sum_{j=1}^{m} \frac{u_j}{m}$
X_i	=	Value of a quality characteristic by measurement on an item i
$\bar{X}$	=	Sample mean
$\bar{X}_d$	=	Means of discarded groups
$\bar{X}_j$	=	Mean of the subgroup j $= \sum_{i=1}^{n} \frac{X_i}{n}$
$\bar{\bar{X}}$	=	Mean or average of subgroup means $= \sum_{j=1}^{m} \frac{\bar{X}_j}{m}$
$\bar{X}'$	=	Process (population) value of sample mean $\bar{X}$
σ'_c	=	Process (population) standard deviation of c
$\hat{\sigma}_c$	=	Estimate of population standard deviation of c
σ'_p	=	Process (population) standard deviation of p
$\hat{\sigma}_p$	=	Estimate of population standard deviation of p
σ'_R	=	Process (population) standard deviation of R
$\hat{\sigma}_R$	=	Estimate of population standard deviation of R $= \frac{d_3 \bar{R}}{d_2}$
σ'_u	=	Process (population) standard deviation of u
$\hat{\sigma}_u$	=	Estimate of population standard deviation of u
σ'_X	=	Process (population) standard deviation of X
$\sigma'_{\bar{x}}$	=	Process (population) standard deviation of $\bar{X} = \frac{\sigma'_x}{\sqrt{n}}$
$\hat{\sigma}_x$	=	Estimate of process (population) standard deviation of X
$\hat{\sigma}_{\bar{x}}$	=	Estimate of process (population) standard deviation of $\bar{x}$ $= \frac{\bar{R}}{d_2 \sqrt{n}}$

The program XBARR

XBARR can be used to develop $\bar{X}$ and R charts to control a process. The process is measured by some quality characteristics - e.g.,length, weight, or some measurable unit. It can handle the cases when the standard values of the quality characteristic are known or unknown. When the standard values are known,the program merely calculates the control limits for immediate use. When the standard values are unknown, all observations by group must be entered. The $\bar{X}$ and R charts should be examined simultaneously during the development of the control limits. In this program, if a point is discarded, it will be ignored in the calculations of both charts.

INPUT
-----
1. The name of the quality characteristic
2. Select the known or unknown standard value option
3. The number of observations per group (>=2)
4. If the standard values are known:
   - the mean of the entire observed groups and
   - the standard deviation of the observations
5. If the standard values are unknown:
   - the number of observed groups and
   - all observation values

OUTPUT

After the above inputs are entered, outputs from the program can be generated via commands. For example, the XBAR command will produce the $\bar{X}$ control chart, and the R command generates the R control chart.

EXAMPLE

Consider the following data pertaining to the length of a part in millimeters used in a manufacturing process. We will illustrate the use of the program with two cases. Assume for the first case that the standard values are known, with the mean length of the part being 760 mm and the standard deviation of the length being 13 mm. In the second case, a number of observations are collected to verify whether the standard set in the first case is reasonable or not. Table 2 contains the 25 groups of observations with five observations per group.

Table 2. Observation Values for Case 2

Group	____ O B S E R V A T I O N ____ N U M B E R ____				
	1	2	3	4	5
1	760	764	761	762	761
2	770	772	769	771	768
3	762	761	765	769	762
4	761	763	762	760	759
5	762	764	762	765	761
6	768	765	764	765	766
7	752	755	761	762	765
8	768	763	762	765	768
9	760	763	762	761	758
10	763	765	762	764	767
11	770	771	772	770	768
12	764	765	761	758	759
13	760	762	765	763	768
14	768	764	763	764	766
15	774	772	769	765	767
16	768	764	765	765	767
17	766	764	765	765	767
18	762	764	761	760	759
19	754	758	753	758	759
20	761	765	762	763	762
21	762	768	769	765	766
22	771	759	763	761	764
23	764	765	768	762	760
24	770	759	763	766	765
25	761	766	768	768	762

Please see example terminal session for the results of the first case. In the second case, after the $\bar{X}$ and R charts are examined for the first time, we assume that the following points may be discarded due to assignable causes: Groups 2, 11, 15 and 19. Although Group 7 is below the lower control limit we assume that no assignable cause can be traced.

Note that the last $\bar{X}$ chart is still called a "TRIAL" control chart because of the convention stated earlier. From the example terminal session, the results of the second case indicate that perhaps the control limits obtained from the first case can be modified.

```
XBARR EXAMPLE TERMINAL SESSION
==============================

X-BAR, R-CHARTS (CONTROL CHARTS FOR VARIABLES)
----------------------------------------------

NEED INTRODUCTION (Y OR N) ... !N

ENTER NAME OF QUALITY CHARACTERISTIC
!LENGTH IN MILLIMETERS

DO YOU KNOW THE STANDARD VALUES (Y OR N)  !Y

ENTER NUMBER OF OBSERVATIONS PER GROUP (>=2)  !5

ENTER THE KNOWN MEAN VALUE   !760
ENTER THE KNOWN STANDARD DEVIATION   !13

PLEASE ENTER COMMAND OR TYPE HELP

COMMAND --> !HELP

AVAILABLE COMMANDS
==================
HELP      - PRINT THIS MESSAGE
XBAR      - GENERATE X-BAR CONTROL CHART
R         - GENERATE R CONTROL CHART
LIST      - LIST OBSNS BY GROUP NUMBER
CHANGE    - CHANGE OBSERVATION IN A GROUP
DISCARD   - DISCARD A GROUP OF OBSERVATIONS
            DUE TO ASSIGNABLE CAUSE
REINSTATE - REINSTATE A DISCARDED GROUP
RESTART   - START A NEW PROBLEM
EXIT      - EXIT FROM THE PROGRAM

NOTE : THE COMMANDS - LIST,CHANGE, DISCARD
AND REINSTATE ARE INACTIVE WHEN STANDARD
VALUES ARE KNOWN.

COMMAND --> !XBAR


X - B A R    C O N T R O L    C H A R T
QUALITY CHARACTERISTIC : LENGTH IN MILLIMETERS
UPPER CONTROL LIMIT = 777.4413302245
CENTRAL VALUE       = 760
LOWER CONTROL LIMIT = 742.5586697755

COMMAND --> !R


R - C O N T R O L    C H A R T
QUALITY CHARACTERISTIC : LENGTH IN MILLIMETERS
UPPER CONTROL LIMIT = 63.934
```

```
CENTRAL VALUE       = 30.238
LOWER CONTROL LIMIT = 0

COMMAND --> !RESTART


X-BAR, R-CHARTS (CONTROL CHARTS FOR VARIABLES)
----------------------------------------------

NEED INTRODUCTION (Y OR N) ... !N

ENTER NAME OF QUALITY CHARACTERISTIC
!LENGTH IN MILLIMETERS

DO YOU KNOW THE STANDARD VALUES (Y OR N)   !N

ENTER NUMBER OF OBSERVATIONS PER GROUP (>=2)  !5
ENTER NUMBER OF OBSERVED GROUPS   !25

ENTER ALL OBSERVATIONS
----------------------

GROUP 1   OBSN  1    !760
GROUP 1   OBSN  2    !764
GROUP 1   OBSN  3    !761
GROUP 1   OBSN  4    !762
GROUP 1   OBSN  5    !761

GROUP 2   OBSN  1    !770
GROUP 2   OBSN  2    !772
GROUP 2   OBSN  3    !769
GROUP 2   OBSN  4    !771
GROUP 2   OBSN  5    !768

GROUP 3   OBSN  1    !762
GROUP 3   OBSN  2    !761
GROUP 3   OBSN  3    !765
GROUP 3   OBSN  4    !769
GROUP 3   OBSN  5    !762

GROUP 4   OBSN  1    !761
GROUP 4   OBSN  2    !763
GROUP 4   OBSN  3    !762
GROUP 4   OBSN  4    !760
GROUP 4   OBSN  5    !759

GROUP 5   OBSN  1    !762
GROUP 5   OBSN  2    !764
GROUP 5   OBSN  3    !762
GROUP 5   OBSN  4    !765
GROUP 5   OBSN  5    !761

GROUP 6   OBSN  1    !768
GROUP 6   OBSN  2    !765
```

```
GROUP  6   OBSN  3   !764
GROUP  6   OBSN  4   !765
GROUP  6   OBSN  5   !766

GROUP  7   OBSN  1   !752
GROUP  7   OBSN  2   !755
GROUP  7   OBSN  3   !761
GROUP  7   OBSN  4   !762
GROUP  7   OBSN  5   !765

GROUP  8   OBSN  1   !768
GROUP  8   OBSN  2   !763
GROUP  8   OBSN  3   !762
GROUP  8   OBSN  4   !765
GROUP  8   OBSN  5   !768

GROUP  9   OBSN  1   !760
GROUP  9   OBSN  2   !763
GROUP  9   OBSN  3   !762
GROUP  9   OBSN  4   !761
GROUP  9   OBSN  5   !758

GROUP 10   OBSN  1   !763
GROUP 10   OBSN  2   !765
GROUP 10   QBSN  3   !762
GROUP 10   OBSN  4   !764
GROUP 10   OBSN  5   !767

GROUP 11   OBSN  1   !770
GROUP 11   OBSN  2   !771
GROUP 11   OBSN  3   !772
GROUP 11   OBSN  4   !770
GROUP 11   OBSN  5   !768

GROUP 12   OBSN  1   !764
GROUP 12   OBSN  2   !765
GROUP 12   OBSN  3   !761
GROUP 12   OBSN  4   !758
GROUP 12   OBSN  5   !759

GROUP 13   OBSN  1   !760
GROUP 13   OBSN  2   !762
GROUP 13   OBSN  3   !765
GROUP 13   OBSN  4   !763
GROUP 13   OBSN  5   !768

GROUP 14   OBSN  1   !768
GROUP 14   OBSN  2   !764
GROUP 14   OBSN  3   !763
GROUP 14   OBSN  4   !764
GROUP 14   OBSN  5   !766

GROUP 15   OBSN  1   !774
GROUP 15   OBSN  2   !772
```

```
GROUP 15   OBSN  3   !769
GROUP 15   OBSN  4   !765
GROUP 15   OBSN  5   !767

GROUP 16   OBSN  1   !768
GROUP 16   OBSN  2   !764
GROUP 16   OBSN  3   !765
GROUP 16   OBSN  4   !765
GROUP 16   OBSN  5   !767

GROUP 17   OBSN  1   !766
GROUP 17   OBSN  2   !764
GROUP 17   OBSN  3   !765
GROUP 17   OBSN  4   !765
GROUP 17   OBSN  5   !767

GROUP 18   OBSN  1   !762
GROUP 18   OBSN  2   !764
GROUP 18   OBSN  3   !761
GROUP 18   OBSN  4   !760
GROUP 18   OBSN  5   !759

GROUP 19   OBSN  1   !754
GROUP 19   OBSN  2   !758
GROUP 19   OBSN  3   !753
GROUP 19   OBSN  4   !758
GROUP 19   OBSN  5   !759

GROUP 20   OBSN  1   !761
GROUP 20   OBSN  2   !765
GROUP 20   OBSN  3   !762
GROUP 20   OBSN  4   !763
GROUP 20   OBSN  5   !762

GROUP 21   OBSN  1   !762
GROUP 21   OBSN  2   !768
GROUP 21   OBSN  3   !769
GROUP 21   OBSN  4   !765
GROUP 21   OBSN  5   !766

GROUP 22   OBSN  1   !771
GROUP 22   OBSN  2   !759
GROUP 22   OBSN  3   !763
GROUP 22   OBSN  4   !761
GROUP 22   OBSN  5   !764

GROUP 23   OBSN  1   !764
GROUP 23   OBSN  2   !765
GROUP 23   OBSN  3   !768
GROUP 23   OBSN  4   !762
GROUP 23   OBSN  5   !760

GROUP 24   OBSN  1   !770
GROUP 24   OBSN  2   !759
```

```
GROUP 24   OBSN  3    !763
GROUP 24   OBSN  4    !766
GROUP 24   OBSN  5    !765

GROUP 25   OBSN  1    !761
GROUP 25   OBSN  2    !766
GROUP 25   OBSN  3    !768
GROUP 25   OBSN  4    !768
GROUP 25   OBSN  5    !762
```

PLEASE ENTER COMMAND OR TYPE HELP

COMMAND --> !XBAR

T R I A L   X - B A R   C O N T R O L   C H A R T
QUALITY CHARACTERISTIC : LENGTH IN MILLIMETERS
UPPER CONTROL LIMIT = 767.5423147632
CENTRAL VALUE   = 763.92
LOWER CONTROL LIMIT = 760.2976852367

```
                          756.000   760.000   764.000   768.000   772.000
GROUP    MEAN ASSIGNABLE CAUSE +---------+---------+---------+---------+
  1    761.600                         :    *    :         :         :
  2    770.000                         :         :         :    *    :
  3    763.800                         :       *:         :         :
  4    761.000                         :  *      :         :         :
  5    762.800                         :       * :         :         :
  6    765.600                         :         :    *    :         :
  7    759.000                  *      :         :         :         :
  8    765.200                         :         :    *    :         :
  9    760.800                         :*        :         :         :
 10    764.200                         :         :*        :         :
 11    770.200                         :         :         :    *    :
 12    761.400                         :  *      :         :         :
 13    763.600                         :       *:         :         :
 14    765.000                         :         :   *     :         :
 15    769.400                         :         :         :       * :
 16    765.800                         :         :     *   :         :
 17    765.400                         :         :    *    :         :
 18    761.200                         : *       :         :         :
 19    756.400             *           :         :         :         :
 20    762.600                         :      *  :         :         :
 21    766.000                         :         :      *  :         :
 22    763.600                         :       *:         :         :
 23    763.800                         :       *:         :         :
 24    764.600                         :        :*         :         :
 25    765.000                         :        :  *       :         :
```

COMMAND --> !R

R E V I S E D   R - C O N T R O L   C H A R T

```
QUALITY CHARACTERISTIC : LENGTH IN MILLIMETERS
UPPER CONTROL LIMIT = 13.27817712812
CENTRAL VALUE       = 6.28
LOWER CONTROL LIMIT = 0

                         -    4.000     1.000     6.000     11.000    16.000
GROUP   MEAN ASSIGNABLE CAUSE +---------+---------+---------+---------+
  1     4.000                          :    *    :         :         :
  2     4.000                          :    *    :         :         :
  3     8.000                          :         :    *    :         :
  4     4.000                          :    *    :         :         :
  5     4.000                          :    *    :         :         :
  6     4.000                          :    *    :         :        *:
  7    13.000                          :         :         :        *:
  8     6.000                          :         :*        :         :
  9     5.000                          :        *:         :         :
 10     5.000                          :        *:         :         :
 11     4.000                          :    *    :         :         :
 12     7.000                          :         :*        :         :
 13     8.000                          :         :    *    :         :
 14     5.000                          :        *:         :         :
 15     9.000                          :         :         *         :
 16     4.000                          :    *    :         :         :
 17     3.000                          :  *      :         :         : .
 18     5.000                          :        *:         :         :
 19     6.000                          :         *         :         :
 20     4.000                          :   *     :         :         :
 21     7.000                          :         :*        :         :
 22    12.000                          :         :         :    *    :
 23     8.000                          :         :    *    :         :
 24    11.000                          :         :         :       * :
 25     7.000                          :         :*        :         :


COMMAND --> !DISCARD

ENTER GROUP NUMBER  !2
ENTER REASON FOR DISCARDING (15 CHARS MAX)   !BAD MATERIAL

COMMAND --> !DISCARD

ENTER GROUP NUMBER  !11
ENTER REASON FOR DISCARDING (15 CHARS MAX)   !BAD MATERIAL

COMMAND --> !DISCARD

ENTER GROUP NUMBER  !15
ENTER REASON FOR DISCARDING (15 CHARS MAX)   !WORKMAN ERROR

COMMAND --> !DISCARD

ENTER GROUP NUMBER  !19
ENTER REASON FOR DISCARDING (15 CHARS MAX)   !MACHINE ERROR
```

```
COMMAND --> !XBAR


TRIAL    X - B A R    C O N T R O L    C H A R T
QUALITY CHARACTERISTIC : LENGTH IN MILLIMETERS
UPPER CONTROL LIMIT = 767.1091156981
CENTRAL VALUE       = 763.4285714286
LOWER CONTROL LIMIT = 759.748027159

                            756.000    760.000    764.000    768.000    772.000
GROUP    MEAN  ASSIGNABLE CAUSE +---------+---------+---------+---------+
  1    761.600                       :       *    :          :
  2    770.000  BAD MATERIAL         :            :          :         D
  3    763.800                       :            *          :
  4    761.000                       :    *       :          :
  5    762.800                       :         *  :          :
  6    765.600                       :            :      *   :
  7    759.000                    *  :            :          :
  8    765.200                       :            :     *    :
  9    760.800                       : *          :          :
 10    764.200                       :            : *        :
 11    770.200  BAD MATERIAL         :            :          :         D
 12    761.400                       :   *        :          :
 13    763.600                       :            *          :
 14    765.000                       :            :    *     :
 15    769.400  WORKMAN ERROR        :            :          :    D
 16    765.800                       :            :      *   :
 17    765.400                       :            :      *   :
 18    761.200                       :    *       :          :
 19    756.400  MACHINE ERROR    D   :            :          :
 20    762.600                       :         *  :          :
 21    766.000                       :            :       *  :
 22    763.600                       :            *          :
 23    763.800                       :            *          :
 24    764.600                       :            : *        :
 25    765.000                       :            :    *     :
```

```
COMMAND --> !R


R E V I S E D    R - C O N T R O L    C H A R T
QUALITY CHARACTERISTIC : LENGTH IN MILLIMETERS
UPPER CONTROL LIMIT = 13.49162674528
CENTRAL VALUE       = 6.380952380952
LOWER CONTROL LIMIT = 0

                          -      4.000     1.000     6.000     11.000    16.000
·GROUP   MEAN ASSIGNABLE CAUSE +---------+---------+---------+---------+
    1    4.000                        :         *         :             :
    2    4.000 BAD MATERIAL           :         D         :             :
    3    8.000                        :         :      *  :             :
    4    4.000                        :         *  :                    :
    5    4.000                        :         *         :             :
    6    4.000                        :         *         :             :
    7   13.000                        :         :         :           *:
    8    6.000                        :         :      *:              :
    9    5.000                        :         :   *  :               :
   10    5.000                        :         :   *  :               :
   11    4.000 BAD MATERIAL           :         D         :            :
   12    7.000                        :         :*                     :
   13    8.000                        :         :      *               :
   14    5.000                        :         :   *  :               :
   15    9.000 WORKMAN ERROR          :         :         :   D        :
   16    4.000                        :      *  :                      :
   17    3.000                        :   *     :                      :
   18    5.000                        :      *  :                      :
   19    6.000 MACHINE ERROR          :         D:                     :
   20    4.000                        :      *  :                      :
   21    7.000                        :         :*                     :
   22   12.000                        :         :         :      *     :
   23    8.000                        :         :      *               :
   24   11.000                        :         :         :         *  :
   25    7.000                        :         :*                     :


COMMAND --> !EXIT
```

```
10 REM    X-BAR, R  CONTROL CHART FOR VARIABLES
20 REM
30 DIM D2(25),D3(25),P$(40),Z$(6)
40 DIM X(30),R(30),R$(30),Z(30),W(30,25)
50 REM    M0 = MAXIMUM NUMBER OF OBSERVED GROUPS
60 REM    N0 = MAXIMUM NUMBER OF OBSERVATIONS PER GROUP
70 M0=30
80 N0=25
90 GOSUB 3900
100 PRINT
110 PRINT "X-BAR, R-CHARTS (CONTROL CHARTS FOR VARIABLES)"
120 PRINT "------------------------------------------------"
130 PRINT
140 PRINT "NEED INTRODUCTION (Y OR N) ... ";
150 INPUT Y$
160 IF Y$ <> "Y" THEN 400
170 PRINT
180 PRINT "THIS PROGRAM DETERMINES THE UPPER AND LOWER"
190 PRINT "CONTROL LIMITS FOR X-BAR AND R-CONTROL CHARTS"
200 PRINT "WITH KNOWN OR UNKNOWN STANDARD VALUES.  IF THE"
210 PRINT "STANDARD VALUES ARE UNKNOWN, THE PROGRAM HAS"
220 PRINT "AN OPTION WHICH ALLOWS THE USER TO DISCARD OR"
230 PRINT "REINSTATE OUT OF CONTROL POINTS."
240 PRINT
250 PRINT "INPUT REQUIREMENTS"
260 PRINT "------------------"
270 PRINT "1. NAME OF QUALITY CHARACTERISTIC"
280 PRINT "2. SELECT KNOWN OR UNKNOWN STANDARD VALUE OPTION"
290 PRINT "3. THE NUMBER OF OBSERVED GROUPS"
300 PRINT "4. IF STANDARD VALUES ARE KNOWN :-"
310 PRINT "     - THE MEAN OF THE OBSERVED GROUPS"
320 PRINT "     - STANDARD DEVIATION OF THE OBSERVED GROUPS"
330 PRINT "5. IF STANDARD VALUES ARE UNKNOWN :-"
340 PRINT "     - NUMBER OF OBSERVATIONS IN A GROUP"
350 PRINT "     - ALL OBSERVATION VALUES"
360 PRINT
370 PRINT "MAXIMUM NUMBER OF OBSERVED GROUPS = ";M0
380 PRINT "MAXIMUM NUMBER OF OBSERVATIONS PER GROUP = ";N0
390 REM *** I N P U T    S E C T I O N ***
400 PRINT
410 PRINT "ENTER NAME OF QUALITY CHARACTERISTIC"
420 INPUT Q$
430 PRINT
440 PRINT "DO YOU KNOW THE STANDARD VALUES (Y OR N)   ";
450 INPUT Y$
460 K9=2
470 IF Y$ <> "Y" THEN 490
480 K9=1
490 PRINT
500 PRINT "ENTER NUMBER OF OBSERVATIONS PER GROUP (>=2)  ";
510 INPUT N
520 N=INT(N)
530 IF N < 2 THEN 490
540 IF N <= N0 THEN 580
```

```
550 PRINT
560 PRINT "*** ERROR ***    MAX ALLOWED = ";N0
570 GOTO 490
580 IF K9=2 THEN 670
590 PRINT
600 PRINT "ENTER THE KNOWN MEAN VALUE   ";
610 INPUT X1
620 PRINT "ENTER THE KNOWN STANDARD DEVIATION   ";
630 INPUT S1
640 IF S1 <= 0 THEN 620
650 GOTO 880
660 REM     STANDARD VALUES UNKNOWN - PROMPT FOR OBSNS
670 PRINT "ENTER NUMBER OF OBSERVED GROUPS   ";
680 INPUT M
690 M=INT(M)
700 IF M < 1 THEN 670
710 IF M <=M0 THEN 750
720 PRINT
730 PRINT "*** ERROR ***    MAX ALLOWED = ";M0
740 GOTO 670
750 PRINT
760 PRINT "ENTER ALL OBSERVATIONS"
770 PRINT "--------------------"
780 FOR I=1 TO M
790 PRINT
800 FOR J=1 TO N
810 PRINT "GROUP ";I;"   OBSN   ";J;"    ";
820 INPUT W(I,J)
830 W(I,J)=ABS(W(I,J))
840 NEXT J
850 GOSUB 3520
860 NEXT I
870 REM    *** ACCEPT COMMANDS HERE ***
880 PRINT
890 PRINT "PLEASE ENTER COMMAND OR TYPE HELP"
900 PRINT
910 PRINT "COMMAND --> ";
920 INPUT C$
930 PRINT
940 IF C$ <> "HELP" THEN 970
950 GOSUB 1220
960 GOTO 900
970 IF C$ <> "XBAR" THEN 1000
980 GOSUB 1960
990 GOTO 900
1000 IF C$ <> "R" THEN 1030
1010 GOSUB 1990
1020 GOTO 900
1030 IF C$ <> "LIST" THEN 1060
1040 GOSUB 1400
1050 GOTO 900
1060 IF C$ <> "CHANGE" THEN 1090
1070 GOSUB 1550
1080 GOTO 900
```

```
1090 IF C$ <> "DISCARD" THEN 1120
1100 GOSUB 1790
1110 GOTO 900
1120 IF C$ <> "REINSTATE" THEN 1150
1130 GOSUB 1790
1140 GOTO 900
1150 IF C$ <> "RESTART" THEN 1180
1160 GOSUB 4090
1170 GOTO 100
1180 IF C$ = "EXIT" THEN 4140
1190 PRINT
1200 PRINT "*** ERROR ***    PLEASE TRY AGAIN !"
1210 GOTO 900
1220 REM    HELP COMMAND
1230 PRINT "AVAILABLE COMMANDS"
1240 PRINT "=================="
1250 PRINT "HELP      - PRINT THIS MESSAGE"
1260 PRINT "XBAR      - GENERATE X-BAR CONTROL CHART"
1270 PRINT "R         - GENERATE R CONTROL CHART"
1280 PRINT "LIST      - LIST OBSNS BY GROUP NUMBER"
1290 PRINT "CHANGE    - CHANGE OBSERVATION IN A GROUP"
1300 PRINT "DISCARD   - DISCARD A GROUP OF OBSERVATIONS"
1310 PRINT "            DUE TO ASSIGNABLE CAUSE"
1320 PRINT "REINSTATE - REINSTATE A DISCARDED GROUP"
1330 PRINT "RESTART   - START A NEW PROBLEM"
1340 PRINT "EXIT      - EXIT FROM THE PROGRAM"
1350 PRINT
1360 PRINT "NOTE : THE COMMANDS - LIST,CHANGE, DISCARD"
1370 PRINT "AND REINSTATE ARE INACTIVE WHEN STANDARD"
1380 PRINT "VALUES ARE KNOWN."
1390 RETURN
1400 REM    LIST COMMAND
1410 IF K9=1 THEN 1540
1420 PRINT "ENTER GROUP NUMBER   ";
1430 INPUT G9
1440 FOR I=1 TO M
1450 IF G9 <> I THEN 1520
1460 FOR J=1 TO N
1470 PRINT "OBSN ";J;"  VALUE = ";W(I,J)
1480 NEXT J
1490 PRINT "GROUP MEAN = ";X(I)
1500 PRINT "GROUP RANGE = ";R(I)
1510 GOTO 1540
1520 NEXT I
1530 GOSUB 1760
1540 RETURN
1550 REM    CHANGE COMMAND
1560 IF K9=1 THEN 1750
1570 PRINT "ENTER GROUP NUMBER   ";
1580 INPUT G9
1590 FOR I=1 TO M
1600 IF G9 <> I THEN 1730
1610 PRINT "ENTER OBSN NUMBER IN THIS GROUP   ";
1620 INPUT G9
```

```
1630 FOR J=1 TO N
1640 IF G9 <> J THEN 1690
1650 PRINT "OLD VALUE = ";W(I,J);"   ENTER NEW VALUE   ";
1660 INPUT W(I,J)
1670 W(I,J)=ABS(W(I,J))
1680 GOTO 1710
1690 NEXT J
1700 GOTO 1740
1710 GOSUB 3520
1720 GOTO 1750
1730 NEXT I
1740 GOSUB 1760
1750 RETURN
1760 PRINT
1770 PRINT "*** ERROR ***   COMMAND ABORTED !!!"
1780 RETURN
1790 REM    DISCARD/REINSTATE   COMMANDS
1800 IF K9 = 1 THEN 1950
1810 PRINT "ENTER GROUP NUMBER   ";
1820 INPUT G9
1830 FOR I=1 TO M
1840 IF G9 <> I THEN 1930
1850 IF C$ = "REINSTATE" THEN 1900
1860 PRINT "ENTER REASON FOR DISCARDING (15 CHARS MAX)   ";
1870 INPUT R$(I)
1880 Z(I)=1
1890 GOTO 1950
1900 R$(I)=" "
1910 Z(I)=0
1920 GOTO 1950
1930 NEXT I
1940 GOSUB 1760
1950 RETURN
1960 REM    XBAR COMMAND
1970 M9=1
1980 GOTO 2020
1990 REM    R  COMMAND
2000 M9=2
2010 REM
2020 REM    GENERATE CONTROL CHART
2030 IF K9 = 2 THEN 2150
2040 REM    KNOWN STANDARD VALUES
2050 IF M9 = 2 THEN 2100
2060 C1=X1
2070 U1=C1+3*S1/SQR(N)
2080 L1=C1-3*S1/SQR(N)
2090 GOTO 2130
2100 C1=D2(N)*S1
2110 U1=C1+3*D3(N)*S1
2120 L1=C1-3*D3(N)*S1
2130 GOSUB 3660
2140 GOTO 2690
2150 REM    STANDARD VALUES UNKNOWN
2160 IF M9=2 THEN 2280
```

```
2170 X1=0
2180 A0=1.E30
2190 A1=-A0
2200 FOR I=1 TO M
2210 IF X(I) >= A0 THEN 2230
2220 A0=X(I)
2230 IF X(I) <= A1 THEN 2250
2240 A1=X(I)
2250 IF Z(I) <> 0 THEN 2270
2260 X1=X1+X(I)
2270 NEXT I
2280 K=0
2290 R1=0
2300 FOR I=1 TO M
2310 IF Z(I) <> 0 THEN 2340
2320 K=K+1
2330 R1=R1+R(I)
2340 NEXT I
2350 IF K > 0 THEN 2400
2360 PRINT
2370 PRINT "!!!!!   ALL GROUPS DISCARDED   !!!!!"
2380 GOSUB 1760
2390 GOTO 2690
2400 X1=X1/K
2410 R1=R1/K
2420 IF M9=2 THEN 2560
2430 C1=X1
2440 U1=C1+3*R1/D2(N)/SQR(N)
2450 L1=C1-3*R1/D2(N)/SQR(N)
2460 IF L1 >= A0 THEN 2480
2470 A0=L1
2480 IF U1 <= A1 THEN 2500
2490 A1=U1
2500 GOSUB 3660
2510 GOSUB 3140
2520 IF L5 > 0 THEN 2540
2530 GOSUB 1760
2540 GOSUB 2700
2550 GOTO 2690
2560 REM     R-CHART OUTPUT
2570 A0=1.E30
2580 A1=-A0
2590 FOR I=1 TO M
2600 IF R(I) >= A0 THEN 2620
2610 A0=R(I)
2620 IF R(I) <= A1 THEN 2640
2630 A1=R(I)
2640 NEXT I
2650 C1=R1
2660 U1=C1+3*R1*D3(N)/D2(N)
2670 L1=C1-3*R1*D3(N)/D2(N)
2680 GOTO 2460
2690 RETURN
2700 REM     PLOTTING ROUTINE
```

```
2710 REM PRINT AXIS
2720 PRINT
2730 PRINT "                           ";
2740 FOR J=L0 TO U0 STEP H0
2750 PRINT USING Z$(5),J;
2760 NEXT J
2770 PRINT
2780 PRINT "GROUP    MEAN ASSIGNABLE CAUSE +";
2790 FOR J=1 TO 4
2800 PRINT "---------+";
2810 NEXT J
2820 PRINT
2830 H=H0/10
2840 J0=INT((L1-L0)/H+0.5)
2850 J1=INT((U1-L0)/H+0.5)
2860 J2=INT((C1-L0)/H+0.5)
2870 FOR I=1 TO M
2880 FOR J=0 TO 40
2890 P$(J)=" "
2900 NEXT J
2910 P$(J0)=":"
2920 P$(J1)=":"
2930 P$(J2)=":"
2940 J9=J1
2950 IF M9=2 THEN 2990
2960 J3=INT((X(I)-L0)/H+0.5)
2970 V=X(I)
2980 GOTO 3010
2990 J3=INT((R(I)-L0)/H+0.5)
3000 V=R(I)
3010 P$(J3)="*"
3020 IF Z(I)=0 THEN 3040
3030 P$(J3)="D"
3040 IF J3 <= J9 THEN 3060
3050 J9=J3
3060 PRINT USING Z$(6),I,V,R$(I);
3070 FOR J=0 TO J9
3080 PRINT P$(J);
3090 NEXT J
3100 PRINT
3110 NEXT I
3120 PRINT
3130 RETURN
3140 REM    S C A L I N G    R O U T I N E
3150 REM    ------------------------------
3160 L5=0
3170 Y0=A0
3180 Y1=A1
3190 A2=(A1-A0)/4
3200 IF A2 <= 0 THEN 3510
3210 L5=L5+1
3220 A3=LOG(A2)/LOG(10)
3230 REM    ROUND DOWN A3 TO A4
3240 A4=INT(A3)
```

```
3250 IF A4 <= A3 THEN 3270
3260 A4=A4-1
3270 A5=10**A4
3280 A6=A2/A5
3290 REM    ROUND UP A6 TO A7
3300 A7=INT(A6+1)
3310 H0=A5*A7
3320 REM    ROUND DOWN L0
3330 L0=INT(A0/H0)
3340 IF L0 <= A0/H0 THEN 3360
3350 L0=L0-1
3360 L0=L0*H0
3370 U0=L0+5*H0
3380 IF U0 >= A1 THEN 3410
3390 A1=A1+A2/2
3400 GOTO 3190
3410 IF L0 <= A0 THEN 3440
3420 A0=A0-A2/2
3430 GOTO 3190
3440 IF U0-H0 <= Y1 THEN 3470
3450 U0=U0-H0
3460 GOTO 3440
3470 IF L0+H0 >= Y0 THEN 3500
3480 L0=L0+H0
3490 GOTO 3470
3500 H0=(U0-L0)/4
3510 RETURN
3520 REM    CALCULATE THE MEAN AND THE RANGE OF A GROUP
3530 A0=1.E30
3540 A1=-A0
3550 A2=0
3560 FOR J=1 TO N
3570 A2=A2+W(I,J)
3580 IF W(I,J) >= A0 THEN 3600
3590 A0=W(I,J)
3600 IF W(I,J) <= A1 THEN 3620
3610 A1=W(I,J)
3620 NEXT J
3630 X(I)=A2/N
3640 R(I)=A1-A0
3650 RETURN
3660 REM    PRINT CONTROL LIMITS
3670 PRINT
3680 IF L1 > 0 THEN 3700
3690 L1=0
3700 IF K9=1 THEN 3840
3710 I0=4
3720 FOR I=1 TO M
3730 IF Z(I) <> 0 THEN 3810
3740 IF M9=2 THEN 3770
3750 V=X(I)
3760 GOTO 3780
3770 V=R(I)
3780 IF V > U1 THEN 3800
```

```
3790 IF V >= L1 THEN 3810
3800 I0=3
3810 NEXT I
3820 PRINT Z$(I0);Z$(M9)
3830 GOTO 3850
3840 PRINT Z$(M9)
3850 PRINT "QUALITY CHARACTERISTIC : ";Q$
3860 PRINT "UPPER CONTROL LIMIT = ";U1
3870 PRINT "CENTRAL VALUE        = ";C1
3880 PRINT "LOWER CONTROL LIMIT = ";L1
3890 RETURN
3900 REM      I N I T I A L I Z A T I O N
3910 Z$(1)="X - B A R     C O N T R O L       C H A R T"
3920 Z$(2)="R - C O N T R O L       C H A R T"
3930 Z$(3)="T R I A L    "
3940 Z$(4)="R E V I S E D    "
3950 Z$(5)="-#####.###"
3960 Z$(6)="### #####.### ###############    "
3970 REM      READ TABLE M OF DUNCAN BOOK FOR D2 AND D3
3980 DATA 1.128,0.853,1.693,0.888,2.059,0.880
3990 DATA 2.326,0.864,2.534,0.848,2.704,0.833
4000 DATA 2.847,0.820,2.970,0.808,3.078,0.797
4010 DATA 3.173,0.787,3.258,0.778,3.336,0.770
4020 DATA 3.407,0.762,3.472,0.755,3.532,0.749
4030 DATA 3.588,0.743,3.640,0.738,3.689,0.733
4040 DATA 3.735,0.729,3.778,0.724,3.819,0.720
4050 DATA 3.858,0.716,3.859,0.712,3.931,0.709
4060 FOR I=2 TO N0
4070 READ D2(I),D3(I)
4080 NEXT I
4090 FOR I=1 TO M0
4100 R$(I)="  "
4110 Z(I)=0
4120 NEXT I
4130 RETURN
4140 END
```

Source for Values in Statements 3980 to 4050: A.J. Duncan, QUALITY CONTROL AND INDUSTRIAL STATISTICS, 4th ed. (Homewood, Illinois: Richard D. Irwin, 1974), Table M.

The Program PUC

PUC can be used to develop p- or u- or c-charts to control a process. The process is measured by the fraction defective (p) or the average number of defects per unit (u) or the number of defects for a single unit (c). It can handle the cases when the standard value of p or u or c is known or unknown. When the standard value is known, the program calculates the standard control limits for immediate use. When the standard value is unknown, appropriate data must be entered for three types of control charts. This is described in the INPUT section below. Note that in this program the terms "group," "unit," "observation" are used interchangeably.

INPUT
-----
1. The type of the control chart (p or u or c)
2. The name of the inspection station
3. Select the known or unknown standard value option
4. If the standard value is known:
      for p-chart   - the size of an observed group
                     - the average fraction defective
      for c-chart   - the average defects per single unit
      for u-chart   - the size of an observed group
                     - the average defects per unit
5. If the standard value is unknown:
      for p-chart   - the number of observed groups
                     - the size of an observed group
                     - the number of defectives in each group
      for c-chart   - the number of single units observed
                     - the number of defects in each unit
      for u-chart   - the number of observed groups
                     - the size of an observed group
                     - the number of defects in each group

OUTPUT

After the above inputs are entered, the user can use the CHART command to generate the type of control chart specified in the input section. In addition, the LIST, CHANGE, DISCARD, and REINSTATE commands will prompt the user for the appropriate information based on the the type of control chart selected. These commands only affect the output for the case when the standard value is unknown and raw data are entered for the purpose of developing the control limits.

EXAMPLES

We will illustrate the use of the PUC program by giving a detail terminal session for the development of the p- and the u-charts only. In practice, these two charts are also more popular than the c-chart. The user can consult the examples in the references for the c-chart.

Consider an example of developing a p-chart at inspection station 10. Two cases are considered. First, the standard value of the average fraction defective is known to be 0.07 and the size of an observed group is 60 items and we are interested in computing the upper and lower control limits. In the second case, 25 groups of data are collected for the number of defectives in each group of 60 items and we are interested in developing the control limits and the central value. Table 3 contains the number of defective items in a group.

Table 3. Number of Defectives in a Group of 60

GROUP	NO. OF DEFECTIVES	GROUP	NO. OF DEFECTIVES
1	4	14	5
2	3	15	6
3	5	16	2
4	10	17	1
5	6	18	5
6	4	19	19
7	2	20	11
8	7	21	4
9	18	22	5
10	9	23	7
11	6	24	9
12	22	25	4
13	10		

As opposed to the p-chart, which classifies an item as being either good or defective, the u-chart records the average number of defects for a group of items. The item in this case could be, for example, 100 yards of carpet from a manufacturing process. If the total number of defective spots in 10 obervations of 100 yards of carpet is 22, then the average number of defects per item or unit is 2.2.

The following data will be used to develop control limits for a u-chart at inspection station number 20. In the first case, the standard value of the average number of defects per unit is 1.8, and the station has been collecting data for 10 units at a time when inspecting. In the second case, we assume that 25 groups of 10 units have been inspected and the total number of defects for each group is as shown in Table 4 below. We wish to use the program PUC to set up the u-control limits at the station.

Table 4. Number of Defects in a Group of 10 Units

GROUP	NO. OF DEFECTS	GROUP	NO. OF DEFECTS
1	21	14	33
2	18	15	19
3	16	16	12
4	22	17	15
5	28	18	11
6	18	19	16
7	12	20	22
8	14	21	28
9	29	22	16
10	9	23	14
11	11	24	11
12	8	25	10
13	25		

We begin the example terminal session by selecting the p-chart for inspection station 10 with known standard value option -i.e., the first case. The size of the observed group is 60, and the known fraction defective 0.07. Next the command HELP is entered to show all available commands, and the CHART command is selected to compute the upper and lower control limits which are 0.1688 and 0, respectively. The second case for the p-chart is initiated using the RESTART command. This time we select the unknown standard value option, and the data of Table 3 are entered. At the command level, the CHART command is given. The resulting p-chart is recognized by the program to be a trial p-control chart because there are three groups of observations that are out of control. They are groups 9, 12, and 19. We have decided to discard these groups using the DISCARD command. The next CHART command produces the revised p-control chart with upper control limit of 0.2081, the central value of 0.0947, and the lower control limit of 0.

The RESTART command is entered next to develop the u-chart control limits for inspection station 20. As with the p-chart case, we select the known standard value for the first case. The size of the observed group or the number of units in a group is 10, and the known average defect per unit is 1.8. The command CHART produces the upper and lower control limits of 3.0728 and 0.5272, respectively. A RESTART command is then entered to develop the u-chart control limits for the second case. After the data of Table 4 are entered, the CHART command produces a trial u-control chart in which group number 14 is indicated as being out of control. Assuming that we have determined its assignable cause as due to machine error, we proceed to use the DISCARD command and obtain a revised u-control chart using the CHART command again. The upper control limit, the central value, and the lower control limit are 2.9199, 1.6875, and 0.4551, respectively.

```
PUC EXAMPLE TERMINAL SESSION
============================

P U C  CHARTS (CONTROL CHARTS FOR ATTRIBUTES)
---------------------------------------------

NEED INTRODUCTION (Y OR N) ... !N

ENTER TYPE OF CHART ( P , U , C ) --> !P

ENTER NAME OF THE INSPECTION STATION
!INSPECTION STATION 10

DO YOU KNOW THE STANDARD VALUE ( Y OR N ) ... !Y

ENTER SIZE OF AN OBSERVED GROUP !60

ENTER THE KNOWN FRACTION DEFECTIVE !0.07

PLEASE ENTER COMMAND OR TYPE HELP

COMMAND --> !HELP

AVAILABLE COMMANDS
==================
HELP      - PRINT THIS MESSAGE
CHART     - GENERATE P OR U OR C CONTROL CHARTS
LIST      - LIST OBSNS BY GROUP OR UNIT NUMBER
CHANGE    - CHANGE OBSNS IN A GROUP OR UNIT
DISCARD   - DISCARD A GROUP OR UNIT OF OBSNS
            DUE TO ASSIGNABLE CAUSE
REINSTATE - REINSTATE A DISCARDED GROUP OR UNIT
RESTART   - START A NEW PROBLEM
EXIT      - EXIT FROM THE PROBLEM

NOTE : THE COMMANDS -LIST,CHANGE,DISCARD
AND REINSTATE ARE INACTIVE WHEN STANDARD
VALUES ARE KNOWN.

COMMAND --> !CHART


P - C O N T R O L    C H A R T
INSPECTION STATION : INSPECTION STATION 10
UPPER CONTROL LIMIT = .168818014552
CENTRAL VALUE       = .07
LOWER CONTROL LIMIT = 0

COMMAND --> !RESTART

ENTER TYPE OF CHART ( P , U , C ) --> !P
```

```
ENTER NAME OF THE INSPECTION STATION
!INSPECTION STATION 10

DO YOU KNOW THE STANDARD VALUE ( Y OR N ) ... !N

ENTER SIZE OF AN OBSERVED GROUP !60

ENTER NUMBER OF OBSERVED GROUP  !25

ENTER ALL OBSERVATIONS
----------------------
GROUP  1   NO.OF DEFECTIVES   =  !4
GROUP  2   NO.OF DEFECTIVES   =  !3
GROUP  3   NO.OF DEFECTIVES   =  !5
GROUP  4   NO.OF DEFECTIVES   =  !10
GROUP  5   NO.OF DEFECTIVES   =  !6
GROUP  6   NO.OF DEFECTIVES   =  !4
GROUP  7   NO.OF DEFECTIVES   =  !2
GROUP  8   NO.OF DEFECTIVES   =  !7
GROUP  9   NO.OF DEFECTIVES   =  !18
GROUP 10   NO.OF DEFECTIVES   =  !9
GROUP 11   NO.OF DEFECTIVES   =  !6
GROUP 12   NO.OF DEFECTIVES   =  !22
GROUP 13   NO.OF DEFECTIVES   =  !10
GROUP 14   NO.OF DEFECTIVES   =  !5
GROUP 15   NO.OF DEFECTIVES   =  !6
GROUP 16   NO.OF DEFECTIVES   =  !2
GROUP 17   NO.OF DEFECTIVES   =  !1
GROUP 18   NO.OF DEFECTIVES   =  !5
GROUP 19   NO.OF DEFECTIVES   =  !19
GROUP 20   NO.OF DEFECTIVES   =  !11
GROUP 21   NO.OF DEFECTIVES   =  !4
GROUP 22   NO.OF DEFECTIVES   =  !5
GROUP 23   NO.OF DEFECTIVES   =  !7
GROUP 24   NO.OF DEFECTIVES   =  !9
GROUP 25   NO.OF DEFECTIVES   =  !4

PLEASE ENTER COMMAND OR TYPE HELP

COMMAND --> !CHART
```

```
TRIAL     P - CONTROL    CHART
INSPECTION STATION : INSPECTION STATION 10
UPPER CONTROL LIMIT = .2497215104678
CENTRAL VALUE       = .1226666666667
LOWER CONTROL LIMIT = 0

                                   .000      .100      .200      .300      .400
OBSVN   MEAN  ASSIGNABLE CAUSE    +---------+---------+---------+---------+
  1     .067                      :    *    :         :         :         :
  2     .050                      :  *      :         :         :         :
  3     .083                      :      *  :         :         :         :
  4     .167                      :         :      *  :         :         :
  5     .100                      :        *:         :         :         :
  6     .067                      :    *    :         :         :         :
  7     .033                      : *       :         :         :         :
  8     .117                      :         *         :         :         :
  9     .300                      :         :         :         :    *    :
 10     .150                      :         :  *      :         :         :
 11     .100                      :        *:         :         :         :
 12     .367                      :         :         :         :        *:
 13     .167                      :         :      *  :         :         :
 14     .083                      :      *  :         :         :         :
 15     .100                      :        *:         :         :         :
 16     .033                      : *       :         :         :         :
 17     .017                      :*        :         :         :         :
 18     .083                      :      *  :         :         :         :
 19     .317                      :         :         :         :     *   :
 20     .183                      :         :        *:         :         :
 21     .067                      :    *    :         :         :         :
 22     .083                      :      *  :         :         :         :
 23     .117                      :         *         :         :         :
 24     .150                      :         :  *      :         :         :
 25     .067                      :    *    :         :         :         :

COMMAND --> !DISCARD

ENTER GROUP OR UNIT NUMBER !9
ENTER REASON FOR DISCARDING (15 CHARS MAX)    !BAD MATERIAL

COMMAND --> !DISCARD

ENTER GROUP OR UNIT NUMBER !12
ENTER REASON FOR DISCARDING (15 CHARS MAX)    !WORKER ERROR

COMMAND --> !DISCARD

ENTER GROUP OR UNIT NUMBER !19
ENTER REASON FOR DISCARDING (15 CHARS MAX)    !MACHINE ERROR

COMMAND --> !CHART
```

```
R E V I S E D      P - C O N T R O L      C H A R T
INSPECTION STATION : INSPECTION STATION 10
UPPER CONTROL LIMIT = .2080963437873
CENTRAL VALUE       = .09469696969696
LOWER CONTROL LIMIT = 0

                                  .000       .100       .200       .300       .400
    OBSVN    MEAN  ASSIGNABLE CAUSE +---------+---------+---------+---------+
      1      .067                 :     *   :         :         :
      2      .050                 :   *     :         :         :
      3      .083                 :      *  :         :         :
      4      .167                 :         :      *  :         :
      5      .100                 :         :*        :         :
      6      .067                 :     *   :         :         :
      7      .033                 : *       :         :         :
      8      .117                 :         : *       :         :
      9      .300  BAD MATERIAL   :         :         :         D
     10      .150                 :         :      *  :         :
     11      .100                 :         :*        :         :
     12      .367  WORKER ERROR   :         :         :         :         D
     13      .167                 :         :       * :         :
     14      .083                 :      *  :         :         :
     15      .100                 :         :*        :         :
     16      .033                 :  *      :         :         :
     17      .017                 : *       :         :         :
     18      .083                 :      *  :         :         :
     19      .317  MACHINE ERROR  :         :         :         :   D
     20      .183                 :         :         :*        :
     21      .067                 :     *   :         :         :
     22      .083                 :      *  :         :         :
     23      .117                 :         :  *      :         :
     24      .150                 :         :      *  :         :
     25      .067                 :     *   :         :         :

COMMAND --> !RESTART

ENTER TYPE OF CHART ( P , U , C ) --> !U

ENTER NAME OF THE INSPECTION STATION
!INSPECTION STATION 20

DO YOU KNOW THE STANDARD VALUE ( Y OR N ) ... !Y

ENTER SIZE OF AN OBSERVED GROUP   !10

ENTER THE KNOWN AVERAGE DEFECTS PER UNIT   !1.8

PLEASE ENTER COMMAND OR TYPE HELP

COMMAND --> !CHART
```

```
U - CONTROL   CHART
INSPECTION STATION : INSPECTION STATION 20
UPPER CONTROL LIMIT = 3.072792206136
CENTRAL VALUE       = 1.8
LOWER CONTROL LIMIT = .5272077938642

COMMAND --> !RESTART


ENTER TYPE OF CHART ( P , U , C ) --> !U

ENTER NAME OF THE INSPECTION STATION
!INSPECTION STATION 20

DO YOU KNOW THE STANDARD VALUE ( Y OR N ) ... !N

ENTER SIZE OF AN OBSERVED GROUP  !10

ENTER NUMBER OF OBSERVED GROUPS  !25

ENTER ALL OBSERVATIONS
----------------------
GROUP  1   NUMBER OF DEFECTS  =  !21
GROUP  2   NUMBER OF DEFECTS  =  !18
GROUP  3   NUMBER OF DEFECTS  =  !16
GROUP  4   NUMBER OF DEFECTS  =  !22
GROUP  5   NUMBER OF DEFECTS  =  !28
GROUP  6   NUMBER OF DEFECTS  =  !18
GROUP  7   NUMBER OF DEFECTS  =  !12
GROUP  8   NUMBER OF DEFECTS  =  !14
GROUP  9   NUMBER OF DEFECTS  =  !29
GROUP  10  NUMBER OF DEFECTS  =  !9
GROUP  11  NUMBER OF DEFECTS  =  !11
GROUP  12  NUMBER OF DEFECTS  =  !8
GROUP  13  NUMBER OF DEFECTS  =  !25
GROUP  14  NUMBER OF DEFECTS  =  !33
GROUP  15  NUMBER OF DEFECTS  =  !19
GROUP  16  NUMBER OF DEFECTS  =  !12
GROUP  17  NUMBER OF DEFECTS  =  !15
GROUP  18  NUMBER OF DEFECTS  =  !11
GROUP  19  NUMBER OF DEFECTS  =  !16
GROUP  20  NUMBER OF DEFECTS  =  !22
GROUP  21  NUMBER OF DEFECTS  =  !28
GROUP  22  NUMBER OF DEFECTS  =  !16
GROUP  23  NUMBER OF DEFECTS  =  !14
GROUP  24  NUMBER OF DEFECTS  =  !11
GROUP  25  NUMBER OF DEFECTS  =  !10

PLEASE ENTER COMMAND OR TYPE HELP

COMMAND --> !CHART
```

```
T R I A L    U - C O N T R O L    C H A R T
INSPECTION STATION : INSPECTION STATION 20
UPPER CONTROL LIMIT = 3.007706972187
CENTRAL VALUE       = 1.752
LOWER CONTROL LIMIT = .4962930278125

                                      .000       1.000      2.000      3.000      4.000
  OBSVN     MEAN  ASSIGNABLE CAUSE  +---------+---------+---------+---------+
    1       2.100                   :         :         : *       :
    2       1.800                   :         :        *:         :
    3       1.600                   :         :     *   :         :
    4       2.200                   :         :         :  *      :
    5       2.800                   :         :         :       * :
    6       1.800                   :         :        *:         :
    7       1.200                   :         :  *      :         :
    8       1.400                   :         :     *   :         :
    9       2.900                   :         :         :        *:
   10        .900                   :      *  :         :         :
   11       1.100                   :        *:         :         :
   12        .800                   :   *     :         :         :
   13       2.500                   :         :         :   *     :
   14       3.300                   :         :         :         :  *
   15       1.900                   :         :        :*         :
   16       1.200                   :         : *       :         :
   17       1.500                   :         :      *  :         :
   18       1.100                   :        *:         :         :
   19       1.600                   :         :    *    :         :
   20       2.200                   :         :         : *       :
   21       2.800                   :         :         :      *  :
   22       1.600                   :         :   *     :         :
   23       1.400                   :         :  *      :         :
   24       1.100                   :        *:         :         :
   25       1.000                   :       * :         :         :

COMMAND --> !DISCARD

ENTER GROUP OR UNIT NUMBER !14
ENTER REASON FOR DISCARDING (15 CHARS MAX)   !MACHINE ERROR
```

```
COMMAND --> !CHART


R E V I S E D      U - C O N T R O L     C H A R T
INSPECTION STATION : INSPECTION STATION 20
UPPER CONTROL LIMIT = 2.919875754386
CENTRAL VALUE       = 1.6875
LOWER CONTROL LIMIT = .4551242456133

                                  .000      1.000     2.000     3.000     4.000
OBSVN    MEAN ASSIGNABLE CAUSE   +-.--------+---------+---------+---------+
  1     2.100                    :          :     *   :
  2     1.800                    :          :*        :
  3     1.600                    :        *:          :
  4     2.200                    :          :      *  :
  5     2.800                    :          :         :  *:
  6     1.800                    :          :*        :
  7     1.200                    :      *   :         :
  8     1.400                    :        * :         :
  9     2.900                    :          :         :   *
 10      .900                    :   *      :         :
 11     1.100                    :      *   :         :
 12      .800                    : *        :         :
 13     2.500                    :          :         *:
 14     3.300 MACHINE ERROR      :          :         :       D
 15     1.900                    :          :  *      :
 16     1.200                    :      *   :         :
 17     1.500                    :        * :         :
 18     1.100                    :    *     :         :
 19     1.600                    :         *:         :
 20     2.200                    :          :      *  :
 21     2.800                    :          :         :  *:
 22     1.600                    :         *:         :
 23     1.400                    :        * :         :
 24     1.100                    :     *    :         :
 25     1.000                    :       *  :         :

COMMAND --> !EXIT
```

```
10 REM       P, U, C CONTROL CHART FOR ATTRIBUTES
20 REM
30 DIM W(30),X(30),Z(30),R$(30)
40 DIM P$(40),Z$(7)
50 REM       M0 = MAXIMUM NUMBER OF OBSERVED GROUPS
60 M0=30
70 GOSUB 4590
80 PRINT
90 PRINT "P U C CHARTS (CONTROL CHARTS FOR ATTRIBUTES)"
100 PRINT "----------------------------------------"
110 PRINT
120 PRINT "NEED INTRODUCTION (Y OR N) ... ";
130 INPUT Y$
140 IF Y$<>"Y" THEN 460
150 PRINT
160 PRINT "THIS PROGRAM DETERMINES THE UPPER AND LOWER"
170 PRINT "CONTROL LIMITS FOR P, U AND C CONTROL CHARTS"
180 PRINT "WITH KNOWN OR UNKNOWN STANDARD VALUES. IF THE"
190 PRINT "STANDARD VALUES ARE UNKNOWN, THE PROGRAM HAS"
200 PRINT "AN OPTION WHICH ALLOWS THE USER TO DISCARD OR"
210 PRINT "REINSTATE OUT OF CONTROL POINTS."
220 PRINT
230 PRINT "INPUT REQUIREMENTS"
240 PRINT "------------------"
250 PRINT "1. SELECT TYPE OF CHART ( P, U OR C )"
260 PRINT "2. NAME OF THE INSPECTION STATION"
270 PRINT "3. SELECT KNOWN OR UNKNOWN STANDARD VALUE OPTION"
280 PRINT "4. IF THE STANDARD VALUE IS KNOWN :-"
290 PRINT "    FOR P-CHART, -SIZE OF AN OBSERVED GROUP"
300 PRINT "                -FRACTION DEFECTIVE VALUE"
310 PRINT "    FOR C-CHART, -VALUE OF DEFECTS PER SINGLE UNIT"
320 PRINT "    FOR U-CHART, -SIZE OF THE SAMPLE"
330 PRINT "                -AVERAGE DEFECTS PER UNIT"
340 PRINT "5. IF THE STANDARD VALUE IS UNKNOWN :-"
350 PRINT "    FOR P-CHART, -NUMBER OF OBSERVED GROUPS"
360 PRINT "                -SIZE OF AN OBSERVED GROUP"
370 PRINT "                -NUMBER OF DEFECTIVE IN EACH GROUP"
380 PRINT "    FOR C-CHART, -NUMBER OF SINGLE UNIT OBSERVED"
390 PRINT "                -NUMBER OF DEFECTS IN EACH UNIT"
400 PRINT "    FOR U-CHART, -NUMBER OF OBSERVED GROUPS"
410 PRINT "                -SIZE OF AN OBSERVED GROUP"
420 PRINT "                -NUMBER OF DEFECTS IN EACH GROUP"
430 PRINT
440 PRINT "MAXIMUM NUMBER OF OBSERVED GROUPS OR UNITS = ";M0
450 REM *** I N P U T    S E C T I O N ***
460 PRINT
470 PRINT "ENTER TYPE OF CHART ( P , U , C ) --> ";
480 INPUT F$
490 IF F$<>"P" THEN 520
500 M9=1
510 GOTO 610
520 IF F$<>"C" THEN 550
530 M9=2
540 GOTO 610
```

```
550 IF F$<>"U" THEN 580
560 M9=3
570 GOTO 610
580 PRINT
590 PRINT "*** ERROR ***    PLEASE TRY AGAIN ! "
600 GOTO 460
610 PRINT
620 PRINT "ENTER NAME OF THE INSPECTION STATION"
630 INPUT Q$
640 PRINT
650 PRINT "DO YOU KNOW THE STANDARD VALUE ( Y OR N ) ... ";
660 INPUT Y$
670 K9=2
680 IF Y$<>"Y" THEN 700
690 K9=1
700 IF M9<>1 THEN 1040
710 REM P-CHART INPUT
720 PRINT
730 PRINT "ENTER SIZE OF AN OBSERVED GROUP ";
740 INPUT S
750 S = INT(S)
760 IF S<2 THEN 720
770 IF K9=2 THEN 840
780 PRINT
790 PRINT "ENTER THE KNOWN FRACTION DEFECTIVE ";
800 INPUT D1
810 D1 = ABS(D1)
820 GOTO 1570
830 REM STANDARD VALUE UNKNOWN  -PROMPT FOR OBSNS
840 PRINT
850 PRINT "ENTER NUMBER OF OBSERVED GROUP   ";
860 INPUT M
870 GOSUB 3360
880 IF E1=0 THEN 900
890 GOTO 840
900 PRINT
910 PRINT "ENTER ALL OBSERVATIONS"
920 PRINT "----------------------"
930 FOR I = 1 TO M
940 PRINT "GROUP ";I;"  NO.OF DEFECTIVES  = ";
950 INPUT D
960 D=ABS(D)
970 X(I)=D/S
980 IF X(I)<=1 THEN 1020
990 PRINT
1000 PRINT "*** ERROR *** FRACTION DEFECTIVE EXCEEDS ONE !!!"
1010 GOTO 940
1020 NEXT I
1030 GOTO 1570
1040 IF M9<>2 THEN 1280
1050 REM C-CHART INPUT
1060 IF K9=2 THEN 1130
1070 PRINT
1080 PRINT "ENTER THE KNOWN DEFECTS PER SINGLE UNIT   ";
```

```
1090 INPUT D1
1100 D1 = ABS(D1)
1110 GOTO 1570
1120 REM STANDARD VALUE UNKNOWN -PROMPT FOR OBSNS
1130 PRINT
1140 PRINT "ENTER NUMBER OF SINGLE UNITS OBSERVED ";
1150 INPUT M
1160 GOSUB 3360
1170 IF E1=0 THEN 1190
1180 GOTO 1130
1190 PRINT
1200 PRINT "ENTER ALL OBSERVATIONS"
1210 PRINT "---------------------"
1220 FOR I = 1 TO M
1230 PRINT "UNIT   ";I;"  NUMBER OF DEFECTS   = ";
1240 INPUT X(I)
1250 X(I) = ABS(X(I))
1260 NEXT I
1270 GOTO 1570
1280 REM U-CHART INPUT
1290 PRINT
1300 PRINT "ENTER SIZE OF AN OBSERVED GROUP   ";
1310 INPUT S
1320 S = INT(S)
1330 IF S<2 THEN 1290
1340 IF K9=2 THEN 1410
1350 PRINT
1360 PRINT "ENTER THE KNOWN AVERAGE DEFECTS PER UNIT   ";
1370 INPUT D1
1380 D1 = ABS(D1)
1390 GOTO 1570
1400 REM STANDARD VALUE UNKNOWN  -PROMPT FOR OBSNS
1410 PRINT
1420 PRINT "ENTER NUMBER OF OBSERVED GROUPS   ";
1430 INPUT M
1440 GOSUB 3360
1450 IF E1=0 THEN 1470
1460 GOTO 1410
1470 PRINT
1480 PRINT "ENTER ALL OBSERVATIONS"
1490 PRINT "---------------------"
1500 FOR I = 1 TO M
1510 PRINT "GROUP   ";I;"  NUMBER OF DEFECTS   = ";
1520 INPUT D
1530 D = ABS(D)
1540 X(I) = D/S
1550 NEXT I
1560 REM  *** ACCEPT COMMANDS HERE ***
1570 PRINT
1580 PRINT "PLEASE ENTER COMMAND OR TYPE HELP"
1590 PRINT
1600 PRINT "COMMAND --> ";
1610 INPUT C$
1620 PRINT
```

```
1630 IF C$<>"HELP" THEN 1660
1640 GOSUB 1880
1650 GOTO 1590
1660 IF C$<>"CHART" THEN 1690
1670 GOSUB 2650
1680 GOTO 1590
1690 IF C$<>"LIST" THEN 1720
1700 GOSUB 2050
1710 GOTO 1590
1720 IF C$<>"CHANGE" THEN 1750
1730 GOSUB 2240
1740 GOTO 1590
1750 IF C$<>"DISCARD" THEN 1780
1760 GOSUB 2480
1770 GOTO 1590
1780 IF C$<>"REINSTATE" THEN 1810
1790 GOSUB 2480
1800 GOTO 1590
1810 IF C$<>"RESTART" THEN 1840
1820 GOSUB 4670
1830 GOTO 460
1840 IF C$="EXIT" THEN 4720
1850 PRINT
1860 PRINT "*** ERROR *** PLEASE TRY AGAIN !!! "
1870 GOTO 1590
1880 REM HELP COMMAND
1890 PRINT "AVAILABLE COMMANDS"
1900 PRINT "=================="
1910 PRINT "HELP      - PRINT THIS MESSAGE"
1920 PRINT "CHART     - GENERATE P OR U OR C CONTROL CHARTS"
1930 PRINT "LIST      - LIST OBSNS BY GROUP OR UNIT NUMBER"
1940 PRINT "CHANGE    - CHANGE OBSNS IN A GROUP OR UNIT"
1950 PRINT "DISCARD   - DISCARD A GROUP OR UNIT OF OBSNS"
1960 PRINT "            DUE TO ASSIGNABLE CAUSE"
1970 PRINT "REINSTATE - REINSTATE A DISCARDED GROUP OR UNIT"
1980 PRINT "RESTART   - START A NEW PROBLEM"
1990 PRINT "EXIT      - EXIT FROM THE PROBLEM"
2000 PRINT
2010 PRINT "NOTE : THE COMMANDS -LIST,CHANGE,DISCARD"
2020 PRINT "AND REINSTATE ARE INACTIVE WHEN STANDARD"
2030 PRINT "VALUES ARE KNOWN."
2040 RETURN
2050 REM LIST COMMAND
2060 IF K9 = 1 THEN 2230
2070 PRINT "ENTER GROUP OR UNIT NUMBER   ";
2080 INPUT G9
2090 FOR I = 1 TO M
2100 IF G9<>I THEN 2210
2110 IF M9<>1 THEN 2150
2120 PRINT "GROUP ";I;"  NO. OF DEFECTIVES = ";X(I)*S
2130 PRINT "FRACTION DEFECTIVE =  ";X(I)
2140 GOTO 2230
2150 IF M9<>2 THEN 2180
2160 PRINT "UNIT  ";I;"  NUMBER OF DEFECTS = ";X(I)
```

```
2170 GOTO 2230
2180 PRINT "GROUP ";I;"  NUMBER OF DEFECTS = ";X(I)*S
2190 PRINT "AVERAGE DEFECTS PER UNIT   = ";X(I)
2200 GOTO 2230
2210 NEXT I
2220 GOSUB 3330
2230 RETURN
2240 REM CHANGE COMMAND
2250 IF K9 = 1 THEN 2470
2260 PRINT "ENTER GROUP OR UNIT NUMBER  ";
2270 INPUT G9
2280 FOR I = 1 TO M
2290 IF G9<>I THEN 2450
2300 IF M9=2 THEN 2410
2310 PRINT "OLD VALUE = ";X(I)*S;"  ENTER NEW VALUE  ";
2320 INPUT D
2330 D = ABS(D)
2340 X(I)=D/S
2350 IF M<>1 THEN 2400
2360 IF X(I)<=1 THEN 2400
2370 PRINT
2380 PRINT "*** ERROR *** FRACTION DEFECTIVE EXEEDS ONE !!!"
2390 GOTO 2310
2400 GOTO 2470
2410 PRINT "OLD VALUE = ";X(I);"  ENTER NEW VALUE   ";
2420 INPUT X(I)
2430 X(I) = ABS(X(I))
2440 GOTO 2470
2450 NEXT I
2460 GOSUB 3330
2470 RETURN
2480 REM DISCARD/REINSTATE COMMANDS
2490 IF K9 = 1 THEN 2640
2500 PRINT "ENTER GROUP OR UNIT NUMBER ";
2510 INPUT G9
2520 FOR I = 1 TO M
2530 IF G9<>I THEN 2620
2540 IF C$="REINSTATE" THEN 2590
2550 PRINT "ENTER REASON FOR DISCARDING (15 CHARS MAX)   ";
2560 INPUT R$(I)
2570 Z(I) = 1
2580 GOTO 2640
2590 R$(I)=" "
2600 Z(I) = 0
2610 GOTO 2640
2620 NEXT I
2630 GOSUB 3330
2640 RETURN
2650 REM CHART COMMAND
2660 IF M9<>1 THEN 2970
2670 IF K9 = 2 THEN 2780
2680 REM P-CHART KNOWN STD.VALUE
2690 C1 = D1
2700 U1 = D1+(3*(SQR(D1*(1-D1)/S)))
```

```
2710 L1 = D1-(3*(SQR(D1*(1-D1)/S)))
2720 IF U1<1 THEN 2740
2730 U1 = 1
2740 IF L1>0 THEN 2760
2750 L1 = 0
2760 GOSUB 4400
2770 GOTO 3320
2780 REM P-CHART UNKNOWN STD.VALUE
2790 GOSUB 3450
2800 C1 = X1/(K*S)
2810 U1 = C1+(3*(SQR(C1*(1-C1)/S)))
2820 L1 = C1-(3*(SQR(C1*(1-C1)/S)))
2830 IF U1<1 THEN 2850
2840 U1=1
2850 IF L1>0 THEN 2870
2860 L1=0
2870 IF L1>=A0 THEN 2890
2880 A0=L1
2890 IF U1<=A1 THEN 2910
2900 A1=U1
2910 GOSUB 4400
2920 GOSUB 4020
2930 IF L5>0 THEN 2950
2940 GOSUB 3330
2950 GOSUB 3630
2960 GOTO 3320
2970 IF M9<>2 THEN 3150
2980 IF K9=2 THEN 3070
2990 REM C-CHART KNOWN STD. VALUE
3000 C1=D1
3010 U1=D1+3*(SQR(D1))
3020 L1=D1-3*(SQR(D1))
3030 IF L1>0 THEN 3050
3040 L1=0
3050 GOSUB 4400
3060 GOTO 3320
3070 REM C-CHART UNKNOWN STD.VALUE
3080 GOSUB 3450
3090 C1=X1/K
3100 U1=C1+3*(SQR(C1))
3110 L1=C1-3*(SQR(C1))
3120 IF L1>0 THEN 3140
3130 L1=0
3140 GOTO 2870
3150 IF K9=2 THEN 3240
3160 REM U-CHART KNOWN STD.VALUE
3170 C1=D1
3180 U1=D1+3*(SQR(D1/S))
3190 L1=D1-3*(SQR(D1/S))
3200 IF L1>0 THEN 3220
3210 L1=0
3220 GOSUB 4400
3230 GOTO 3320
3240 REM U-CHART UNKNOWN STD.VALUE
```

```
3250 GOSUB 3450
3260 C1=X1/(K*S)
3270 U1=C1+3*(SQR(C1/S))
3280 L1=C1-3*(SQR(C1/S))
3290 IF L1>0 THEN 3310
3300 L1=0
3310 GOTO 2870
3320 RETURN
3330 PRINT
3340 PRINT "*** ERROR ***    COMMAND ABORTED !!!"
3350 RETURN
3360 REM MAX. OBSNS. TEST
3370 E1=0
3380 M = INT(M)
3390 IF M<=1 THEN 3430
3400 IF M<=M0 THEN 3440
3410 PRINT
3420 PRINT "*** ERROR ***   MAX. ALLOWED =   ";M0
3430 E1=1
3440 RETURN
3450 REM SUBROUTINE TO FIND SUM,MAX,MIN
3460 K=0
3470 X1=0
3480 A0=1.E30
3490 A1=-A0
3500 FOR I = 1 TO M
3510 IF X(I)>=A0 THEN 3530
3520 A0=X(I)
3530 IF X(I)<=A1 THEN 3550
3540 A1=X(I)
3550 IF Z(I)<>0 THEN 3610
3560 K=K+1
3570 IF M9=2 THEN 3600
3580 X1=X1+X(I)*S
3590 GOTO 3610
3600 X1=X1+X(I)
3610 NEXT I
3620 RETURN
3630 REM   PLOTTING ROUTINE
3640 REM PRINT AXIS
3650 PRINT
3660 PRINT TAB(26);
3670 FOR J=L0 TO U0 STEP H0
3680 PRINT USING Z$(6),J;
3690 NEXT J
3700 PRINT
3710 PRINT "OBSVN    MEAN ASSIGNABLE CAUSE +";
3720 FOR J=1 TO 4
3730 PRINT "---------+";
3740 NEXT J
3750 PRINT
3760 H=H0/10
3770 J0=INT((L1-L0)/H+0.5)
3780 J1=INT((U1-L0)/H+0.5)
```

```
3790 J2=INT((C1-L0)/H+0.5)
3800 FOR I = 1 TO M
3810 FOR J = 0 TO 40
3820 P$(J)=" "
3830 NEXT J
3840 P$(J0)=":"
3850 P$(J1)=":"
3860 P$(J2)=":"
3870 J9=J1
3880 J3=INT((X(I)-L0)/H+0.5)
3890 V=X(I)
3900 P$(J3)="*"
3910 IF Z(I)=0 THEN 3930
3920 P$(J3)="D"
3930 IF J3<= J9 THEN 3950
3940 J9=J3
3950 PRINT USING Z$(7),I,V,R$(I);
3960 FOR J=0 TO J9
3970 PRINT P$(J);
3980 NEXT J
3990 PRINT
4000 NEXT I
4010 RETURN
4020 REM     S C A L I N G      R O U T I N E
4030 REM     -------------------------------
4040 L5=0
4050 Y0=A0
4060 Y1=A1
4070 A2=(A1-A0)/4
4080 IF A2<=0 THEN 4390
4090 L5=L5+1
4100 A3=LOG(A2)/LOG(10)
4110 REM    ROUND DOWN A3 TO A4
4120 A4=INT(A3)
4130 IF A4<=A3 THEN 4150
4140 A4=A4-1
4150 A5=10**A4
4160 A6=A2/A5
4170 REM ROUND UP A6 TO A7
4180 A7=INT(A6+1)
4190 H0=A5*A7
4200 REM    ROUND DOWN L0
4210 L0=INT(A0/H0)
4220 IF L0<=A0/H0 THEN 4240
4230 L0=L0-1
4240 L0=L0*H0
4250 U0=L0+5*H0
4260 IF U0 >=A1 THEN 4290
4270 A1=A1+A2/2
4280 GOTO 4070
4290 IF L0 <= A0 THEN 4320
4300 A0=A0-A2/2
4310 GOTO 4070
4320 IF U0-H0 <= Y1 THEN 4350
```

```
4330 U0=U0-H0
4340 GOTO 4320
4350 IF L0+H0 >= Y0 THEN 4380
4360 L0=L0+H0
4370 GOTO 4350
4380 H0=(U0-L0)/4
4390 RETURN
4400 REM   PRINT CONTROL LIMITS
4410 PRINT
4420 IF K9=1 THEN 4530
4430 I0=5
4440 FOR I = 1 TO M
4450 IF Z(I)<>0 THEN 4500
4460 V=X(I)
4470 IF V>U1 THEN 4490
4480 IF V>=L1 THEN 4500
4490 I0=4
4500 NEXT I
4510 PRINT Z$(I0);Z$(M9)
4520 GOTO 4540
4530 PRINT Z$(M9)
4540 PRINT "INSPECTION STATION : ";Q$
4550 PRINT "UPPER CONTROL LIMIT = ";U1
4560 PRINT "CENTRAL VALUE       = ";C1
4570 PRINT "LOWER CONTROL LIMIT = ";L1
4580 RETURN
4590 REM    I N I T I A L I Z A T I O N
4600 Z$(1)="P - C O N T R O L    C H A R T"
4610 Z$(2)="C - C O N T R O L    C H A R T"
4620 Z$(3)="U - C O N T R O L    C H A R T"
4630 Z$(4)="T R I A L         "
4640 Z$(5)="R E V I S E D      "
4650 Z$(6)="-#####.###"
4660 Z$(7)="### #####.### ###############   "
4670 FOR I = 1 TO M0
4680 R$(I)="  "
4690 Z(I)=0
4700 NEXT I
4710 RETURN
4720 END
```

II. Product Control

To ensure acceptable quality of manufactured items in modern industry, some procedure is needed to determine whether these items conform to desired quality standards. Either 100 percent inspection or sampling inspection can be used. If feasible, 100 percent inspection is generally considered more reliable. However, there is evidence to suggest that this procedure is not foolproof unless some automatic or mechanical method is used. Sampling inspection requires the inspection of some fraction of the items supplied. The items are inspected more completely and with more precision than is possible under 100 percent inspection. All sampling inspection schemes presuppose management's willingness to accept some defective products, since no sampling system will guarantee 100 percent perfect products.

Lot-by-Lot Single-Sampling Acceptance Plans by Attributes

In general, lot-by-lot single-sampling acceptance by attributes is the classical method of ensuring acceptance quality by sampling in industry. Because it is easy to understand and easy to apply, it has become the most widely used method today. Under this method, a sample is selected from a group of identical items called a lot made under uniform conditions. The sample units are classified as defective or nondefective with respect to some quality characteristic. The lot is accepted or rejected according to how the proportion of defectives in the sample compares with a predetermined ratio. Lot-by-lot single-sampling acceptance plans by attributes are generally of two types -- with or without rectifying inspection. Acceptance sampling plans without rectifying inspection generally protect the consumer from accepting individual bad lots and give incentive to the supplier to produce acceptable quality. The inspection process has no effect on the quality of the product. Rectifying inspection plans, developed by Dodge and Romig [2], call for 100 percent inspection of rejected lots. The rectifying sampling plans chosen guarantee that the average outgoing quality (AOQ) will not exceed a specified value, referred to as the average outgoing quality limit (AOQL). The following are examples of lot-by-lot single-sampling plans by attributes.

a. Lot-by-Lot Rectifying Single-Sampling Plans by Attributes for a Specified AOQL Based on Minimum Average Total Inspection (ATI)

ATI, average total inspection, refers to the number of items actually sampled over a long period of time. It is an expected value from our statistical model.

From every lot of size N a sample of n items is inspected. The lot is accepted if c or fewer of the inspected items are defective. The value c is called the acceptance number. If the lot is rejected, all items are tested and the defects are replaced with items of acceptable quality. We assume that p' is the process average or the probability that an item is defective. This is the value at which we seek minimum ATI for the designated AOQL.

Using the Poisson approximation to the binomial distribution, the probability of acceptaing a lot is P_a, where

$$P_a = \sum_{x=0}^{c} \frac{e^{-np'}(np')^x}{x!}$$

= Prob(number of defectives ⩽ c)

The statistical model implies

$$AOQL = \frac{y}{n}(1 - \frac{n}{N})$$

where $y = P_a \cdot n \cdot p_m'$ and p_m' is the process average that yields the given AOQL as the AOQ. Therefore,

$$n = \frac{y N}{N \cdot AOQL + y} .$$

The values of y are dependent on c and are available in Dodge and Romig [2, Table 2-3]. The planner chooses an arbitrary c value and computes n and P_a as shown above. Since

$$ATI = n + (1-P_a)(N-n) ,$$

the corresponding value of ATI can be computed for each c. By calculating the ATI for consecutive values of c, the planner can find the c and n values that yield the lowest ATI, thus obtaining the desired plan. A comprehensive treatment of the lot-by-lot rectifying single-sampling plans by attributes is given in Dodge and Romig [2].

b. Lot-by-Lot Nonrectifying Single-Sampling Plans by Attributes for Specified Producer's and Consumer's Risks

Since any sampling plan is defined by its operating characteristic curve, we can design our plan by specifying two points on this curve. If (p', α) is a point on the OC curve, then we know that the probability of accepting a lot with a fraction defective of p' will be α. Let our two points be (AQL, 1-α) and (LTFD, β) where α is the producer's risk, β is the consumer's risk, AQL is the acceptable quality level p'_1, and LTFD is the lot tolerance fraction defective p'_2.

The corresponding parameters n and c of the plan can be found by using the appropriate probability model. The sampling distribution of defectives is modeled as binomial if the sampling is from an infinite population -i.e., when we are considering a "stream" of lots. When the lots are isolated and when individual lot quality is important, the appropriate probability model is the hypergeometric distribution. In either case, the Poisson approximation is appropriate.

For a single-sampling plan with points (p'_1, 1-α) and (p'_2, β) on the OC curve, n and c must then satisfy the expressions (1) and (2):

(1) $$1-\alpha = \sum_{x=0}^{c} \binom{n}{x} p'^x_1 (1-p'_1)^{n-x}$$

and

(2) $$\beta = \sum_{x=0}^{c} \binom{n}{x} p'^x_2 (1-p'_2)^{n-x} .$$

If np' is small, we may use the Poisson approximation to the binomial. Hence:

$$1-\alpha = \sum_{x=0}^{c} \frac{e^{-np'_1} (np'_1)^x}{x!}$$

and

$$\beta = \sum_{x=0}^{c} \frac{e^{-np'_2} (np'_2)^x}{x!}$$

The exact solution to these equations cannot usually be obtained because n and c must be integers. An approximate solution to this problem is given by Duncan [3].

Two computer programs dealing with single-sampling plans by attributes will now be presented. The first program, LBL.AOQL, can be used to design a lot-by-lot rectifying single-sampling plan for a specified value of AOQL based on minimum ATI. The second program, AQL.LTFD, can be used for designing lot-by-lot nonrectifying single-sampling plans for specified Producer's and Consumer's Risks of 0.05 and 0.10, respectively.

The Program LBL.AOQL

LBL.AOQL can be used to develop a lot-by-lot rectifying single-sampling plan by attributes for a specified value of average outgoing quality limit (AOQL) based on minimum average total inspection (ATI).

INPUT
-----
1. Name of the inspection station
2. Lot size (N)
3. Average outgoing quality limit (AOQL)
4. Process average fraction defective (p')
5. Plotting increment of the fraction defective when the commands PLOTOC and PLOTAOQ are activated

OUTPUT

After the above inputs (1 through 4) are entered, the optimal sampling plan can be obtained using the PLAN command. The commands PLOTOC and PLOTAOQ are used to plot the OC curve, and the AOQ curve for the plan, respectively.

EXAMPLE

Design a lot-by-lot rectifying single-sampling plan by attributes for the final inspection of a product line that will have an AOQL of 0.015 and minimizes ATI when the process average fraction defective is 0.025. The lot size is 1000.

The example terminal session on the next page shows that the optimal plan is to randomly sample 175 items from the lot and accept the lot if the number of defective items is less than or equal to 5, otherwise the lot is rejected. Because of the tight requirement on the average outgoing quality limit, the average total number of items inspected is 400 items according to this plan.

The OC curve and the AOQ curve for this plan are next generated via the commands PLOTOC and PLOTAOQ, respectively, using a plotting increment of 0.005.

```
LBL.AOQL EXAMPLE TERMINAL SESSION
=================================

LOT-BY-LOT RECTIFYING SINGLE-SAMPLING PLANS BY ATTRIBUTES
FOR A SPECIFIED A O Q L BASED ON AVERAGE TOTAL INSPECTION
---------------------------------------------------------

NEED INTRODUCTION (Y OR N) ... !N

ENTER NAME OF INSPECTION STATION    !FINAL INSPECTION

ENTER LOT SIZE ... !1000

ENTER   A  O  Q  L  ... !.015

ENTER PROCESS AVERAGE FRACTION DEFECTIVE   !.025

PLEASE ENTER COMMAND OR TYPE HELP

COMMAND --> !HELP

AVAILABLE COMMANDS
==================
HELP       - PRINT THIS MESSAGE
PLAN       - GENERATE THE SAMPLING PLAN
PLOTOC     - PLOT THE OC CURVE
PLOTAOQ    - PLOT THE AOQ CURVE
RESTART    - START A NEW PROBLEM
EXIT       - EXIT FROM THE PROGRAM

NOTE : FOR PLOTTING OC AND AOQ CURVES THE PROGRAM
WILL INFORM THE USER OF THE APPLICABLE RANGE OF
FRACTION DEFECTIVE.  THEREFORE THE USER CAN SPECIFY
THE APPROPRIATE PLOTTING INCREMENT.

COMMAND --> !PLAN

LOT-BY-LOT RECTIFYING SINGLE-SAMPLING PLAN BY ATTRIBUTES
--------------------------------------------------------

SAMPLING AT INSPECTION STATION  : FINAL INSPECTION
GIVEN :
LOT SIZE = 1000
A O Q L  = .015
PROCESS AVERAGE FRACTION DEFECTIVE = .025

THE OPTIMAL PLAN IS :
  - SAMPLE SIZE                175
  - ACCEPTANCE NUMBER            5
  - AVERAGE TOTAL INSPECTION   400

---------------------------------------------------------
```

COMMAND -->!PLOTOC

P L O T T I N G   R A N G E   IS   0 < P < .08857142857143
ENTER PLOTTING INCREMENT OF FRACTION DEFECTIVE P !.005

****
THE OPERATING CHARACTERISTIC CURVE
****

SAMPLING AT INSPECTION STATION : FINAL INSPECTION
GIVEN :
LOT SIZE = 1000
A O Q L = .015
PROCESS AVERAGE FRACTION DEFECTIVE = .025

THE OPTIMAL PLAN IS :
 - SAMPLE SIZE   175
 - ACCEPTANCE NUMBER   5
 - AVERAGE TOTAL INSPECTION 400

```
                                  PROB(ACCEPTANCE)
                         .0000    .2500    .5000    .7500   1.0000
PROC.DEF.  PROB(ACCEPT)  +--------+--------+--------+--------+
  .000       1.0000      I                                   *
  .005        .9997      I                                   *
  .010        .9909      I                                   *
  .015        .9491      I                                 *
  .020        .8576      I                            *
  .025        .7241      I                      *
  .030        .5722      I                *
  .035        .4258      I          *
  .040        .3007      I      *
  .045        .2030      I   *
  .050        .1317      I *
  .055        .0827      I *
  .060        .0504      I *
  .065        .0299      I*
  .070        .0174      I*
  .075        .0099      *
  .080        .0055      *
  .085        .0030      *
```

COMMAND -->!PLOTAOQ

P L O T T I N G   R A N G E   IS   0 < P < .08857142857143
ENTER PLOTTING INCREMENT OF FRACTION DEFECTIVE P !.005

```
***************************************
*THE AVERAGE OUTGOING QUALITY CURVE*
***************************************

SAMPLING AT INSPECTION STATION  : FINAL INSPECTION
GIVEN :
LOT SIZE = 1000
A O Q L  = .015
PROCESS AVERAGE FRACTION DEFECTIVE = .025

THE OPTIMAL PLAN IS :
 - SAMPLE SIZE              175
 - ACCEPTANCE NUMBER          5
 - AVERAGE TOTAL INSPECTION 400

                                   AVERAGE OUTGOING QUALITY
                           .0000      .0037     .0075     .0112      .0150
PROC.DEF.    A   O   Q     +---------+---------+---------+---------+
 .000       .0000          *                                        :
 .005       .0041       I       *                                   :
 .010       .0082       I                 *                         :
 .015       .0117       I                          *                :
 .020       .0142       I                                    *      :
 .025       .0149       I                                        *
 .030       .0142       I                                    *      :
 .035       .0123       I                               *           :
 .040       .0099       I                      *                    :
 .045       .0075       I                 *                         :
 .050       .0054       I            *                              :
 .055       .0038       I       *                                   :
 .060       .0025       I   *                                       :
 .065       .0016       I *                                         :
 .070       .0010       I *                                         :
 .075       .0006       I*                                          :
 .080       .0004       I*                                          :
 .085       .0002       I*                                          :

COMMAND -->!EXIT
```

365

```
10 REM     LBL.AOQL
20 REM
30 DIM P$(40),Y(40),F$(8)
40 GOSUB 2290
50 PRINT
60 PRINT "LOT-BY-LOT RECTIFYING SINGLE-SAMPLING PLANS BY ATTRIBUTES"
70 PRINT "FOR A SPECIFIED A O Q L BASED ON AVERAGE TOTAL INSPECTION"
80 PRINT "----------------------------------------------------------"
90 PRINT
100 PRINT "NEED INTRODUCTION (Y OR N) ... ";
110 INPUT Y$
120 IF Y$<>"Y" THEN 300
130 PRINT
140 PRINT "THIS PROGRAM DETERMINES LOT-BY-LOT RECTIFYING"
150 PRINT "SAMPLING PLAN BY ATTRIBUTES GIVEN THE AVERAGE"
160 PRINT "OUTGOING QUALITY LIMIT (A O Q L) WHICH HAS"
170 PRINT "THE MINIMUM AVERAGE TOTAL INSPECTION (A T I)."
180 PRINT "THE PLAN WILL CONSIST OF THE OPTIMAL ACCEPTANCE"
190 PRINT "NUMBER, SAMPLE SIZE AND THE ATI ACHIEVED FROM"
200 PRINT "THE PLAN."
210 PRINT
220 PRINT "INPUT REQUIREMENTS"
230 PRINT "------------------"
240 PRINT " 1. NAME OF INSPECTION STATION"
250 PRINT " 2. LOT SIZE"
260 PRINT " 3. A O Q L VALUE"
270 PRINT " 4. THE PROCESS AVERAGE FRACTION DEFECTIVE"
280 PRINT
290 REM  *** I N P U T   S E C T I O N ***
300 PRINT
310 PRINT "ENTER NAME OF INSPECTION STATION    ";
320 INPUT S$
330 PRINT
340 PRINT "ENTER LOT SIZE ... ";
350 INPUT N
360 N=ABS(N)
370 PRINT
380 PRINT "ENTER    A  O  Q  L  ... ";
390 INPUT L
400 L=ABS(L)
410 PRINT
420 PRINT "ENTER PROCESS AVERAGE FRACTION DEFECTIVE   ";
430 INPUT P
440 P=ABS(P)
450 IF P<1 THEN 490
460 PRINT
470 PRINT "*** ERROR *** PLEASE TRY AGAIN !!!"
480 GOTO 410
490 REM ** C A L C U L A T I O N    S E C T I O N **
500 A0=1E+30
510 FOR I = 0 TO 40
520 N1=(Y(I)*N)/(N*L+Y(I))
530 P3 = N1*P
540 P2 = EXP(-P3)
```

```
550 P1=P2
560 IF I=0 THEN 610
570 FOR J = 1 TO I
580 P2=P2*P3/J
590 P1=P1+P2
600 NEXT J
610 A1=N1+(1-P1)*(N-N1)
620 IF A1>A0 THEN 660
630 A0=A1
640 Z1=I
650 Z3=INT(A1+1)
660 NEXT I
670 Z2=INT((Y(Z1)*N)/(N*L+Y(Z1))+1)
680 REM *** ACCEPT COMMANDS HERE ***
690 PRINT
700 PRINT "PLEASE ENTER COMMAND OR TYPE HELP"
710 PRINT
720 PRINT "COMMAND -->";
730 INPUT C$
740 PRINT
750 IF C$ <> "HELP" THEN 780
760 GOSUB 930
770 GOTO 710
780 IF C$ <> "PLAN" THEN 810
790 GOSUB 1080
800 GOTO 710
810 IF C$ <> "PLOTOC" THEN 840
820 GOSUB 1160
830 GOTO 710
840 IF C$ <> "PLOTAOQ" THEN 870
850 GOSUB 1160
860 GOTO 710
870 IF C$ <> "RESTART" THEN 890
880 GOTO 290
890 IF C$="EXIT" THEN 2480
900 PRINT
910 PRINT " *** ERROR *** PLEASE TRY AGAIN ! "
920 GOTO 690
930 REM     *** HELP   C O M M A N D ***
940 PRINT "AVAILABLE COMMANDS"
950 PRINT "=================="
960 PRINT "HELP      - PRINT THIS MESSAGE"
970 PRINT "PLAN      - GENERATE THE SAMPLING PLAN"
980 PRINT "PLOTOC    - PLOT THE OC CURVE"
990 PRINT "PLOTAOQ   - PLOT THE AOQ CURVE"
1000 PRINT "RESTART   - START A NEW PROBLEM"
1010 PRINT "EXIT      - EXIT FROM THE PROGRAM"
1020 PRINT
1030 PRINT "NOTE : FOR PLOTTING OC AND AOQ CURVES THE PROGRAM"
1040 PRINT "WILL INFORM THE USER OF THE APPLICABLE RANGE OF"
1050 PRINT "FRACTION DEFECTIVE.  THEREFORE THE USER CAN SPECIFY"
1060 PRINT "THE APPROPRIATE PLOTTING INCREMENT."
1070 RETURN
1080 REM *** PLAN   C O M M A N D ***
```

```
1090 PRINT
1100 PRINT "LOT-BY-LOT RECTIFYING SINGLE-SAMPLING PLAN BY ATTRIBUTES"
1110 PRINT "--------------------------------------------------------"
1120 GOSUB 1920
1130 PRINT "--------------------------------------------------------"
1140 RETURN
1150 REM      ***  PLOTOC/PLOTAOQ    C O M M A N D S  ***
1160 K9=2
1170 IF C$ <> "PLOTOC" THEN 1190
1180 K9=1
1190 U1=1
1200 L1=Z1/Z2
1210 I0=.03
1220 GOSUB 2060
1230 U1 = T
1240 I0=(U1-L1)/2
1250 GOSUB 2060
1260 IF P1 < U1 THEN 1280
1270 U1 = T
1280 L1=0
1290 PRINT "P L O T T I N G     R A N G E     IS    0 < P < ";U1
1300 PRINT "ENTER PLOTTING INCREMENT OF FRACTION DEFECTIVE   P ";
1310 INPUT I0
1320 I0=ABS(I0)
1330 IF I0 < U1 THEN 1360
1340 PRINT "*** INCREMENT NOT WITHIN PLOTTING RANGE *** TRY AGAIN !"
1350 GOTO 1290
1360 PRINT
1370 PRINT
1380 PRINT
1390 PRINT "**********************************"
1400 PRINT "*";F$(K9);"*"
1410 PRINT "**********************************"
1420 GOSUB 1920
1430 PRINT
1440 PRINT TAB(36);F$(2+K9)
1450 L0=0
1460 U0=L
1470 IF K9=2 THEN 1490
1480 U0=1
1490 H0=U0/4
1500 GOSUB 1520
1510 RETURN
1520 REM     PLOTTING ROUTINE
1530 REM     PRINT AXIS
1540 PRINT TAB(20);
1550 FOR J=L0 TO U0 STEP H0
1560 PRINT USING F$(7),J;
1570 NEXT J
1580 PRINT
1590 PRINT F$(4+K9);
1600 FOR J= 1 TO 4
1610 PRINT "---------+";
1620 NEXT J
```

```
1630 PRINT
1640 J1 = 40
1650 FOR J = 1 TO J1
1660 P$(J)=" "
1670 NEXT J
1680 H=H0/10
1690 FOR I = L1 TO U1 STEP I0
1700 GOSUB 2170
1710 IF J1=0 THEN 1730
1720 P$(J1) =" "
1730 P$(0)= "I"
1740 IF K9=2 THEN 1790
1750 J1 = INT((P1-L0)/H+.5)
1760 V =P1
1770 J9=J1
1780 GOTO 1830
1790 J1 =INT((A1-L0)/H+.5)
1800 J9=INT((L-L0)/H+.5)
1810 V=A1
1820 P$(J9)=":"
1830 P$(J1)="*"
1840 PRINT USING F$(8),I,V;
1850 FOR J=0 TO J9
1860 PRINT P$(J);
1870 NEXT J
1880 PRINT
1890 NEXT I
1900 PRINT
1910 RETURN
1920 REM     ROUTINE TO PRINT OPTIMAL SAMPLING PLAN
1930 PRINT
1940 PRINT "SAMPLING AT INSPECTION STATION  : ";S$
1950 PRINT "GIVEN :"
1960 PRINT "LOT SIZE = ";N
1970 PRINT "A O Q L  = ";L
1980 PRINT "PROCESS AVERAGE FRACTION DEFECTIVE = ";P
1990 PRINT
2000 PRINT "THE OPTIMAL PLAN IS :"
2010 PRINT "  - SAMPLE SIZE              ";Z2
2020 PRINT "  - ACCEPTANCE NUMBER        ";Z1
2030 PRINT "  - AVERAGE TOTAL INSPECTION ";Z3
2040 PRINT
2050 RETURN
2060 REM     ROUTINE TO SEARCH FOR RANGE OF OC CURVE
2070 FOR I = L1 TO U1 STEP I0
2080 T = I
2090 GOSUB 2170
2100 IF K9=2 THEN 2130
2110 IF P1 < .0125 THEN 2160
2120 GOTO 2150
2130 IF I < (Z1/Z2) THEN 2150
2140 IF A1 < (L/80) THEN 2160
2150 NEXT I
2160 RETURN
```

```
2170 REM     ROUTINE TO FIND PROB/AOQ
2180 P3=I*Z2
2190 P2=EXP(-P3)
2200 P1=P2
2210 IF Z1=0 THEN 2270
2220 IF I=0 THEN 2270
2230 FOR J=1 TO Z1
2240 P2=P2*P3/J
2250 P1=P1+P2
2260 NEXT J
2270 A1=((N-Z2)*I*P1)/N
2280 RETURN
2290 REM     I N I T I A L I Z A T I O N
2300 F$(1)="THE OPERATING CHARACTERISTIC CURVE"
2310 F$(2)="THE AVERAGE OUTGOING QUALITY CURVE"
2320 F$(3)="    PROB(ACCEPTANCE)"
2330 F$(4)="AVERAGE OUTGOING QUALITY"
2340 F$(5)="PROC.DEF.   PROB(ACCEPT) +"
2350 F$(6)="PROC.DEF.    A   O   Q    +"
2360 F$(7)="#####.####"
2370 F$(8)="###.###       ###.####    "
2380 DATA .3679,.84,1.371,1.942,2.544,3.168,3.812
2390 DATA 4.472,5.146,5.831,6.528,7.223,7.948,8.67
2400 DATA 9.398,10.13,10.88,11.62,12.37,13.13,13.89
2410 DATA 14.66,15.43,16.2,16.98,17.76,18.54,19.33
2420 DATA 20.12,20.91,21.7,22.5,23.3,24.1,24.9
2430 DATA 25.71,26.52,27.33,28.14,28.96,29.77
2440 FOR I = 0 TO 40
2450 READ Y(I)
2460 NEXT I
2470 RETURN
2480 END
```

Source for Values in Statements 2380-2430: Harold F. Dodge and Harry G. Romig, SAMPLING INSPECTION TABLES - SINGLE AND DOUBLE SAMPLING, 2nd ed. (New York: John Wiley, 1959), p.39. Reprinted by permission, Bell Telephone Laboratories, Inc.

The Program AQL.LTFD

AQL.LTFD can be used to develop a lot-by-lot nonrectifying single-sampling plan by attributes for the producer's risk (ALPHA) of 0.05 and the consumer's risk (BETA) of 0.10 as described in Duncan [3, p.167]. As mentioned earlier, the exact solution to this problem cannot be obtained due to the fact that the sample size (n) and the acceptance number (c) must be integers. Therefore, up to four possible sampling plans with closest approximations to ALPHA of 0.05 and BETA of 0.10 are produced by the program. The selection of one of these plans should be made as a result of the trade-offs made among the risk levels involved.

INPUT
-----
1. The name of the inspection station
2. The lot size
3. Acceptable Quality Limit (AQL)
4. Lot Tolerance Fraction Defective (LTFD)

Note that the lot size is not used in the calculation, but is inputted to show its size relative to the sample size.

OUTPUT

There are five output generating commands in this program. The command PLANS produces a table showing four possible sampling plans. These plans are usually unique except when the ratio of AQL to LTFD exceeds 45.1 or when the ratio is less than 2.12. Under these conditions the program will generate two sets of identical plans. The commands OC1, OC2, OC3, and OC4 generate the OC curves for plans 1 through 4, respectively. The program also requests the plotting increment of the fraction defective when these commands are entered.

EXAMPLE

Consider the design of such a sampling plan at inspection station number 12. Assume that the lot size is 1000 items and AQL is 0.009 and LTFD is 0.078. The example terminal session on the next page shows that after the data are entered, the HELP command is given to display all available commands. Next the PLANS command is entered and the program generates four possible sampling plans. Note that plans 1 and 3 yield ALPHA values close to 0.05, whereas plans 2 and 4 yield BETA values close to 0.10. The OC3 command is entered next to produce the OC curve for plan 3 (n=91, c=2). The selected plotting increment of the fraction defective is 0.005.

371

```
AQL.LTFD EXAMPLE TERMINAL SESSION
=================================

LOT-BY-LOT NONRECTIFYING SINGLE-SAMPLING PLANS BY ATTRIBUTES
FOR PRODUCER'S RISK OF 0.05 AND CONSUMER'S RISK OF 0.10
------------------------------------------------------------

NEED INTRODUCTION (Y OR N) ... !N

ENTER NAME OF INSPECTION STATION   !STATION 12

ENTER LOT SIZE ...  !1000

ENTER ACCEPTABLE QUALITY LEVEL (AQL) ...  !.009

ENTER LOT TOLERANCE FRACTION DEFECTIVE (LTFD) ...  !.078

PLEASE ENTER COMMAND OR TYPE HELP

COMMAND -->!HELP


AVAILABLE COMMANDS
==================
HELP    - PRINT THIS MESSAGE
PLANS   - GENERATE FOUR POSSIBLE SAMPLING PLANS
OC1     - GENERATE OC CURVE FOR PLAN 1
OC2     - GENERATE OC CURVE FOR PLAN 2
OC3     - GENERATE OC CURVE FOR PLAN 3
OC4     - GENERATE OC CURVE FOR PLAN 4
RESTART - START A NEW PROBLEM
EXIT    - EXIT FROM THE PROGRAM

COMMAND -->!PLANS


*******************************************
*  NONRECTIFYING SINGLE-SAMPLING PLANS   *
*******************************************

I N S P E C T I O N   S T A T I O N : STATION 12
         LOT SIZE  = 1000
         A Q L     = .009
         LTFD      = .078

-----------------------------------------------------
 PLAN NO.      PLAN           ALPHA        BETA
-----------------------------------------------------
    1      N=  40   C=  1      .051        .182
    2      N=  50   C=  1      .075        .099
    3      N=  91   C=  2      .050        .028
    4      N=  69   C=  2      .025        .096
-----------------------------------------------------
```

```
COMMAND -->!OC3

PLOTTING RANGE OF OC CURVE IS   0<P< .108
ENTER PLOTTING INCREMENT OF FRAC. DEFECT. P   !.005


OPERATING CHARACTERISTIC CURVE
------------------------------

I N S P E C T I O N   S T A T I O N : STATION 12
      LOT SIZE = 1000
      A Q L    = .009
      LTFD     = .078

          PLAN NO.        PLAN            ALPHA           BETA
             3      N=  91  C=  2         .050            .028

                                      PROB(ACCEPTANCE)
                              .0000    .2500    .5000    .7500   1.0000
  FRAC.DEF.   PROB(ACCEPT)  +---------+---------+---------+---------+
    .000        1.0000      I                                        *
    .005         .9888      I                                        *
    .010         .9355      I                                     *
    .015         .8419      I                                  *
    .020         .7253      I                            *
    .025         .6027      I                       *
    .030         .4863      I                  *
    .035         .3830      I              *
    .040         .2957      I          *
    .045         .2245      I       *
    .050         .1680      I     *
    .055         .1242      I   *
    .060         .0909      I  *
    .065         .0659      I *
    .070         .0474      I *
    .075         .0338      I*
    .080         .0240      I*
    .085         .0169      I*
    .090         .0119      *


COMMAND -->!EXIT
```

```
10  REM     AQL.LTFD
20  REM
30  DIM K(15,2),R(15),N(4),C(4),A(8),P(2),P$(40)
40  GOSUB 2350
50  PRINT
60  PRINT "LOT-BY-LOT NONRECTIFYING SINGLE-SAMPLING PLANS BY ATTRIBUTES"
70  PRINT "FOR PRODUCER'S RISK OF 0.05 AND CONSUMER'S RISK OF 0.10"
80  PRINT "-------------------------------------------------------------"
90  PRINT
100 PRINT "NEED INTRODUCTION (Y OR N) ... ";
110 INPUT Y$
120 IF Y$<>"Y" THEN 280
130 PRINT
140 PRINT "THIS PROGRAM DETERMINES LOT-BY-LOT NONRECTIFYING"
150 PRINT "SINGLE-SAMPLING PLANS BY ATTRIBUTES GIVEN THE VALUE"
160 PRINT "OF PRODUCER'S RISK (ALPHA) OF APPROXIMATELY 0.05"
170 PRINT "AND CONSUMER'S RISK (BETA) OF APPROXIMATELY 0.10."
180 PRINT
190 PRINT "INPUT REQUIREMENTS"
200 PRINT "------------------"
210 PRINT "1. NAME OF INSPECTION STATION"
220 PRINT "2. LOT SIZE"
230 PRINT "3. ACCEPTANCE QUALITY LIMIT (AQL)"
240 PRINT "4. LOT TOLERANCE FRACTION DEFECTIVE (LTFD)"
250 PRINT "5. INCREMENTAL VALUE OF PROCESS AVERAGE"
260 PRINT "     FOR PLOTTING OC CURVE"
270 PRINT
280 REM  I N P U T    S E C T I O N
290 PRINT
300 PRINT "ENTER NAME OF INSPECTION STATION  ";
310 INPUT S$
320 PRINT
330 PRINT "ENTER LOT SIZE ... ";
340 INPUT N0
350 N0=ABS(N0)
360 PRINT
370 PRINT "ENTER ACCEPTABLE QUALITY LEVEL (AQL) ... ";
380 INPUT P(1)
390 P(1)=ABS(P(1))
400 IF P(1) < 1 THEN 430
410 GOSUB 2520
420 GOTO 360
430 PRINT
440 PRINT "ENTER LOT TOLERANCE FRACTION DEFECTIVE (LTFD) ... ";
450 INPUT P(2)
460 P(2)=ABS(P(2))
470 IF P(2) > P(1) THEN 500
480 GOSUB 2520
490 GOTO 430
500 REM *** FIND POSSIBLE SAMPLING PLANS ***
510 M=0
520 Y=P(2)/P(1)
530 IF Y >= R(0) THEN 660
540 IF Y < R(15) THEN 680
```

```
550 FOR I=1 TO 15
560 IF Y < R(I) THEN 650
570 FOR J=I-1 TO I
580 FOR L=1 TO 2
590 M=M+1
600 C(M)=J
610 N(M)=INT(K(J,L)/P(L)+0.999)
620 NEXT L
630 NEXT J
640 GOTO 760
650 NEXT I
660 M1=0
670 GOTO 690
680 M1=15
690 FOR J=1 TO 2
700 FOR L=1 TO 2
710 M=M+1
720 C(M)=M1
730 N(M)=INT(K(M1,L)/P(L)+0.999)
740 NEXT L
750 NEXT J
760 REM *** FIND ALPHA AND BETA FOR ALL PLANS ***
770 FOR M=1 TO 4
780 FOR L=1 TO 2
790 V1=C(M)
800 I=P(L)
810 V2=N(M)
820 GOSUB 2250
830 L1=(L-1)*4+M
840 A(L1)=P7
850 IF L=2 THEN 880
860 A(L1)=1-A(L1)
870 NEXT L
880 NEXT M
890 REM *** ACCEPT COMMANDS HERE ***
900 PRINT
910 PRINT "PLEASE ENTER COMMAND OR TYPE HELP"
920 PRINT
930 PRINT "COMMAND -->";
940 INPUT C$
950 PRINT
960 IF C$<>"HELP" THEN 990
970 GOSUB 1190
980 GOTO 920
990 IF C$ <> "PLANS" THEN 1020
1000 GOSUB 1320
1010 GOTO 920
1020 IF C$ <> "OC1" THEN 1050
1030 M=1
1040 GOTO 1130
1050 IF C$ <> "OC2" THEN 1080
1060 M=2
1070 GOTO 1130
1080 IF C$ <> "OC3" THEN 1110
```

```
1090 M=3
1100 GOTO 1130
1110 IF C$ <> "OC4" THEN 1150
1120 M=4
1130 GOSUB 1560
1140 GOTO 920
1150 IF C$ = "RESTART" THEN 280
1160 IF C$ = "EXIT" THEN 2540
1170 GOSUB 2520
1180 GOTO 920
1190 REM     *** HELP COMMAND ***
1200 PRINT
1210 PRINT "AVAILABLE COMMANDS"
1220 PRINT "=================="
1230 PRINT "HELP     - PRINT THIS MESSAGE"
1240 PRINT "PLANS    - GENERATE FOUR POSSIBLE SAMPLING PLANS"
1250 PRINT "OC1      - GENERATE OC CURVE FOR PLAN 1"
1260 PRINT "OC2      - GENERATE OC CURVE FOR PLAN 2"
1270 PRINT "OC3      - GENERATE OC CURVE FOR PLAN 3"
1280 PRINT "OC4      - GENERATE OC CURVE FOR PLAN 4"
1290 PRINT "RESTART  - START A NEW PROBLEM"
1300 PRINT "EXIT     - EXIT FROM THE PROGRAM"
1310 RETURN
1320 REM     *** PLANS COMMAND ***
1330 PRINT
1340 PRINT "*****************************************"
1350 PRINT "*   NONRECTIFYING SINGLE-SAMPLING PLANS    *"
1360 PRINT "*****************************************"
1370 PRINT
1380 GOSUB 1500
1390 PRINT
1400 PRINT F$(1)
1410 PRINT F$(2)
1420 PRINT F$(1)
1430 FOR M=1 TO 4
1440 PRINT USING F$(3),M,N(M),C(M),A(M),A(M+4)
1450 NEXT M
1460 PRINT F$(1)
1470 PRINT
1480 PRINT
1490 RETURN
1500 REM     *** ROUTINE TO PRINT DATA ***
1510 PRINT "I N S P E C T I O N    S T A T I O N : ";S$
1520 PRINT TAB(10);"LOT SIZE  = ";N0
1530 PRINT TAB(10);"A Q L     = ";P(1)
1540 PRINT TAB(10);"LTFD      = ";P(2)
1550 RETURN
1560 REM     *** OC1,OC2,OC3,OC4 COMMANDS ***
1570 S9=0.03
1580 V1=C(M)
1590 V2=N(M)
1600 V3=P(2)
1610 V3=V3+S9
1620 I=V3
```

```
1630 GOSUB 2250
1640 IF P7 <= 0.0125 THEN 1660
1650 GOTO 1610
1660 PRINT "PLOTTING RANGE OF OC CURVE IS  0<P< ";V3
1670 PRINT "ENTER PLOTTING INCREMENT OF FRAC. DEFECT. P  ";
1680 INPUT D
1690 D=ABS(D)
1700 IF D <= V3 THEN 1740
1710 GOSUB 2520
1720 PRINT
1730 GOTO 1660
1740 FOR J=1 TO 3
1750 PRINT
1760 NEXT J
1770 PRINT "OPERATING CHARACTERISTIC CURVE"
1780 PRINT "-----------------------------"
1790 PRINT
1800 GOSUB 1500
1810 PRINT
1820 PRINT TAB(9);F$(2)
1830 PRINT TAB(9);
1840 PRINT USING F$(3),M,N(M),C(M),A(M),A(M+4)
1850 PRINT
1860 PRINT
1870 PRINT TAB(36);"PROB(ACCEPTANCE)"
1880 L0=0
1890 L9=1
1900 H0 = L9/4
1910 REM PRINT AXIS
1920 PRINT TAB(20);
1930 FOR J = L0 TO L9 STEP H0
1940 PRINT USING F$(4),J;
1950 NEXT J
1960 PRINT
1970 PRINT F$(5);
1980 FOR J = 1 TO 4
1990 PRINT "---------+";
2000 NEXT J
2010 PRINT
2020 J1=40
2030 P$(0)="I"
2040 FOR J=1 TO J1
2050 P$(J)=" "
2060 NEXT J
2070 H = H0/10
2080 FOR I=0 TO V3 STEP D
2090 P$(J1)=" "
2100 IF I <> 0 THEN 2130
2110 P7=1
2120 GOTO 2140
2130 GOSUB 2250
2140 J1= INT((P7-L0)/H+0.5)
2150 P$(J1)="*"
2160 PRINT USING F$(6),I,P7;
```

```
2170 FOR J = 0 TO J1
2180 PRINT P$(J);
2190 NEXT J
2200 PRINT
2210 IF P7 < 0.0125 THEN 2230
2220 NEXT I
2230 PRINT
2240 RETURN
2250 REM    ROUTINE TO FIND PROB(ACCEPT)
2260 P9=V2*I
2270 P8=EXP(-P9)
2280 P7=P8
2290 IF V1<1 THEN 2340
2300 FOR J = 1 TO V1
2310 P8=P8*P9/J
2320 P7=P7+P8
2330 NEXT J
2340 RETURN
2350 REM    I N I T I A L I Z A T I O N
2360 F$(1)="-------------------------------------------------"
2370 F$(2)=" PLAN NO.      PLAN         ALPHA        BETA"
2380 F$(3)="    ##      N=####  C=###     #.###       #.###"
2390 F$(4)="#####.####"
2400 F$(5)="FRAC.DEF.   PROB(ACCEPT) +"
2410 F$(6)="###.###      ###.####    "
2420 DATA 0.051,2.3,45.1,0.355,3.89,10.96,0.818,5.32,6.5
2430 DATA 1.366,6.68,4.89,1.97,7.99,4.06,2.613,9.28,3.55
2440 DATA 3.285,10.53,3.21,3.981,11.77,2.96,4.695,12.99
2450 DATA 2.77,5.425,14.21,2.62,6.169,15.41,2.5,6.924
2460 DATA 16.6,2.4,7.69,17.78,2.31,8.464,18.96,2.24
2470 DATA 9.246,20.13,2.18,10.04,21.29,2.12
2480 FOR I = 0 TO 15
2490 READ K(I,1),K(I,2),R(I)
2500 NEXT I
2510 RETURN
2520 PRINT "*** ERROR ***   PLEASE TRY AGAIN !!!"
2530 RETURN
2540 END
```

Source for Values in Statements 2420 to 2470: F.E. Grubbs, "On Designing Single Sampling Inspection Plans," ANNALS OF MATHEMATICAL STATISTICS, XX (1949), 256.

Lot-by-Lot Single-Sampling Acceptance Plans by Variables

A variable is a measure of some quality characteristic of the product, such as the diameter of an opening or the percentage concentration of a chemical. Variables sampling plans are based on sample statistics of the probability model of the measured variable. The normal distribution is the model usually employed. The decision to be made in this instance is whether we should accept or reject a lot based on the measurements obtained from a sample of size n.

Let X be the normal random variable representing the measure of quality characteristic, $\bar{X}'$ be the actual mean of X, and σ' be the known standard deviation of X. Also let L and U be the lower and upper specification limits, respectively. Then, for either single-limit plans (lower or upper) or double-limit plans (lower and upper), two parameters n and k are required to specify the sampling plan.

When the standard deviation is known, if a lot is to be accepted, the sample mean must be greater than or equal to $L+k\sigma'$ for the lower-limit case; or it must be less than or equal to $U-k\sigma'$ for the upper-limit case; or it must lie within $L+k\sigma'$ and $U-k\sigma'$ for the double-limit case. When the standard deviation is unknown and for single-limit cases, we replace the known standard deviation with the sample standard deviation in the above statements. However, the derivations of n and k values for each case differ. We will show the derivation of n and k for the simplest case of lower limit with known standard deviation. Other cases may be found in Duncan [3].

Since the OC curve of any sampling plan indicates the degree of risk as a function of the possible fraction defective (p') in the lot, we can determine the values n and k by specifying the points $(p_1', 1-\alpha)$ and (p_2', β) on the desired OC curve; where α is the producer's risk, β is the consumer's risk, and p_1' and p_2' are the corresponding fraction defectives. At the point $(p_1', 1-\alpha)$ on the desired OC curve, we know that the proportion of nondefective items is

$$(3) \quad P(X \geq L) = P\left(\frac{X-\bar{X}'}{\sigma'} \geq \frac{L-\bar{X}'}{\sigma'}\right) = P\left(z \geq \frac{L-\bar{X}'}{\sigma'}\right)$$

$$= P(z \geq z_{1-p_1'}) = 1-p_1' \quad .$$

Also note that z_τ is the standard normal deviate for which the upper-tail probability is τ, and $z_{1-\tau} = -z_\tau$ due to symmetry of the normal distribution.

If $\bar{X}$ represents the sample mean, then the standard deviation of the sample mean is $\sigma'/\sqrt{n}$. The condition of the decision rule at the point $(p_1', 1-\alpha)$ on the OC curve is

$$P(\bar{X} \geq L + k\sigma') = 1-\alpha$$

or $$P\left(\frac{X-\bar{X}'}{\sigma'/\sqrt{n}} \geq \frac{L+k\sigma'-\bar{X}'}{\sigma'/\sqrt{n}}\right) = 1-\alpha .$$

This implies $$\frac{L+k\sigma'-\bar{X}'}{\sigma'/\sqrt{n}} = (-z_{p_1'}+k)\sqrt{n} = z_{1-\alpha} = -z_\alpha$$

and $$P[z \geq (-z_{p_1'}+k)\sqrt{n}] = 1-\alpha .$$

Hence

(4) $$k = z_{p_1'} - z_\alpha/\sqrt{n}.$$

Similarly, at the point (p_2', β) on the OC curve, we obtain

(5) $$k = z_{p_2'} + z_\beta/\sqrt{n}.$$

Equating (4) and (5) and solving for n, we get

(6) $$n = \left(\frac{z_\alpha + z_\beta}{z_{p_1'} - z_{p_2'}}\right)^2$$

After n is found, k can be determined using either (4) or (5).

We will now present the program LBL.SDV, which can be used to design variables sampling plans for single and double specification limits with known and unknown standard deviation. However, for the unknown standard deviation case, only the single-limit case can be handled.

The Program LBL.SDV

There are essentially three possible computational procedures in LBL.SDV. Each one can be considered a separate subprogram. They are

(i) Single specification limit with known standard deviation of the quality characteristic (SSKNOWN)

(ii) Double specification limits with known standard deviation (DSKNOWN)

(iii) Single specification limit with unknown standard deviation (SSUNKNOWN)

The program also detects, under certain conditions of input data, whether or not the process is unacceptable and can be rejected without sampling. If a sampling procedure is required, the program provides a detailed sampling procedure that is simple to understand and completely eliminates the need of having to look up nomographs or perform unnecessary calculations except the calculation of the sample mean.

All computational procedures used in this program are based on Duncan [3]. It is important to recognize in general that since the sample size needs to be integer value, the OC curve cannot be forced to pass through the desired values of the producer's risk (ALPHA) and the consumer's risk (BETA) exactly.

INPUT

Depending on the subprogram selected -i.e., either SSKNOWN or DSKNOWN or SSUNKNOWN - the program will prompt the user with the appropriate data as follows:

1. Name of the quality characteristic
2. Acceptable Quality Level (AQL)
3. Producer's risk (ALPHA value)
4. Lot Tolerance Fraction Defective (LTFD)
5. Consumer's risk (BETA value)
6. The value of the standard deviation if it is known.
   If it is unknown:
   - the number of observations and
   - all observation values
7. The specification limit(s), which can be either an upper or a lower limit or both
8. For double specification limits
   - allowable maximum fraction defective

OUTPUT

Once a subprogram is selected, output can be obtained using either the PLAN or the PLANOC command. The first provides a short summary of some of the input data plus a sampling procedure. If the latter is entered, an OC curve will also be generated based on the plotting increment of the process fraction defective which must be specified by the user. The program will inform the user of the appropriate range. Two additional commands (LIST and CHANGE) are provided to allow the user to edit the observation values. The two commands are only active when the subprogram SSUNKNOWN is in effect.

EXAMPLE

We give one example terminal session which illustrates the use of the three subprograms in LBL.SDV. In all three cases, we set AQL=0.01, ALPHA=0.05, LTFD=0.08, and BETA=0.10.

Suppose, in the first case, we are interested in designing a single-sampling plan by variable with known standard deviation (SSKNOWN). The quality characteristic is the life of a type of electric motor in hours. The known standard deviation of the life of the motor is 800 hours. The lower specification limit is 17,000 hours.

In the second case (SSUNKNOWN), we assume that the standard deviation of the life of the motor is unknown but 10 observations have been collected, and on the basis of the data we wish to obtain a sampling procedure that will help us control the quality of the motors. Table 5 contains the 10 observation values.

Table 5. Observations for SSUNKNOWN

Observation	Life in Hours
1	17,250
2	17,345
3	17,260
4	17,190
5	16,955
6	17,300
7	17,220
8	17,248
9	17,050
10	17,100

The third case (DSKNOWN) illustrates the design of a sampling procedure for a quality characteristic with double specification limits. Suppose the resistance of a type of resistor in some electronic device requires that the maximum allowable variation be between 0.87 and 0.88 ohms. The known standard deviation of the resistance is 0.0005 ohms.

The example terminal session shows that the first command that is entered is the HELP command which displays all available commands. Next the SSKNOWN command is entered for the first case in our example. After the data are entered, the PLANOC command is selected together with a plotting increment of 0.01 for the OC curve. The sampling procedure to be used is to randomly choose a sample of size 10 and calculate the mean life of the motors. If the mean exceeds 18,446.86 hours the lot should be accepted; otherwise the lot should be rejected. The OC curve shows that at process fraction defective of 0.01 (AQL), the probability of acceptance is 0.9494, which corresponds to the ALPHA value of 0.0506, and at LTFD=0.08 the value of BETA is 0.1011. These values are, of course, very close to the desired values of 0.05 and 0.10, respectively.

The SSUNKNOWN command is entered next for the second case and the PLANOC command produces a sampling procedure that requires a random sample of 27 motors. The lot should be rejected if the mean of the sample falls below 17,218.88 hours.

The DSKNOWN command initiates the third case. After all data are entered, the PLANOC command tells us to randomly choose 10 resistors and measure their resistance. If the mean resistance of the sample is between 0.8709 and 0.8790, the lot should be accepted. Otherwise the lot should be rejected.

```
LBL.SDV EXAMPLE TERMINAL SESSION
================================

    LOT-BY-LOT ACCEPTANCE SINGLE-SAMPLING PLANS BY
    VARIABLES FOR SINGLE OR DOUBLE SPECIFICATION LIMIT(S)
    -----------------------------------------------------

NEED INTRODUCTION (Y OR N) ... !N

PLEASE ENTER COMMAND OR TYPE HELP

    COMMAND -->!HELP

    AVAILABLE COMMANDS
    ==================
    SSKNOWN   - PROGRAM FOR SINGLE SPEC.KNOWN STANDARD DEV.
    DSKNOWN   - PROGRAM FOR DOUBLE SPEC.KNOWN STANDARD DEV.
    SSUNKNOWN - PROGRAM FOR SINGLE SPEC.UNKNOWN STANDARD DEV.
    HELP      - PRINT THIS MESSAGE
    PLAN      - GENERATE VARIABLE PLAN WITHOUT OC CURVE
    PLANOC    - GENERATE VARIABLE PLAN WITH OC CURVE
    LIST      - LIST ONE OBSERVATION AT A TIME
    CHANGE    - CHANGE ONE OBSERVATION AT A TIME
    EXIT      - EXIT FROM THE PROGRAM

    NOTE : SPECIFY THE TYPE OF PROGRAM FIRST BEFORE USING
           THE PLAN, PLANOC, LIST OR CHANGE COMMANDS.
           LIST AND CHANGE COMMANDS ARE INACTIVE WHEN
           THE STANDARD DEVIATION IS KNOWN.

    COMMAND -->!SSKNOWN

    ENTER NAME OF QUALITY CHARACTERISTIC    !LIFE OF ELECTRIC MOTOR

    ENTER ACCEPTABLE QUALITY LIMIT (AQL) ... !0.01
    ENTER PRODUCER'S RISK (ALPHA VALUE)    ... !0.05

    ENTER LOT TOLERANCE FRACTION DEFECTIVE (LTFD) ... !0.08
    ENTER CONSUMER'S RISK (BETA VALUE)    ... !0.10

    ENTER THE KNOWN STANDARD DEVIATION  .. !800

    SINGLE SPECIFICATION : L=LOWER , U=UPPER
    ENTER TYPE OF SPECIFICATION (L OR U) ---> !L

    ENTER THE SPECIFICATION LIMIT   ... !17000

    COMMAND -->!PLANOC

    PLOTTING RANGE OF OC CURVE IS    0 <= P <= .15
    ENTER PLOTTING INCREMENT OF FRAC. DEFECT. P !0.01
```

```
****************************
*THE VARIABLE-SAMPLING PLAN*
****************************
FOR SINGLE SPECIFICATION WITH KNOWN STANDARD DEVIATION

QUALITY CHARACTERISTIC  : LIFE OF ELECTRIC MOTOR

AQL = .01    ALPHA = .05    LTFD = .08    BETA = .10
LOWER SPECIFICATION LIMIT   = 17000
STANDARD DEVIATION          = 800


S A M P L I N G     P R O C E D U R E

1. RANDOMLY CHOOSE A SAMPLE OF SIZE 10
2. CALCULATE THE MEAN OF THE SAMPLE.
3. ACCEPT THE LOT IF THE MEAN EXCEEDS   18446.86571175
4. OTHERWISE, REJECT THE LOT.

                                    PROB(ACCEPTANCE)
                          .0000    .2500    .5000    .7500    1.0000
FRAC.DEF.   PROB(ACCEPT)  +---------+---------+---------+---------+
 .000         1.0000      I                                       *
 .010          .9494      I                                     *
 .020          .7813      I                                *
 .030          .5908      I                        *
 .040          .4278      I                  *
 .050          .3027      I            *
 .060          .2114      I       *
 .070          .1465      I    *
 .080          .1011      I  *
 .090          .0696      I *
 .100          .0479      I *
 .110          .0329      I*
 .120          .0226      I*
 .130          .0155      I*


COMMAND -->!SSUNKNOWN

ENTER NAME OF QUALITY CHARACTERISTIC     !LIFE OF ELECTRIC MOTOR

ENTER ACCEPTABLE QUALITY LIMIT (AQL) ...   !.01
ENTER PRODUCER'S RISK (ALPHA VALUE)      ...  !.05

ENTER LOT TOLERANCE FRACTION DEFECTIVE (LTFD) ... !.08
ENTER CONSUMER'S RISK (BETA VALUE)   ...  !.10

SINGLE SPECIFICATION : L=LOWER , U=UPPER
ENTER TYPE OF SPECIFICATION (L OR U) ---> !L

ENTER THE SPECIFICATION LIMIT   ...  !17000
```

```
ENTER NUMBER OF OBSERVATIONS    ... !10

ENTER ALL OBSERVATIONS
----------------------

OBSERVATION NO.  1   117250
OBSERVATION NO.  2   117345
OBSERVATION NO.  3   117260
OBSERVATION NO.  4   117190
OBSERVATION NO.  5   116955
OBSERVATION NO.  6   117300
OBSERVATION NO.  7   117220
OBSERVATION NO.  8   117248
OBSERVATION NO.  9   117050
OBSERVATION NO. 10   117100

COMMAND --> !PLANOC

PLOTTING RANGE OF OC CURVE IS    0 <= P <= .15
ENTER PLOTTING INCREMENT OF FRAC. DEFECT. P  !.01

*****************************
*THE VARIABLE-SAMPLING PLAN*
*****************************
FOR SINGLE SPECIFICATION WITH UNKNOWN STANDARD DEVIATION

QUALITY CHARACTERISTIC  :  LIFE OF ELECTRIC MOTOR

AQL =  .01    ALPHA =  .05    LTFD =  .08    BETA =  .10
LOWER SPECIFICATION LIMIT  = 17000


S A M P L I N G     P R O C E D U R E

1. RANDOMLY CHOOSE A SAMPLE OF SIZE 27
2. CALCULATE THE MEAN OF THE SAMPLE.
3. ACCEPT THE LOT IF THE MEAN EXCEEDS  17218.87983283
4. OTHERWISE, REJECT THE LOT.
```

```
                                  PROB(ACCEPTANCE)
                           .0000    .2500    .5000    .7500   1.0000
FRAC.DEF.   PROB(ACCEPT)  +---------+---------+---------+---------+
  .000        1.0000      I                                        *
  .010         .9513      I                                     *
  .020         .7838      I                              *
  .030         .5916      I                    *
  .040         .4267      I               *
  .050         .3003      I         *
  .060         .2084      I      *
  .070         .1434      I    *
  .080         .0983      I  *
  .090         .0671      I *
  .100         .0458      I *
  .110         .0312      I*
  .120         .0213      I*
  .130         .0145      I*
```

COMMAND --> !DSKNOWN

ENTER NAME OF QUALITY CHARACTERISTIC   !RESISTANCE OF A RESISTOR

ENTER ACCEPTABLE QUALITY LIMIT (AQL) ... !.01
ENTER PRODUCER'S RISK (ALPHA VALUE) ... !.05

ENTER LOT TOLERANCE FRACTION DEFECTIVE (LTFD) ... !.08
ENTER CONSUMER'S RISK (BETA VALUE) ... !.10

ENTER THE KNOWN STANDARD DEVIATION .. !.0005

ENTER THE LOWER SPECIFICATION LIMIT ... !.87
ENTER THE UPPER SPECIFICATION LIMIT ... !.88

ENTER MAXIMUM ALLOWABLE FRACTION DEFECTIVE ... !.01

```
COMMAND -->!PLANOC

PLOTTING RANGE OF OC CURVE IS     0 <= P <= .15
ENTER PLOTTING INCREMENT OF FRAC. DEFECT. P !.01

****************************
*THE VARIABLE-SAMPLING PLAN*
****************************
FOR DOUBLE SPECIFICATION WITH KNOWN STANDARD DEVIATION

QUALITY CHARACTERISTIC  :  RESISTANCE OF A RESISTOR

AQL =  .01     ALPHA =  .05     LTFD =  .08     BETA =  .10
LOWER SPECIFICATION LIMIT    = .87
UPPER SPECIFICATION LIMIT    = .88
MAXIMUM ALLOWABLE FRACTION DEFECTIVE = .01
STANDARD DEVIATION           = .0005


S A M P L I N G     P R O C E D U R E

  1. RANDOMLY CHOOSE A SAMPLE OF SIZE 10
  2. CALCULATE THE MEAN OF THE SAMPLE.
  3. ACCEPT THE LOT IF .8709042910698 <= MEAN <= .8790957089302
  4. OTHERWISE, REJECT THE LOT.
```

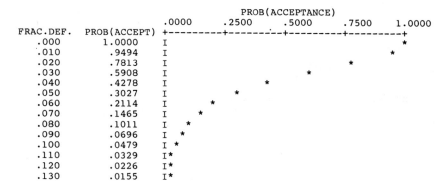

```
COMMAND -->!EXIT
```

```
10  REM    LBL.SDV
20  REM
30  DIM X(30),P$(40),F$(7)
40  REM M0 = MAXIMUM NUMBER OF OBSERVATIONS
50  M0=30
60  GOSUB 4220
70  PRINT
80  PRINT "   LOT-BY-LOT ACCEPTANCE SINGLE-SAMPLING PLANS BY"
90  PRINT "VARIABLES FOR SINGLE OR DOUBLE SPECIFICATION LIMIT(S)"
100 PRINT "--------------------------------------------------"
110 PRINT
120 PRINT "NEED INTRODUCTION (Y OR N) ... ";
130 INPUT Y$
140 IF Y$<>"Y" THEN 460
150 PRINT
160 PRINT "THIS PROGRAM DETERMINES THE LOT-BY-LOT ACCEPTANCE"
170 PRINT "SAMPLING PLAN BY VARIABLES FOR SINGLE OR DOUBLE"
180 PRINT "SPECIFICATION(S) WHICH ARE :"
190 PRINT "   1. ";F$(1)
200 PRINT "   2. ";F$(2)
210 PRINT "   3. ";F$(3)
220 PRINT
230 PRINT "INPUT REQUIREMENTS"
240 PRINT "------------------"
250 PRINT " 1. NAME OF QUALITY CHARACTERISTIC"
260 PRINT " 2. ACCEPTABLE QUALITY LEVEL (AQL)"
270 PRINT " 3. PRODUCER'S RISK (ALPHA VALUE)"
280 PRINT " 4. LOT TOLERANCE FRACTION DEFECTIVE (LTFD)"
290 PRINT " 5. CONSUMER'S RISK (BETA VALUE)"
300 PRINT " 6. IF STANDARD DEVIATION KNOWN :"
310 PRINT "        -STANDARD DEVIATION VALUE"
320 PRINT "     IF STANDARD DEVIATION UNKNOWN :"
330 PRINT "        -NUMBER OF OBSERVATIONS"
340 PRINT "        -OBSERVATION VALUE "
350 PRINT " 7. SPECIFICATION LIMIT(S)"
360 PRINT " 8. FOR DOUBLE SPECIFICATIONS :"
370 PRINT "      THE ALLOWABLE MAXIMUM FRACTION DEFECTIVE"
380 PRINT
390 PRINT "MAXIMUM NUMBER OF OBSERVATIONS = ";M0
400 PRINT
410 PRINT "NOTE : FOR PLOTTING OC CURVE THE PROGRAM WILL INFORM"
420 PRINT "       THE USER OF THE APPLICABLE RANGE OF THE FRACTION"
430 PRINT "       DEFECTIVE. HENCE, THE USER CAN SPECIFY THE"
440 PRINT "       APPROPRIATE PLOTTING INCREMENT."
450 PRINT
460 REM *** ACCEPT COMMANDS HERE ***
470 PRINT
480 PRINT "PLEASE ENTER COMMAND OR TYPE HELP"
490 PRINT
500 PRINT "COMMAND -->";
510 INPUT C$
520 PRINT
530 IF C$<>"SSKNOWN" THEN 580
540 K9=1
```

```
550 X8=0
560 GOSUB 980
570 GOTO 490
580 IF C$<>"DSKNOWN" THEN 630
590 K9=2
600 X8 = 0
610 GOSUB 980
620 GOTO 490
630 IF C$<>"SSUNKNOWN" THEN 680
640 K9=3
650 X8=0
660 GOSUB 980
670 GOTO 490
680 IF C$<>"HELP" THEN 710
690 GOSUB 2570
700 GOTO 490
710 IF C$<>"PLAN" THEN 840
720 IF K9=0 THEN 490
730 IF X8<>1 THEN 780
740 PRINT
750 PRINT "PROCESS IS UNACCEPTABLE !   REJECT WITHOUT SAMPLING !"
760 PRINT "===================================================="
770 GOTO 490
780 IF C$="PLAN" THEN 810
790 X7=1
800 GOTO 820
810 X7=0
820 GOSUB 2760
830 GOTO 490
840 IF C$<>"PLANOC" THEN 860
850 GOTO 720
860 IF C$<>"LIST" THEN 900
870 IF K9=0 THEN 490
880 GOSUB 3070
890 GOTO 490
900 IF C$<>"CHANGE" THEN 940
910 IF K9=0 THEN 490
920 GOSUB 3180
930 GOTO 490
940 IF C$="EXIT" THEN 4310
950 PRINT
960 PRINT "*** ERROR ***   PLEASE TRY AGAIN ! "
970 GOTO 490
980 REM INPUT ROUTINE
990 PRINT "ENTER NAME OF QUALITY CHARACTERISTIC        ";
1000 INPUT S$
1010 PRINT
1020 PRINT "ENTER ACCEPTABLE QUALITY LIMIT (AQL) ... ";
1030 INPUT P5
1040 P5=ABS(P5)
1050 IF P5>1 THEN 1010
1060 PRINT "ENTER PRODUCER'S RISK (ALPHA VALUE)        ... ";
1070 INPUT A5
1080 A5=ABS(A5)
```

```
1090 IF A5>1 THEN 1010
1100 PRINT
1110 PRINT "ENTER LOT TOLERANCE FRACTION DEFECTIVE (LTFD) ... ";
1120 INPUT P6
1130 P6=ABS(P6)
1140 IF P6>1 THEN 1100
1150 PRINT "ENTER CONSUMER'S RISK (BETA VALUE)  ... ";
1160 INPUT A6
1170 A6=ABS(A6)
1180 IF A6>1 THEN 1100
1190 PRINT
1200 IF K9=3 THEN 1270
1210 PRINT
1220 PRINT "ENTER THE KNOWN STANDARD DEVIATION  .. ";
1230 INPUT S
1240 S = ABS(S)
1250 IF K9=2 THEN 1350
1260 PRINT
1270 PRINT "SINGLE SPECIFICATION : L=LOWER , U=UPPER"
1280 PRINT "ENTER TYPE OF SPECIFICATION (L OR U) ---> ";
1290 INPUT Y3$
1300 PRINT
1310 PRINT "ENTER THE SPECIFICATION LIMIT   ... ";
1320 INPUT S2
1330 IF K9=3 THEN 1470
1340 GOTO 1620
1350 PRINT
1360 PRINT "ENTER THE LOWER SPECIFICATION LIMIT  ... ";
1370 INPUT S3
1380 S3=ABS(S3)
1390 PRINT "ENTER THE UPPER SPECIFICATION LIMIT  ... ";
1400 INPUT S4
1410 S4=ABS(S4)
1420 PRINT
1430 PRINT "ENTER MAXIMUM ALLOWABLE FRACTION DEFECTIVE ... ";
1440 INPUT W0
1450 W0=ABS(W0)
1460 GOTO 1620
1470 PRINT
1480 PRINT "ENTER NUMBER OF OBSERVATIONS  ... ";
1490 INPUT M
1500 IF M <= M0 THEN 1530
1510 PRINT " *** ERROR *** MAX.ALLOWED = ";M0
1520 GOTO 1470
1530 PRINT
1540 PRINT "ENTER ALL OBSERVATIONS "
1550 PRINT "---------------------- "
1560 PRINT
1570 FOR I = 1 TO M
1580 PRINT "OBSERVATION NO. ";I;"   ";
1590 INPUT X(I)
1600 X(I)=ABS(X(I))
1610 NEXT I
1620 REM
```

```
1630 P = 1-P5
1640 GOSUB 2240
1650 Z1=ABS(Z)
1660 P=1-A5
1670 GOSUB 2240
1680 Z2=ABS(Z)
1690 P=P6
1700 GOSUB 2240
1710 Z3=ABS(Z)
1720 P=A6
1730 GOSUB 2240
1740 Z4=ABS(Z)
1750 IF K9<>3 THEN 1780
1760 GOSUB 2110
1770 GOTO 2230
1780 IF K9=1 THEN 2040
1790 REM     CHECK IF THE PROCESS CAN BE REJECTED WITHOUT SAMPLING
1800 Z9=(S4-S3)/(2*S)
1810 IF Z9 > 3 THEN 2040
1820 GOSUB 2420
1830 W1 = 2*(1-P)
1840 IF W1<= W0 THEN 1870
1850 X8 =1
1860 GOTO 2230
1870 REM     FIND ROOT OF EQUATION USING NEWTON'S METHOD
1880 X9=S3
1890 S5 =(S4-S3)/10000
1900 Z9 =(X9-S3)/S
1910 GOSUB 2420
1920 P1 =(1-P)
1930 R1 =Z7
1940 Z9 =(S4-X9)/S
1950 GOSUB 2420
1960 P2 =(1-P)
1970 R2 =Z7
1980 G1 = P1 + P2 -P5
1990 G2 =(R2-R1)/S
2000 X9 =X9 - (G1/G2)
2010 IF ABS(G1/G2)<S5 THEN 2030
2020 GOTO 1900
2030 Z1 = (X9-S3)/S
2040 N=INT((((Z2+Z4)/(Z1-Z3))**2)+.5)
2050 K=((Z3+Z4/SQR(N))+(Z1-Z2/SQR(N)))/2
2060 IF K9=1 THEN 2230
2070 Z9 =K*(SQR(N/(N-1)))
2080 GOSUB 2420
2090 M2=1-P
2100 GOTO 2230
2110 M4 = 0
2120 S6 = 0
2130 FOR I=1 TO M
2140 M4=M4+X(I)
2150 S6=S6+X(I)*X(I)
2160 NEXT I
```

```
2170 M4=M4/M
2180 S=0
2190 IF M <=1 THEN 2210
2200 S=SQR((S6-M4*M4*M)/(M-1))
2210 K=(Z2*Z3 + Z4*Z1)/(Z2+Z4)
2220 N=INT((1+K*K/2)*(((Z2+Z4)/(Z1-Z3))**2)+0.5)
2230 RETURN
2240 REM SUBROUTINE FINDING NORMAL VARIATE
2250 C0=2.515517
2260 C1=0.802853
2270 C2=0.010328
2280 D1=1.432788
2290 D2=0.189269
2300 D3=0.001308
2310 E=0
2320 IF P<=0.5 THEN 2350
2330 E=1
2340 P=1-P
2350 T = SQR(LOG(1/(P*P)))
2360 Y = T-((C2*T+C1)*T+C0)/(((D3*T+D2)*T+D1)*T+1)
2370 IF E=1 THEN 2400
2380 Z=-Y
2390 GOTO 2410
2400 Z=Y
2410 RETURN
2420 REM SUBROUTINE FINDING NORMAL PROBABILITY
2430 A1=0.4361836
2440 A2=-0.1201676
2450 A3=0.9372980
2460 H=0.33267
2470 E=0
2480 IF Z9>=0 THEN 2510
2490 Z9=-Z9
2500 E=1
2510 Z7=0.3989423*EXP(-Z9*Z9/2)
2520 T=1/(1+(H*Z9))
2530 P=1-Z7*(((A3*T+A2)*T+A1)*T)
2540 IF E<>1 THEN 2560
2550 P = 1-P
2560 RETURN
2570 REM HELP COMMAND
2580 PRINT "AVAILABLE COMMANDS"
2590 PRINT "=================="
2600 PRINT "SSKNOWN    - PROGRAM FOR SINGLE SPEC.KNOWN STANDARD DEV."
2610 PRINT "DSKNOWN    - PROGRAM FOR DOUBLE SPEC.KNOWN STANDARD DEV."
2620 PRINT "SSUNKNOWN  - PROGRAM FOR SINGLE SPEC.UNKNOWN STANDARD DEV."
2630 PRINT "HELP       - PRINT THIS MESSAGE"
2640 PRINT "PLAN       - GENERATE VARIABLE PLAN WITHOUT OC CURVE"
2650 PRINT "PLANOC     - GENERATE VARIABLE PLAN WITH OC CURVE"
2660 PRINT "LIST       - LIST ONE OBSERVATION AT A TIME"
2670 PRINT "CHANGE     - CHANGE ONE OBSERVATION AT A TIME"
2680 PRINT "EXIT       - EXIT FROM THE PROGRAM"
2690 PRINT
2700 PRINT "NOTE : SPECIFY THE TYPE OF PROGRAM FIRST BEFORE USING"
```

```
2710 PRINT "          THE PLAN, PLANOC, LIST OR CHANGE COMMANDS."
2720 PRINT "          LIST AND CHANGE COMMANDS ARE INACTIVE WHEN"
2730 PRINT "          THE STANDARD DEVIATION IS KNOWN."
2740 PRINT
2750 RETURN
2760 REM    *** PLAN/PLANOC COMMANDS ***
2770 IF X7=0 THEN 2790
2780 GOSUB 3350
2790 PRINT
2800 PRINT "*****************************"
2810 PRINT "*THE VARIABLE-SAMPLING PLAN*"
2820 PRINT "*****************************"
2830 PRINT "FOR ";F$(K9)
2840 PRINT
2850 GOSUB 4060
2860 PRINT
2870 IF K9=2 THEN 2940
2880 IF Y3$="U" THEN 2920
2890 Y5=S2+(K*S)
2900 G$="EXCEEDS"
2910 GOTO 2940
2920 Y5=S2-(K*S)
2930 G$="FALLS BELOW"
2940 PRINT
2950 PRINT "S A M P L I N G     P R O C E D U R E"
2960 PRINT
2970 PRINT " 1. RANDOMLY CHOOSE A SAMPLE OF SIZE ";N
2980 PRINT " 2. CALCULATE THE MEAN OF THE SAMPLE."
2990 IF K9=2 THEN 3020
3000 PRINT " 3. ACCEPT THE LOT IF THE MEAN ";G$;"    ";Y5
3010 GOTO 3030
3020 PRINT " 3. ACCEPT THE LOT IF ";K*S+S3;" <= MEAN <= ";S4-K*S
3030 PRINT " 4. OTHERWISE, REJECT THE LOT."
3040 IF X7 = 0 THEN 3060
3050 GOSUB 3520
3060 RETURN
3070 REM    *** LIST COMMAND ***
3080 IF K9<>3 THEN 3170
3090 PRINT "ENTER OBSERVATION NUMBER ";
3100 INPUT G9
3110 FOR I = 1 TO M
3120 IF G9<>I THEN 3150
3130 PRINT "OBSERVATION NO. ";I;"   VALUE = ";X(I)
3140 GOTO 3170
3150 NEXT I
3160 GOSUB 3320
3170 RETURN
3180 REM    *** CHANGE COMMAND ***
3190 IF K9<>3 THEN 3310
3200 PRINT "ENTER OBSERVATION NUMBER ";
3210 INPUT G9
3220 FOR I = 1 TO M
3230 IF G9<>I THEN 3290
3240 PRINT "OLD VALUE = ";X(I);" ,ENTER NEW VALUE ";
```

```
3250 INPUT X(I)
3260 X(I)=ABS(X(I))
3270 GOSUB 1970
3280 GOTO 3310
3290 NEXT I
3300 GOSUB 3320
3310 RETURN
3320 PRINT
3330 PRINT " *** ERROR *** COMMAND ABORTED !"
3340 RETURN
3350 REM     *** ROUTINE TO FIND RANGE OF OC CURVE ***
3360 U1=1
3370 U0=0
3380 I0=0.03
3390 FOR I=U0 TO U1 STEP I0
3400 GOSUB 3910
3410 IF P7<0.0125 THEN 3430
3420 NEXT I
3430 U1=I
3440 PRINT "PLOTTING RANGE OF OC CURVE IS    0 <= P <= ";U1
3450 PRINT "ENTER PLOTTING INCREMENT OF FRAC. DEFECT. P   ";
3460 INPUT I0
3470 I0=ABS(I0)
3480 IF I0<U1 THEN 3510
3490 PRINT " ** INCREMENT NOT IN PLOTTING RANGE ** PLEASE TRY AGAIN !"
3500 GOTO 3440
3510 RETURN
3520 REM     *** ROUTINE TO PLOT OC CURVE ***
3530 PRINT
3540 PRINT
3550 PRINT TAB(38);"PROB(ACCEPTANCE)"
3560 L0=0
3570 L9=1
3580 H0=L9/4
3590 REM PRINT AXIS
3600 PRINT TAB(20);
3610 FOR J=L0 TO L9 STEP H0
3620 PRINT USING F$(4),J;
3630 NEXT J
3640 PRINT
3650 PRINT F$(5);
3660 FOR J = 1 TO 4
3670 PRINT "---------+";
3680 NEXT J
3690 PRINT
3700 J1=40
3710 P$(0)="I"
3720 FOR J = 1 TO J1
3730 P$(J)=" "
3740 NEXT J
3750 H1=H0/10
3760 FOR I = U0 TO U1 STEP I0
3770 GOSUB 3910
3780 IF P7<0.0125 THEN 3890
```

```
3790 P$(J1)=" "
3800 J1=INT((P7-L0)/H1+0.5)
3810 V=P7
3820 P$(J1)="*"
3830 PRINT USING F$(6),I,V;
3840 FOR J = 0 TO J1
3850 PRINT P$(J);
3860 NEXT J
3870 PRINT
3880 NEXT I
3890 PRINT
3900 RETURN
3910 REM    ROUTINE TO FIND PROB(ACCEPT)
3920 IF I <>0 THEN 3950
3930 P7=1
3940 GOTO 4050
3950 P=I
3960 GOSUB 2240
3970 IF K9=3 THEN 4020
3980 Z9=(K - ABS(Z))*SQR(N)
3990 GOSUB 2420
4000 P7= 1-P
4010 GOTO 4050
4020 Z9=(K-ABS(Z))/SQR((1/N)+(K*K/(2*N)))
4030 GOSUB 2420
4040 P7=1-P
4050 RETURN
4060 REM    ROUTINE TO PRINT DATA
4070 PRINT "QUALITY CHARACTERISTIC   : ";S$
4080 PRINT
4090 PRINT USING F$(7),P5,A5,P6,A6
4100 IF K9=2 THEN 4160
4110 IF Y3$="U" THEN 4140
4120 PRINT "LOWER SPECIFICATION LIMIT   = ";S2
4130 GOTO 4190
4140 PRINT "UPPER SPECIFICATION LIMIT   = ";S2
4150 GOTO 4190
4160 PRINT "LOWER SPECIFICATION LIMIT   = ";S3
4170 PRINT "UPPER SPECIFICATION LIMIT   = ";S4
4180 PRINT "MAXIMUM ALLOWABLE FRACTION DEFECTIVE = ";W0
4190 IF K9=3 THEN 4210
4200 PRINT "STANDARD DEVIATION      = ";S
4210 RETURN
4220 REM I N I T I A L I Z A T I O N
4230 F$(1)= "SINGLE SPECIFICATION WITH KNOWN STANDARD DEVIATION"
4240 F$(2)= "DOUBLE SPECIFICATION WITH KNOWN STANDARD DEVIATION"
4250 F$(3)= "SINGLE SPECIFICATION WITH UNKNOWN STANDARD DEVIATION"
4260 F$(4)= "#####.####"
4270 F$(5)= "FRAC.DEF.    PROB(ACCEPT) +"
4280 F$(6)= "###.###       ###.####      "
4290 F$(7)="AQL = #.##    ALPHA = #.##    LTFD = #.##    BETA = #.##"
4300 RETURN
4310 END
```

III. Published Sampling Plans

Published sampling plans contained in military standards MIL-STD-105D [6] and MIL-STD-414 [7], and Dodge Romig Inspection Tables [2] permit a wide choice of sampling schemes. In the Dodge-Romig tables the sampling plans are cataloged by Average Outgoing Quality Limit (AOQL) or by Lot Tolerance Percent Defective (LTPD). Some Limiting Quality (LQ) plans are also contained in MIL-STD-105D under the heading "Limiting Quality Protection."

MIL-STD-105D provides single, double, and multiple sampling inspection plans by attributes. The plans provided fall into two classes:

a. Those that ensure lot quality (AQL or LQ)

b. Those that ensure average quality (AOQL)

MIL-STD-414 provides sampling inspection plans by variables based on unknown and known variability. These plans can be used for single and double specification limits. A more recent publication approved by the American National Standard Institute, Inc. related to MIL-STD-414 is the ANSI/ASQC Z1.9-1980, which is titled "Sampling Procedures and Tables for Inspection by Variables for Percent Nonconforming." It is a revision of the MIL-STD-414 that roughly matches the MIL-STD-105D in terms of the protection given by the plans for stated values of AQL and Inspection Levels.

IV. Dimensional Specifications

Program XBARR

Let m and n be the maximum number of observed groups and the maximum number of observations per group. Currently, m=30 and n=25. The value n=25 represents the maximum possible for this program because the tables that we use from Duncan's book contain values up to n=25. We note that n must be at least 2. Considerable savings in memory storage can be achieved if m and n are adjusted to fit an application using proper values in the following statements:

```
Line No.
-------------------------------------------------------------
30 DIM D2(n),D3(n),P$(40),Z$(6)
40 DIM X(m),R(m),R$(m),Z(m),W(m,n)
70 M0=m
80 N0=n
-------------------------------------------------------------
```

Program PUC

The maximum number of subgroups, m, which is currently set at 30, can be increased or decreased by adjusting these statements:

Line No.
----
30 DIM W(m),X(m),Z(m),R$(m)
60 M0=m

Program LBL.AOQL

LBL.AOQL does not require any dimensional modification. However, it should be noted that the maximum acceptance number for a sampling plan is 40. This is the maximum value from Table 2-3 in [2]. In practice, an acceptance number of 40 will be very rare due to the large sample size the plan will generate.

Program AQL.LTFD

Similar to the LBL.AOQL, the program AQL.LFTD does not require redimensioning of any array. Since the program only handles the case when ALPHA=0.05 and BETA=0.10, the table from Duncan [3] allows us to generate plans with a maximum acceptance number of 15. For conditions that require the acceptance number to be greater than 15, the program will force it to be 15.

Program LBL.SDV

LBL.SDV dimensional requirement is due to the SSUNKNOWN option for which observation values are required to develop the single specification variable-sampling plan when the standard deviation is unknown. Let m be the maximum number of observations which is currently set at 30. The following statements must be modified if there is a need to change m:

Line No.
----
30 DIM X(m),P$(40),F$(7)
50 M0=m

V. REFERENCES

1. Besterfield, Dale H., QUALITY CONTROL, Prentice-Hall, Englewood Cliffs, N.J., 1979.

2. Dodge, Harold F., and Harry G. Romig, SAMPLING INSPECTION TABLES - SINGLE AND DOUBLE SAMPLING, 2nd ed., John Wiley, New York, 1959.

3. Duncan, Acheson J., QUALITY CONTROL AND INDUSTRIAL STATISTICS, 4th ed., Richard D. Irwin, Homewood, Ill., 1974.

4. Grant, E.L., and R.L. Leavenworth, STATISTICAL QUALITY CONTROL, 5th ed., McGraw-Hill, New York, 1980.

5. Grubbs, F.E., "On Designing Single Sampling Plans," ANNALS OF MATHEMATICAL STATISTICS, 20 (1949), 256.

6. MILITARY STANDARD 105D: SAMPLING PROCEDURES AND TABLES FOR INSPECTION BY ATTRIBUTES, Superintendent of Documents, Government Printing Office, Washington, D.C., 1963 (the ABC Standard).

7. MILITARY STANDARD 414: SAMPLING PROCEDURES AND TABLES FOR INSPECTION BY VARIABLES FOR PERCENT DEFECTIVE, Superintendent of Documents, Government Printing Office, Washington, D.C., 1957.

CHAPTER 10

WAITING LINE ANALYSIS

I. Model Notation

II. Performance Statistics of Waiting Line Models

III. Mathematical Relationships of Waiting Line Models

IV. The Program QUEUE
- Input
- Output
- Examples
- QUEUE Example Terminal Session 1
- QUEUE Example Terminal Session 2
- QUEUE Program Listing

V. Dimensional Specifications

VI. References

Waiting lines are a common phenomenon in life and are often an unavoidable and unpleasant experience. A waiting line is formed when a person or a machine requiring service arrives at a service facility but finds it busy and decides to wait for the service. Thus, when the service facility is busy, a queue is formed and the new arrival joins this queue. The new customers are generated over time from an input source called an input population. The input population can be characterized as being finite or infinite in size. The service facility may consist of single or multiple servers. When a server becomes available, a member of the queue is selected for service according to some predetermined rule called "service discipline." Figure 1 illustrates a typical waiting line system.

The difference in time between when the service begins and when it ends is defined as the service time, which may vary from customer to customer and may follow some probability distribution. Similarly, the interarrival time of the customers may also follow some distribution.

In general, the behavior of a waiting line system depends on the following characteristics:

1. Arrival and service time distributions

2. The size of the input population, which may be either finite or infinite

3. The number of servers in the service facility -i.e., single server or multiple servers

4. The service discipline - e.g., First-Come, First-Served (FCFS) or some other priority rule

5. The capacity of the queue, which means that no arrival is permitted to enter the system once its capacity is reached

Waiting line models have been under intensive investigation for the past seventy years, and a large body of knowledge exists which permits us to understand their characteristics. In this chapter, we present a computer program called QUEUE which can be used to analyze the nine most commonly known queueing models. These models are described in more detail in many books, such as Allen [1], Gross and Harris [4], and Hillier and Lieberman [5].

Section I describes the standard model representation of queueing systems. In Section II, we define the variables representing the various performance statistics used in

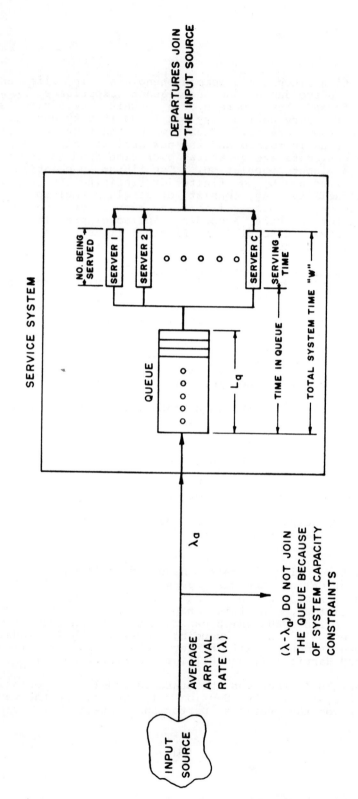

Figure 1. A Typical Waiting Line System

evaluating a queueing system. Section III contains the formulas for each of the models used in the program QUEUE and briefly states the relationships among the nine models. Section IV discusses the program QUEUE usage with two example terminal sessions. Sections V and VI contain the dimensional specifications and list of references, respectively.

I. Model Notation

The Kendall notation, named after David Kendall, is generally used to describe queueing systems and has the form

$$A/B/c/K/m/Z$$

where
- A = Interarrival time distribution,
- B = Service time distribution,
- c = Number of identical servers,
- K = System capacity, which is the number of customers allowed in the system - i.e., the number of customers in the queue plus those being served (if a specific value is not prescribed, K is assumed to be infinite),
- m = Size of the input population or input source (which is also assumed to be infinite if it is not explicitly specified),
- Z = Service discipline.

The symbols used for A and B are

- M = Markovian process - i.e., exponential interarrival or service time distribution,
- D = Deterministic (constant) interarrival or service time distribution,
- E_k = Erlang distribution with shape parameter equal to k for interarrival or service time distribution,
- GI = General independent interarrival time distribution,
- G = General service time distribution.

II. Performance Statistics of Waiting Line Models

In order to provide a consistent set of notation throughout the chapter, we define the following symbols which will be used to describe the formulas of the next section.

Let λ = Mean arrival rate -e.g., average number of customers arriving per hour,

μ = Mean service rate per server -e.g., average number of customers served per hour by a server,

c = Number of identical servers in the service facility.

Let w be the random variable representing the time a customer spends in the system. Also, let q be the random variable representing the time a customer spends in the queue. The following statistics are normally used to evaluate queueing system performances.

W = E(w) = Expected time a customer spends in the system.

W_q = E(q) = Expected time a customer spends in the queue before the service begins.

$W_{q|q>0}$ = Expected time spent in the queue given that the customer must wait.

L = Expected number of customers in the system.

L_q = Expected number of customers in the queue.

ρ = Server utilization.

u = Traffic intensity expressed in Erlang units.

P_n = Steady state probability that there are n customers in the system.

$P_q(r)$ = The rth percentile of the time in the queue. For example, if r=50, and $P_q(50)=2$ minutes, this means that 50 percent of the customers will wait less than 2 minutes.

$P_w(r)$ = The rth percentile of the time spent in the system. For example, 75 percent of the customers will spend less than $P_w(75)$ minutes in the system.

λ_a = Effective arrival rate for those models that assume finite queue length.

III. Mathematical Relationships of Waiting Line Models

In this section, we give the formulas for the steady state performance statistics used in the program QUEUE. All models assume that the interarrival time distribution is exponential. This means that the first character in the Kendall notation model representation will always be M. The service discipline will always be First-Come, First-Served (FCFS) and hence is omitted from the model representation.

1. Single-Server, Exponential Service Time Model (M/M/1)

This is one of the most basic models in the queueing literature. The service time is exponentially distributed. The input population size and the system's capacity are assumed to be infinite. The model is a special case of the M/M/c model for which c=1. The following formulas are relevant:

(1) $$u = \rho = \lambda/\mu < 1 .$$

(2) $$P_n = (1-\rho)\rho^n , \text{ for } n=0,1,2,...$$

(3) $$W_q = \frac{\rho}{(1-\rho)\mu} .$$

(4) $$W = W_q + 1/\mu .$$

(5) $$L = \lambda W .$$

(6) $$L_q = \lambda W_q .$$

(7) $$W_{q|q>0} = W .$$

(8) $$P_q(r) = W \ln(\frac{100\rho}{100-r}) , r > 100(1-\rho) .$$

2. Single-Server, Erlangian Service Model (M/E_k/1)

In this model, the service time distribution is Erlangian with parameter k. The k value must be given as integer. When k=1, this model is identical to the M/M/1 model. The pertinent formulas are

(9) $$u = \rho = \lambda/\mu < 1 .$$

(10) $$W_q = \frac{k+1}{2k} \frac{\rho}{(1-\rho)\mu} = E(q) .$$

(11) $$W = W_q + 1/\mu = E(w) .$$

(12) $$L = \lambda W .$$

(13) $$L_q = \lambda W_q .$$

(14) $$W_{q|q>0} = W_q/\rho .$$

The 90th and 95th percentile of the time a customer spends in the system based on Martin's estimates [7] are, respectively,

(15) $$P_w(90) = E(w) + 1.3 \sigma_w$$

(16) and $$P_w(95) = E(w) + 2.0 \sigma_w$$

where σ_w is the standard deviation of the time spent in the system. It is determined from equations (17), (18), and (19) as follows:

(17) $$E(q^2) = 2(E(q))^2 + \frac{\rho(k+1)(k+2)}{3(1-\rho)k^2\mu^2} .$$

(18) $$E(w^2) = E(q^2) + \frac{k+1}{(1-\rho)k\mu^2} .$$

(19) $$\sigma_w^2 = E(w^2) - (E(w))^2 .$$

3. Single-Server, Constant Service Time Model (M/D/1)

The service time under the $M/E_k/1$ model degenerates to a deterministic or constant value when k approaches infinity. This yields the M/D/1 model. In fact, the formulas given below compare very well with those of the $M/E_k/1$ model:

(20) $$u = \rho \equiv \lambda/\mu < 1 .$$

(21) $$W_q = \frac{\rho}{2(1-\rho)\mu} = E(q) .$$

(22) $$W = W_q + 1/\mu = E(w) .$$

(23) $\quad L = \lambda W$.

(24) $\quad L_q = \lambda W_q$.

(25) $\quad W_{q|q>0} = W_q/\rho$.

(26) $\quad P_w(90) = E(w) + 1.3\, \sigma_w$.

(27) $\quad P_w(95) = E(w) + 2.0\, \sigma_w$.

(28) $\quad E(q^2) = 2[E(q)]^2 + \dfrac{\rho}{3(1-\rho)\mu^2}$.

(29) $\quad E(w^2) = E(q^2) + \dfrac{1}{(1-\rho)\mu^2}$.

(30) $\quad \sigma_w^2 = E(w^2) - [E(w)]^2$.

4. Single-Server, General Service Time Model (M/G/1)

In this model, the service time of the server can be any general distribution. This implies that the steady state probabilities cannot be determined. Hence only a limited number of performance statistics are available. These statistics also depend on the knowledge of the moments of the service time distribution. For example, Var(s) is the variance of the service time distribution. In our program we use the following formulas, which are based on Pollaczek-Klintchine's results:

(31) $\quad u = \rho = \lambda/\mu < 1$.

(32) $\quad L_q = \dfrac{\lambda^2 \text{Var}(s) + \rho^2}{2(1-\rho)}$.

(33) $\quad L = L_q + \rho$.

(34) $\quad W_q = L_q/\lambda$.

(35) $\quad W = L/\lambda$

5. **Single-Server, Exponential Service, Finite Queue Model (M/M/1/K)**

As opposed to the M/M/1 model in which the system capacity is infinite, the M/M/1/K model assumes that the system can accommodate only K customers. Customers are turned away when the system is full. This model is also a special case of the M/M/c/K model to be discussed below. The formulas for the system's performance statistics can be found by letting c=1 in the M/M/c/K model given under model 8.

6. **Single-Server, Exponential Service, Finite Queue, and Source (M/M/1/K/K)**

This is a typical model used in analyzing what is known as the repairman problem. Under this model, the input source o input population is also finite - e.g., the number of machines (K) that will break down in a single-repairman shop. If all K machines break down, the system capacity is also K. The model is a special case of the M/M/c/K/K model in which there are c repairmen servicing the machines. The relevant formulas used i the program QUEUE are given under model 9 below.

7. **Multi-Server, Exponential Service Time Model (M/M/c)**

As mentioned in the first model, the M/M/c model is a generalization of the M/M/1 model in which there are c servers in the service facility. The formulas given below reduce to those given under the M/M/1 model when c is set to 1:

(36) $$u = \lambda/\mu \ .$$

(37) $$\rho = u/c < 1 \ .$$

(38) $$P_0 = \left[\sum_{n=0}^{c-1} \frac{u^n}{n!} + \frac{u^c}{c!(1-\rho)} \right]^{-1}$$

(39) $$P_n = \begin{cases} \frac{u^n}{n!} P_0 & \text{, for } n=0,1,\ldots,c \\ \frac{u^n}{c! c^{n-c}} P_0 & \text{, for } n=c+1,\ldots \ . \end{cases}$$

The probability that all c servers are busy, $C(c,u)$, can be determined from Erlang's C formula of equation (40):

$$(40) \quad C(c,u) = \frac{u^c/c!}{[u^c/c! + (1-\rho) \sum_{n=0}^{c-1} u^n/n!]}$$

$$(41) \quad W_q = \frac{C(c,u)}{c\mu(1-\rho)}$$

$$(42) \quad W = W_q + 1/\mu$$

$$(43) \quad L = \lambda W$$

$$(44) \quad L_q = \lambda W_q$$

$$(45) \quad W_{q|q>0} = \frac{1}{c\mu(1-\rho)}$$

$$(46) \quad P_q(r) = \frac{1}{c\mu(1-\rho)} \ln\left(\frac{100 C(c,u)}{100-r}\right), \; r > 100[1-C(c,u)].$$

8. Multi-Server, Exponential Service, Finite Queue Model (M/M/c/K)

We assume that K is greater than or equal to c. When no queue is allowed to form - i.e., when K=c - this model reduces to the M/G/c/c model. In other words, the service time distribution can be any general distribution [4, p. 110]. The following formulas are pertinent to our program:

$$(47) \quad u = \lambda/\mu$$

$$(48) \quad r = u/c$$

$$(49) \quad P_0 = \begin{cases} \left[\sum_{n=0}^{c-1} \frac{u^n}{n!} + \frac{u^c}{c!} \frac{1-r^{(K-c+1)}}{1-r}\right]^{-1}, & r \neq 1 \\ \left[\sum_{n=0}^{c-1} \frac{u^n}{n!} + \frac{u^c}{c!}(K-c+1)\right]^{-1}, & r = 1 \end{cases}$$

$$(50) \quad P_n = \begin{cases} \frac{u^n}{n!} P_0, & \text{for } n=0,1,\ldots,c \\ \frac{u^n}{c! c^{n-c}} P_0, & \text{for } n=c+1,\ldots,K \end{cases}$$

(51) $$L = \sum_{n=0}^{K} nP_n \;.$$

(52) $$L_q = \sum_{n=c}^{K} (n-c)P_n \;.$$

(53) $$\lambda_a = \lambda(1-P_K) \;.$$

(54) $$W = L/\lambda_a \;.$$

(55) $$W_q = L_q/\lambda_a \;.$$

(56) $$W_{q|q>0} = W_q/(1 - \sum_{n=0}^{c-1} P_n) \;.$$

(57) $$\rho = \frac{\lambda_a}{c\mu} \;.$$

9. Multi-Server, Exponential Service, Finite Queue, and Source (M/M/c/K/K)

We assume that $K \geq c$ in this model. Let $(1/\lambda)$ be the average time that a customer spends outside the system and the time is exponentially distributed. In the machine repairman context, $(1/\lambda)$ is the mean time before failure of a machine. An interesting observation about this model is that when $c=K=1$, the M/M/c/K/K model reduces to the M/M/c/K model. Otherwise the two models are distinctly different. The program QUEUE wil prompt for the mean arrival rate, which must be interpreted as λ. The following formulas are used in the program:

(58) $$u = \lambda/\mu \;.$$

(59) $$P_0 = \left[\sum_{n=0}^{c} \frac{K!}{n!(K-n)!} u^n + \sum_{n=c+1}^{K} \frac{K!}{c!(K-n)!} \frac{u^n}{c^{n-c}} \right]^{-1} \;.$$

(60) $$P_n = \begin{cases} \dfrac{K! \, u^n}{n!(K-n)!} P_0 \;, & n=1,2,\ldots,c \\[2mm] \dfrac{K! u^n}{c!(K-n)!c^{n-c}} P_0 \;, & n=c+1,\ldots,K \end{cases}$$

(61) $$L = \sum_{n=0}^{K} nP_n \;.$$

$$\text{(62)} \qquad L_q = L - c + \sum_{n=0}^{c-1} (c-n) P_n \ .$$

$$\text{(63)} \qquad \lambda_a = \lambda (K-L) \ .$$

$$\text{(64)} \qquad W = L/\lambda_a \ .$$

$$\text{(65)} \qquad W_q = L_q/\lambda_a \ .$$

Let D be the probability that a customer must wait for the service - i.e., the probability that all c servers are busy. Then

$$\text{(66)} \qquad D = \sum_{n=c}^{K} P_n \ .$$

$$\text{(67)} \qquad W_{q \mid q>0} = W_q/D \ .$$

$$\text{(68)} \qquad \rho = \frac{\lambda_a}{c\mu} \ .$$

IV. The Program QUEUE

This program can be used to analyze the queueing models mentioned in the preceding section. The selection of a particular model is accomplished by entering an appropriate command using an abbreviated version of the Kendall notation. For example, the command MM1 represents the M/M/1 model, the command MMCKK means the M/M/c/K/K model, etc. Input data for each model are prompted independently once the appropriate command is activated.

INPUT
-----
1. Problem title (for documentation purposes)
2. Mean arrival rate (λ)
3. Mean service rate per server (μ)
4. Number of identical servers (c) for multi-server systems -i.e., MMC, MMCK and MMCKK commands
5. System capacity (K) for limited capacity systems - i.e., MM1K, MMCK, MM1KK, and MMCKK commands
6. Size of input source for finite population models -i.e., MM1KK and MMCKK commands
7. Erlang shape parameter, k, for ME1 command
8. Variance of the service time distribution for the general service time model -i.e., the MG1 command

OUTPUT

Three types of output are available from the program:

1. The steady state performance statistics, which are applicable to all models. They include statistics such as the expected time in service, the expected time in the queue, the expected time in the system, the average queue length, and the probability that a customer must wait. Please refer to an example terminal session for a complete list.

2. A table of steady state probabilities is given for some models. The table lists the probability that there are n customers in the system, the probability that there are less than or equal to n customers in the system, and the probability that there are more than n customers in the system. They are listed until the cumulative probability exceeds 0.99.

3. The percentile of time spent in the queue or in the system based on Martin's estimates [7]. For the MM1 command and the MMC command, the percentiles of the time spent in the queue are reported. For the ME1 and MD1 commands, the percentiles of the time spent in the system are given. Note that no other model contains these statistics.

EXAMPLES

Because of the large variety of queueing applications even for the set of models contained in this program, we will only provide two example terminal sessions for the purpose of illustrating its usage. The users can consult various books on the subject for possible applications.

The first example terminal session simply illustrates the output results for the following seven problems. Please refer to the first example terminal session for details of the commands and outputs. The input data are as follows:

Problem 1. Command=ME1, $\lambda=10$, $\mu=12$, k=3
Problem 2. Command=MD1, $\lambda=50$, $\mu=80$
Problem 3. Command=MG1, $\lambda=0.85$, $\mu=1.25$, Var(s)=0.04
Problem 4. Command=MM1, $\lambda=7$, $\mu=10$
Problem 5. Command=MMC, $\lambda=7$, $\mu=10$, c=2
Problem 6. Command=MM1K, $\lambda=10$, $\mu=5$, K=4
Problem 7. Command=MMCK, $\lambda=10$, $\mu=5$, c=2, K=4

For the second example terminal session, let us consider the machine repairman problem. Suppose that the management of XYZ company plans to install 6 numerically controlled machines as part of an expansion program in one of its plants. Based on the data supplied by the machine manufacturer, management knows that these machines periodically break down and the time before failure of each machine is exponentially distributed with a mean of 25 days. The time for repairing each machine can also be considered exponentially distributed with a mean of 4 days using a single repairman. Management wishes to know how many repairmen will be required to have at least 5 machines working on the average and the workload of each repairman.

This problem fits the M/M/c/K/K model. Since the mean time before failure of each machine is 25 days, the mean arrival rate to be entered as input to the program is 1/25=0.04. Also, since the mean repair time of each machine is 4 days, the mean service rate of a repairman is 1/4=0.25. Using these data, the MM1KK command shows that, for the one-repairman case, the repairman will be kept busy 71.77 percent of the time. Since the expected number of machines in repair is 1.51, management can expect 4.49 machines to be working on the average. This case is unacceptable because of management performance requirement. Therefore a two-repairman case is explored next.

For the two-repairman case, the MMCKK command indicates that each repairman will be busy only 40.83 percent of the time and the average number of machines in working condition are is 5.10. It appears that, based on this preliminary analysis, two repairmen will be required in order to have 5 machines working on the average.

```
QUEUE EXAMPLE TERMINAL SESSION 1
================================

W A I T I N G    L I N E    M O D E L S
-----------------------------------------

NEED INTRODUCTION  (Y OR N) ... !N

PLEASE ENTER COMMAND OR TYPE HELP

COMMAND --> !HELP

AVAILABLE COMMANDS
==================
HELP   - PRINT THIS MESSAGE
MM1    - SINGLE-SERVER, EXPONENTIAL SERVICE TIME
ME1    - SINGLE-SERVER, ERLANGIAN SERVICE TIME
MD1    - SINGLE-SERVER, CONSTANT SERVICE TIME
MG1    - SINGLE-SERVER, GENERAL SERVICE TIME
MM1K   - SINGLE-SERVER, EXPONENTIAL SERVICE, FINITE QUEUE
MM1KK  - SINGLE-SERVER, EXP. SERVICE,FINITE QUEUE & SOURCE
MMC    - MULTI-SERVER , EXPONENTIAL SERVICE TIME
MMCK   - MULTI-SERVER , EXPONENTIAL SERVICE, FINITE QUEUE
MMCKK  - MULTI-SERVER , EXP. SERVICE,FINITE QUEUE & SOURCE
EXIT   - EXIT FROM THE PROGRAM

COMMAND --> !ME1


ME1    - SINGLE-SERVER, ERLANGIAN SERVICE TIME

ENTER PROBLEM TITLE -->   !PROBLEM 1
ENTER MEAN ARRIVAL RATE (LAMBDA)   !10
ENTER MEAN SERVICE RATE PER SERVER (MU)   !12
ENTER ERLANG SHAPE PARAMETER  K    !3


Q U E U E    P E R F O R M A N C E    S T A T I S T I C S
=========================================================
MEAN ARRIVAL RATE ......................  10.0000
MEAN SERVICE RATE PER SERVER ...........  12.0000
MEAN INTERARRIVAL TIME .................    .1000
MEAN SERVICE TIME PER SERVER ...........    .0833
NUMBER OF SERVERS ......................  1
TRAFFIC INTENSITY (ERLANGS).............    .8333
SERVER UTILIZATION......................  83.3333  %
EXPECTED NUMBER IN THE SYSTEM...........   3.6111
EXPECTED NUMBER IN THE QUEUE ...........   2.7778
EXPECTED NUMBER BEING SERVED ...........    .8333
EXPECTED TIME IN THE SYSTEM ............    .3611
EXPECTED TIME IN THE QUEUE .............    .2778
EXPECTED TIME IN SERVICE ...............    .0833
QUEUE TIME FOR THOSE WHO MUST WAIT .....    .3333
EFFECTIVE ARRIVAL RATE .................  10.0000
```

```
PROBABILITY THAT A CUSTOMER MUST WAIT ...     .8333

90TH PERCENTILE TIME IN THE SYSTEM ......     .7828
95TH PERCENTILE TIME IN THE SYSTEM ......    1.0098

COMMAND --> !MD1


MD1   - SINGLE-SERVER, CONSTANT SERVICE TIME

ENTER PROBLEM TITLE -->  !PROBLEM 2
ENTER MEAN ARRIVAL RATE (LAMBDA)  !50
ENTER MEAN SERVICE RATE PER SERVER (MU)  !80


Q U E U E    P E R F O R M A N C E    S T A T I S T I C S
==========================================================
MEAN ARRIVAL RATE .......................    50.0000
MEAN SERVICE RATE PER SERVER ............    80.0000
MEAN INTERARRIVAL TIME ..................      .0200
MEAN SERVICE TIME PER SERVER ............      .0125
NUMBER OF SERVERS .......................     1
TRAFFIC INTENSITY (ERLANGS)..............      .6250
SERVER UTILIZATION.......................    62.5000  %
EXPECTED NUMBER IN THE SYSTEM............     1.1458
EXPECTED NUMBER IN THE QUEUE ............      .5208
EXPECTED NUMBER BEING SERVED ............      .6250
EXPECTED TIME IN THE SYSTEM .............      .0229
EXPECTED TIME IN THE QUEUE ..............      .0104
EXPECTED TIME IN SERVICE ................      .0125
QUEUE TIME FOR THOSE WHO MUST WAIT ......      .0167
EFFECTIVE ARRIVAL RATE ..................    50.0000
PROBABILITY THAT A CUSTOMER MUST WAIT ...      .6250

90TH PERCENTILE TIME IN THE SYSTEM ......      .0411
95TH PERCENTILE TIME IN THE SYSTEM ......      .0509

COMMAND --> !MG1


MG1   - SINGLE-SERVER, GENERAL SERVICE TIME

ENTER PROBLEM TITLE -->  !PROBLEM 3
ENTER MEAN ARRIVAL RATE (LAMBDA)  !.85
ENTER MEAN SERVICE RATE PER SERVER (MU)  !1.25
ENTER VARIANCE OF SERVICE TIME DISTRIBUTION !0.04


Q U E U E    P E R F O R M A N C E    S T A T I S T I C S
==========================================================
MEAN ARRIVAL RATE .......................     .8500
MEAN SERVICE RATE PER SERVER ............    1.2500
MEAN INTERARRIVAL TIME ..................    1.1765
MEAN SERVICE TIME PER SERVER ............     .8000
```

```
NUMBER OF SERVERS .....................      1
TRAFFIC INTENSITY (ERLANGS)............       .6800
SERVER UTILIZATION.....................     68.0000  %
EXPECTED NUMBER IN THE SYSTEM..........      1.4477
EXPECTED NUMBER IN THE QUEUE ..........       .7677
EXPECTED NUMBER BEING SERVED ..........       .6800
EXPECTED TIME IN THE SYSTEM ...........      1.7031
EXPECTED TIME IN THE QUEUE ............       .9031
EXPECTED TIME IN SERVICE ..............       .8000
QUEUE TIME FOR THOSE WHO MUST WAIT ....      1.3281
EFFECTIVE ARRIVAL RATE ................       .8500
PROBABILITY THAT A CUSTOMER MUST WAIT .       .6800

COMMAND --> !MM1


MM1    - SINGLE-SERVER, EXPONENTIAL SERVICE TIME

ENTER PROBLEM TITLE --> !PROBLEM 4
ENTER MEAN ARRIVAL RATE (LAMBDA)  !7
ENTER MEAN SERVICE RATE PER SERVER (MU)  !10


Q U E U E·  P E R F O R M A N C E    S T A T I S T I C S
===========================================================
MEAN ARRIVAL RATE .....................      7.0000
MEAN SERVICE RATE PER SERVER ..........     10.0000
MEAN INTERARRIVAL TIME ................       .1429
MEAN SERVICE TIME PER SERVER ..........       .1000
NUMBER OF SERVERS .....................      1
TRAFFIC INTENSITY (ERLANGS)............       .7000
SERVER UTILIZATION.....................     70.0000  %
EXPECTED NUMBER IN THE SYSTEM..........      2.3333
EXPECTED NUMBER IN THE QUEUE ..........      1.6333
EXPECTED NUMBER BEING SERVED ..........       .7000
EXPECTED TIME IN THE SYSTEM ...........       .3333
EXPECTED TIME IN THE QUEUE ............       .2333
EXPECTED TIME IN SERVICE ..............       .1000
QUEUE TIME FOR THOSE WHO MUST WAIT ....       .3333
EFFECTIVE ARRIVAL RATE ................      7.0000
PROBABILITY THAT A CUSTOMER MUST WAIT .       .7000

S T E A D Y    S T A T E    P R O B A B I L I T I E S
-----------------------------------------------------
   N        P(N)         P(K <= N)       P(K > N)
---------   ----------   -------------   ------------
   0        .3000         .3000           .7000
   1        .2100         .5100           .4900
   2        .1470         .6570           .3430
   3        .1029         .7599           .2401
   4        .0720         .8319           .1681
   5        .0504         .8824           .1176
   6        .0353         .9176           .0824
   7        .0247         .9424           .0576
```

```
 8     .0173    .9596    .0404
 9     .0121    .9718    .0282
10     .0085    .9802    .0198
11     .0059    .9862    .0138
12     .0042    .9903    .0097
```

P E R C E N T I L E   T I M E   I N   Q U E U E

```
35 PERCENT OF CUSTOMERS WAIT LESS THAN    .0247
40 PERCENT OF CUSTOMERS WAIT LESS THAN    .0514
45 PERCENT OF CUSTOMERS WAIT LESS THAN    .0804
50 PERCENT OF CUSTOMERS WAIT LESS THAN    .1122
55 PERCENT OF CUSTOMERS WAIT LESS THAN    .1473
60 PERCENT OF CUSTOMERS WAIT LESS THAN    .1865
65 PERCENT OF CUSTOMERS WAIT LESS THAN    .2310
70 PERCENT OF CUSTOMERS WAIT LESS THAN    .2824
75 PERCENT OF CUSTOMERS WAIT LESS THAN    .3432
80 PERCENT OF CUSTOMERS WAIT LESS THAN    .4176
85 PERCENT OF CUSTOMERS WAIT LESS THAN    .5135
90 PERCENT OF CUSTOMERS WAIT LESS THAN    .6486
95 PERCENT OF CUSTOMERS WAIT LESS THAN    .8797
```

COMMAND --> !MMC

MMC   - MULTI-SERVER , EXPONENTIAL SERVICE TIME

```
ENTER PROBLEM TITLE -->  !PROBLEM 5
ENTER MEAN ARRIVAL RATE (LAMBDA)    !7
ENTER MEAN SERVICE RATE PER SERVER (MU)  !10
ENTER NUMBER OF IDENTICAL SERVERS (C>=1)  !2
```

Q U E U E   P E R F O R M A N C E   S T A T I S T I C S
===
```
MEAN ARRIVAL RATE ......................    7.0000
MEAN SERVICE RATE PER SERVER ...........   10.0000
MEAN INTERARRIVAL TIME .................     .1429
MEAN SERVICE TIME PER SERVER ...........     .1000
NUMBER OF SERVERS ......................    2
TRAFFIC INTENSITY (ERLANGS).............     .7000
SERVER UTILIZATION......................   35.0000 %
EXPECTED NUMBER IN THE SYSTEM...........     .7977
EXPECTED NUMBER IN THE QUEUE ...........     .0977
EXPECTED NUMBER BEING SERVED ...........     .7000
EXPECTED TIME IN THE SYSTEM ............     .1140
EXPECTED TIME IN THE QUEUE .............     .0140
EXPECTED TIME IN SERVICE ...............     .1000
QUEUE TIME FOR THOSE WHO MUST WAIT .....     .0769
EFFECTIVE ARRIVAL RATE .................    7.0000
PROBABILITY THAT A CUSTOMER MUST WAIT ...    .1815
```

```
STEADY    STATE    PROBABILITIES
-------------------------------------------------------
     N         P(N)         P(K <= N)         P(K > N)
 ---------   ----------   ---------------   -------------
     0         .4815          .4815            .5185
     1         .3370          .8185            .1815
     2         .1180          .9365            .0635
     3         .0413          .9778            .0222
     4         .0145          .9922            .0078


PERCENTILE    TIME    IN    QUEUE
-------------------------------------------------------
85 PERCENT OF CUSTOMERS WAIT LESS THAN     .0147
90 PERCENT OF CUSTOMERS WAIT LESS THAN     .0458
95 PERCENT OF CUSTOMERS WAIT LESS THAN     .0992

COMMAND --> !MM1K


MM1K  - SINGLE-SERVER, EXPONENTIAL SERVICE, FINITE QUEUE

ENTER PROBLEM TITLE --> !PROBLEM 6
ENTER MEAN ARRIVAL RATE (LAMBDA)    !10
ENTER MEAN SERVICE RATE PER SERVER (MU)   !5
ENTER SYSTEM CAPACITY (K)  !4


QUEUE    PERFORMANCE    STATISTICS
============================================================
MEAN ARRIVAL RATE .......................    10.0000
MEAN SERVICE RATE PER SERVER ............     5.0000
MEAN INTERARRIVAL TIME ..................      .1000
MEAN SERVICE TIME PER SERVER ............      .2000
NUMBER OF SERVERS .......................    1
TRAFFIC INTENSITY (ERLANGS)..............     2.0000
SERVER UTILIZATION.......................    96.7742   %
EXPECTED NUMBER IN THE SYSTEM............     3.1613
EXPECTED NUMBER IN THE QUEUE ............     2.1935
EXPECTED NUMBER BEING SERVED ............      .9677
EXPECTED TIME IN THE SYSTEM .............      .6533
EXPECTED TIME IN THE QUEUE ..............      .4533
EXPECTED TIME IN SERVICE ................      .2000
QUEUE TIME FOR THOSE WHO MUST WAIT ......      .4684
EFFECTIVE ARRIVAL RATE ..................     4.8387
PROBABILITY THAT A CUSTOMER MUST WAIT ...      .9677

STEADY    STATE    PROBABILITIES
-------------------------------------------------------
     N         P(N)         P(K <= N)         P(K > N)
 ---------   ----------   ---------------   -------------
     0         .0323          .0323            .9677
     1         .0645          .0968            .9032
     2         .1290          .2258            .7742
```

```
   3        .2581         .4839          .5161
   4        .5161        1.0000          .0000
```

COMMAND --> !MMCK

MMCK - MULTI-SERVER , EXPONENTIAL SERVICE, FINITE QUEUE

```
ENTER PROBLEM TITLE -->   !PROBLEM 7
ENTER MEAN ARRIVAL RATE (LAMBDA)   !10
ENTER MEAN SERVICE RATE PER SERVER (MU)   !5
ENTER NUMBER OF IDENTICAL SERVERS (C>=1)   !2
ENTER SYSTEM CAPACITY (K)  !4
```

QUEUE   PERFORMANCE   STATISTICS
===
```
MEAN ARRIVAL RATE .......................   10.0000
MEAN SERVICE RATE PER SERVER ............    5.0000
MEAN INTERARRIVAL TIME  .................     .1000
MEAN SERVICE TIME PER SERVER ............     .2000
NUMBER OF SERVERS .......................    2
TRAFFIC INTENSITY (ERLANGS)..............    2.0000
SERVER UTILIZATION.......................   77.7778  %
EXPECTED NUMBER IN THE SYSTEM............    2.2222
EXPECTED NUMBER IN THE QUEUE ............     .6667
EXPECTED NUMBER BEING SERVED ............    1.5556
EXPECTED TIME IN THE SYSTEM .............     .2857
EXPECTED TIME IN THE QUEUE ..............     .0857
EXPECTED TIME IN SERVICE ................     .2000
QUEUE TIME FOR THOSE WHO MUST WAIT ......     .1286
EFFECTIVE ARRIVAL RATE ..................    7.7778
PROBABILITY THAT A CUSTOMER MUST WAIT ...     .6667
```

STEADY   STATE   PROBABILITIES

N	P(N)	P(K <= N)	P(K > N)
0	.1111	.1111	.8889
1	.2222	.3333	.6667
2	.2222	.5556	.4444
3	.2222	.7778	.2222
4	.2222	1.0000	.0000

COMMAND --> !EXIT

```
QUEUE EXAMPLE TERMINAL SESSION 2
==================================

W A I T I N G    L I N E    M O D E L S
-----------------------------------------

NEED INTRODUCTION  (Y OR N) ... !N

PLEASE ENTER COMMAND OR TYPE HELP

COMMAND --> !MM1KK


MM1KK - SINGLE-SERVER, EXP. SERVICE,FINITE QUEUE & SOURCE

ENTER PROBLEM TITLE -->   !ONE REPAIRMAN
ENTER MEAN ARRIVAL RATE (LAMBDA)  !.04
ENTER MEAN SERVICE RATE PER SERVER (MU)  !.25
ENTER SYSTEM CAPACITY (K)  !6


Q U E U E    P E R F O R M A N C E    S T A T I S T I C S
==========================================================
MEAN ARRIVAL RATE ......................     .0400
MEAN SERVICE RATE PER SERVER ...........     .2500
MEAN INTERARRIVAL TIME .................   25.0000
MEAN SERVICE TIME PER SERVER ...........    4.0000
NUMBER OF SERVERS ......................    1
TRAFFIC INTENSITY (ERLANGS).............     .1600
SERVER UTILIZATION......................   71.7755  %
EXPECTED NUMBER IN THE SYSTEM...........    1.5140
EXPECTED NUMBER IN THE QUEUE ...........     .7963
EXPECTED NUMBER BEING SERVED ...........     .7178
EXPECTED TIME IN THE SYSTEM ............    8.4376
EXPECTED TIME IN THE QUEUE .............    4.4376
EXPECTED TIME IN SERVICE ...............    4.0000
QUEUE TIME FOR THOSE WHO MUST WAIT .....    6.1826
EFFECTIVE ARRIVAL RATE .................     .1794
PROBABILITY THAT A CUSTOMER MUST WAIT ...    .7178


S T E A D Y    S T A T E    P R O B A B I L I T I E S
-------------------------------------------------------
    N          P(N)        P(K <= N)      P(K > N)
---------   ----------   -------------   -------------
    0         .2822         .2822          .7178
    1         .2710         .5532          .4468
    2         .2168         .7700          .2300
    3         .1387         .9087          .0913
    4         .0666         .9753          .0247
    5         .0213         .9966          .0034

COMMAND --> !MMCKK
```

420

```
MMCKK - MULTI-SERVER , EXP. SERVICE,FINITE QUEUE & SOURCE

ENTER PROBLEM TITLE -->   !TWO REPAIRMEN
ENTER MEAN ARRIVAL RATE (LAMBDA)   !.04
ENTER MEAN SERVICE RATE PER SERVER (MU)   !.25
ENTER NUMBER OF IDENTICAL SERVERS (C>=1)   !2
ENTER SYSTEM CAPACITY (K)  !6


Q U E U E    P E R F O R M A N C E    S T A T I S T I C S
=========================================================
MEAN ARRIVAL RATE ......................    .0400
MEAN SERVICE RATE PER SERVER ...........    .2500
MEAN INTERARRIVAL TIME .................  25.0000
MEAN SERVICE TIME PER SERVER ...........   4.0000
NUMBER OF SERVERS ......................   2
TRAFFIC INTENSITY (ERLANGS).............    .1600
SERVER UTILIZATION......................  40.8347  %
EXPECTED NUMBER IN THE SYSTEM...........    .8957
EXPECTED NUMBER IN THE QUEUE ...........    .0790
EXPECTED NUMBER BEING SERVED ...........    .8167
EXPECTED TIME IN THE SYSTEM ............   4.3868
EXPECTED TIME IN THE QUEUE .............    .3868
EXPECTED TIME IN SERVICE ...............   4.0000
QUEUE TIME FOR THOSE WHO MUST WAIT .....   1.7867
EFFECTIVE ARRIVAL RATE .................    .2042
PROBABILITY THAT A CUSTOMER MUST WAIT ...   .2165


S T E A D Y    S T A T E    P R O B A B I L I T I E S
------------------------------------------------------
   N          P(N)         P(K <= N)       P(K > N)
---------  -----------   -------------   -------------
   0         .3998          .3998           .6002
   1         .3838          .7835           .2165
   2         .1535          .9371           .0629
   3         .0491          .9862           .0138
   4         .0118          .9980           .0020

COMMAND -->  !EXIT
```

```
10 REM    QUEUE  -  WAITING LINE ANALYSIS
20 DIM P(50),P1(50),W(50)
30 DIM R(20),H$(10),F$(18),G$(9)
40 REM    M1=MAXIMUM CAPACITY OF THE SYSTEM.
50 M1=50
60 GOSUB 3910
70 PRINT "W A I T I N G    L I N E    M O D E L S"
80 PRINT "----------------------------------------"
90 PRINT
100 PRINT "NEED INTRODUCTION   (Y OR N) ... ";
110 INPUT A$
120 IF A$ <> "Y" THEN 340
130 PRINT
140 PRINT "FOLLOWING KENDALL NOTATION, THIS PROGRAM CAN BE USED"
150 PRINT "TO ANALYZE THE FOLLOWING QUEUEING MODELS :"
160 PRINT
170 FOR I=1 TO 9
180 PRINT I;".   ";G$(I)
190 NEXT I
200 PRINT
210 PRINT "INPUT REQUIREMENTS"
220 PRINT "------------------"
230 PRINT "1. PROBLEM TITLE"
240 PRINT "2. MEAN ARRIVAL RATE (LAMBDA)"
250 PRINT "3. MEAN SERVICE RATE (MU)"
260 PRINT "4. NUMBER OF IDENTICAL SERVERS FOR MULTI-SERVER SYSTEMS"
270 PRINT "5. MAX SYSTEM CAPACITY FOR LIMITED CAPACITY SYSTEMS"
280 PRINT "6. SIZE OF INPUT SOURCE FOR FINITE POPULATION MODELS"
290 PRINT "7. ERLANG SHAPE PARAMETER, K, FOR ERLANG SERVICE TIME"
300 PRINT "   DISTRIBUTION"
310 PRINT "8. VARIANCE OF SERVICE TIME FOR GENERAL(G) SERVICE"
320 PRINT "   TIME DISTRIBUTION."
330 REM    ***  ACCEPT COMMANDS HERE    ***
340 PRINT
350 PRINT "PLEASE ENTER COMMAND OR TYPE HELP"
360 PRINT
370 PRINT "COMMAND --> ";
380 INPUT C$
390 PRINT
400 IF C$ <> "HELP" THEN 430
410 GOSUB 830
420 GOTO 370
430 IF C$ <> "MM1" THEN 470
440 X9=1
450 GOSUB 930
460 GOTO 370
470 IF C$ <> "ME1" THEN 510
480 X9=2
490 GOSUB 980
500 GOTO 370
510 IF C$ <> "MD1" THEN 550
520 X9=3
530 GOSUB 1070
540 GOTO 370
```

```
550 IF C$ <> "MG1" THEN 590
560 X9=4
570 GOSUB 1140
580 GOTO 370
590 IF C$ <> "MM1K" THEN 630
600 X9=5
610 GOSUB 1350
620 GOTO 370
630 IF C$ <> "MM1KK" THEN 670
640 X9=6
650 GOSUB 1390
660 GOTO 370
670 IF C$ <> "MMC" THEN 710
680 X9=7
690 GOSUB 1430
700 GOTO 370
710 IF C$ <> "MMCK" THEN 750
720 X9=8
730 GOSUB 1980
740 GOTO 370
750 IF C$<> "MMCKK" THEN 790
760 X9=9
770 GOSUB 2420
780 GOTO 370
790 IF C$ = "EXIT" THEN 4300
800 PRINT " ***  ERROR   ***    PLEASE TRY AGAIN  !"
810 GOTO 360
820 REM
830 REM    ***    HELP COMMAND   ***
840 PRINT "AVAILABLE COMMANDS "
850 PRINT "=================="
860 PRINT "HELP  - PRINT THIS MESSAGE"
870 FOR J=1 TO 9
880 PRINT G$(J)
890 NEXT J
900 PRINT "EXIT  - EXIT FROM THE PROGRAM"
910 PRINT
920 RETURN
930 REM ***    MM1 COMMAND    ***
940 C=1
950 GOSUB 2840
960 GOSUB 1460
970 RETURN
980 REM   ***    ME1 COMMAND   ***
990 C=1
1000 GOSUB 2840
1010 PRINT "ENTER ERLANG SHAPE PARAMETER   K    ";
1020 INPUT K
1030 W1=(R*E/(1-R))*((1+1/K)/2)
1040 GOSUB 1230
1050 GOSUB 3470
1060 RETURN
1070 REM   ***    MD1  COMMAND   ***
1080 C=1
```

```
1090 GOSUB 2840
1100 W1=R*E/(2*(1-R))
1110 GOSUB 1230
1120 GOSUB 3470
1130 RETURN
1140 REM      *** MG1 COMMAND    ***
1150 C=1
1160 GOSUB 2840
1170 PRINT "ENTER VARIANCE OF SERVICE TIME DISTRIBUTION ";
1180 INPUT V0
1190 W1=(A*A*V0+R*R)/(2*(1-R))/A
1200 GOSUB 1230
1210 PRINT
1220 RETURN
1230 REM      *** COMMON ROUTINE FOR ME1, MD1, MG1 COMMANDS ***
1240 W=W1+E
1250 L1=A*W1
1260 L=A*W
1270 L2=L-L1
1280 W2=E
1290 E1=W1/R
1300 Z=R*100
1310 A1=A
1320 D=R
1330 GOSUB 3250
1340 RETURN
1350 REM      *** MM1K COMMAND    ***
1360 C=1
1370 GOSUB 2000
1380 RETURN
1390 REM      *** MM1KK COMMAND    ***
1400 C=1
1410 GOSUB 2440
1420 RETURN
1430 REM      *** MMC  COMMAND    ***
1440 C=0
1450 GOSUB 2840
1460 W(0)=1
1470 T1=0
1480 FOR J=1 TO C
1490 T1=T1+W(J-1)
1500 W(J)=W(J-1)*U/J
1510 NEXT J
1520 P(0)=1/(T1+W(C)/(1-R))
1530 FOR J=1 TO C
1540 P(J)=W(J)*P(0)
1550 NEXT J
1560 FOR J=C+1 TO M1
1570 P(J)=P(J-1)*R
1580 NEXT J
1590 P1(0)=P(0)
1600 FOR J=1 TO M1
1610 P1(J)=P1(J-1)+P(J)
1620 IF P1(J) >= 0.99 THEN 1660
```

```
1630 NEXT J
1640 K=M1
1650 GOTO 1680
1660 K=J
1670 REM     F=C(C,U)=ERLANG'S C FORMULA
1680 F=P(C)/(1-R)
1690 L1=U*F/(C*(1-R))
1700 L=L1+U
1710 L2=L-L1
1720 W=L/A
1730 W1=L1/A
1740 W2=W-W1
1750 E1=E/(C*(1-R))
1760 Z=R*100
1770 F1=F*100
1780 REM CALCULATE  PERCENTILE VALUES
1790 FOR J=0 TO 19
1800 R0=J*5
1810 F2=(100-R0)
1820 IF F1 <= F2 THEN 1840
1830 R(J)=E*LOG (100*F/(100-R0))/(C*(1-R))
1840 NEXT J
1850 A1=A
1860 GOSUB 3640
1870 GOSUB 3250
1880 GOSUB 3750
1890 GOSUB 3860
1900 FOR J=0 TO 19
1910 R0=J*5
1920 F2=(100-R0)
1930 IF F1 <= F2 THEN 1950
1940 PRINT USING H$(8),R0,R(J)
1950 NEXT J
1960 PRINT
1970 RETURN
1980 REM ***    MMCK  COMMAND    ***
1990 C = 0
2000 GOSUB 2840
2010 W(0)=1
2020 T1=0
2030 FOR J=1 TO C
2040 T1=T1+W(J-1)
2050 W(J)=W(J-1)*U/J
2060 NEXT J
2070 IF R <> 1 THEN 2100
2080 P(0)=1/(T1+W(C)*(K-C+1))
2090 GOTO 2120
2100 Y1=R**(K-C+1)
2110 P(0)=1/(T1+W(C)/(1-R)*(1-Y1))
2120 FOR J=1 TO C
2130 P(J)=W(J)*P(0)
2140 NEXT J
2150 IF K=C THEN 2190
2160 FOR J=C+1 TO K
```

```
2170 P(J)=P(J-1)*U/C
2180 NEXT J
2190 P1(0)=P(0)
2200 FOR J=1 TO K
2210 P1(J)=P1(J-1)+P(J)
2220 NEXT J
2230 L=0
2240 FOR J=1 TO K
2250 L=L+J*P(J)
2260 NEXT J
2270 L1=0
2280 IF K=C THEN 2320
2290 FOR J=C+1 TO K
2300 L1=L1+(J-C)*P(J)
2310 NEXT J
2320 A1=A*(1-P(K))
2330 W=L/A1
2340 W1=L1/A1
2350 Z=A1/C/M*100
2360 L2=L-L1
2370 W2=W-W1
2380 GOSUB 3640
2390 GOSUB 3250
2400 GOSUB 3750
2410 RETURN
2420 REM ***    MMCKK COMMAND    ***
2430 C = 0
2440 GOSUB 2840
2450 W(0)=1
2460 FOR J=1 TO C
2470 W(J)=W(J-1)*(K-J+1)*U/J
2480 NEXT J
2490 IF K <= C THEN 2530
2500 FOR J=C+1 TO K
2510 W(J)=W(J-1)/C*U*(K-J+1)
2520 NEXT J
2530 T1=0
2540 FOR J=0 TO K
2550 T1=T1+W(J)
2560 NEXT J
2570 REM    CALCULATE STEADY STATE PROB.
2580 P(0)=1/T1
2590 P1(0)=P(0)
2600 FOR J= 1 TO K
2610 P(J)=W(J)*P(0)
2620 P1(J)=P1(J-1)+P(J)
2630 NEXT J
2640 L=0
2650 FOR J=1 TO K
2660 L=L + J*P(J)
2670 NEXT J
2680 L1=0
2690 FOR J = C TO K
2700 L1=L1 + (J-C)*P(J)
```

```
2710 NEXT J
2720 A1=A*(K-L)
2730 W=L/A1
2740 W1=L1/A1
2750 L2=L-L1
2760 W2=W-W1
2770 R=A1*E/C
2780 Z=R*100
2790 GOSUB 3640
2800 GOSUB 3250
2810 PRINT
2820 GOSUB 3750
2830 RETURN
2840 REM      *** I N P U T    R O U T I N E ***
2850 PRINT
2860 PRINT G$(X9)
2870 PRINT
2880 PRINT "ENTER PROBLEM TITLE  -->  ";
2890 INPUT P$
2900 PRINT "ENTER MEAN ARRIVAL RATE (LAMBDA)  ";
2910 INPUT A
2920 PRINT "ENTER MEAN SERVICE RATE PER SERVER (MU)   ";
2930 INPUT M
2940 IF C=1 THEN 3010
2950 PRINT "ENTER NUMBER OF IDENTICAL SERVERS (C>=1)  ";
2960 INPUT C
2970 IF C <= M1 THEN 3010
2980 PRINT "*** ERROR *** MAX ALLOWABLE SERVERS = ";M1
2990 PRINT
3000 GOTO 2950
3010 U=A/M
3020 R=U/C
3030 E=1/M
3040 T=1/A
3050 IF R<1 THEN 3110
3060 IF X9 <= 4 THEN 3080
3070 IF X9 <> 7 THEN 3110
3080 PRINT "*** ERROR *** ARRIVAL RATE MUST BE LESS THAN ";C*M
3090 PRINT "       !!!!!! PLEASE START INPUT AGAIN !!!!!!!!"
3100 GOTO 2850
3110 IF X9 <= 4 THEN 3240
3120 IF X9=7 THEN 3240
3130 PRINT "ENTER SYSTEM CAPACITY (K) ";
3140 INPUT K
3150 IF K <= M1 THEN 3190
3160 PRINT "*** ERROR ***  SYSTEM CAPACITY MUST BE <= ";M1
3170 PRINT
3180 GOTO 3130
3190 IF K >= C THEN 3240
3200 PRINT "*** ERROR ***  SYSTEM CAPACITY CANNOT BE LESS THAN"
3210 PRINT "               THE NUMBER OF SERVERS OF ";C
3220 PRINT
3230 GOTO 3130
3240 RETURN
```

```
3250 REM PRINT ROUTINE
3260 PRINT
3270 PRINT
3280 PRINT F$(1)
3290 PRINT F$(2)
3300 PRINT USING F$(3),A
3310 PRINT USING F$(4),M
3320 PRINT USING F$(5),T
3330 PRINT USING F$(6),E
3340 PRINT USING F$(7),C
3350 PRINT USING F$(8),U
3360 PRINT USING F$(9),Z
3370 PRINT USING F$(10),L
3380 PRINT USING F$(11),L1
3390 PRINT USING F$(12),L2
3400 PRINT USING F$(13),W
3410 PRINT USING F$(14),W1
3420 PRINT USING F$(15),W2
3430 PRINT USING F$(16),E1
3440 PRINT USING F$(17),A1
3450 PRINT USING F$(18),D
3460 RETURN
3470 REM CALCULATE MARTIN'S ESTIMATES
3480 IF X9<>2 THEN 3520
3490 T2=R*E*E*(K+1)*(K+2)/(3*(1-R)*K*K)
3500 T3=E*E*(1+1/K)/(1-R)
3510 GOTO 3540
3520 T2=R*E*E/(3*(1-R))
3530 T3=E*E/(1-R)
3540 T1=2*W1*W1
3550 V=T1+T2+T3-W*W
3560 S=SQR(V)
3570 P0=W+1.3*S
3580 P5=W+2*S
3590 PRINT
3600 PRINT USING H$(9),P0
3610 PRINT USING H$(10),P5
3620 PRINT
3630 RETURN
3640 REM    D = PROB(WAIT), E1=AVE WAIT TIME GIVEN MUST WAIT
3650 D=0
3660 IF C=K THEN 3710
3670 D=1
3680 FOR J=0 TO C-1
3690 D=D-P(J)
3700 NEXT J
3710 E1=0
3720 IF D=0 THEN 3740
3730 E1=W1/D
3740 RETURN
3750 REM    PRINT TABLE OF STEADY STATE PROBABILITIES
3760 PRINT
3770 FOR H9=1 TO 4
3780 PRINT H$(H9)
```

```
3790 NEXT H9
3800 FOR J=0 TO K
3810 PRINT USING H$(5),J,P(J),P1(J),1-P1(J)
3820 IF P1(J) > 0.99 THEN 3840
3830 NEXT J
3840 PRINT
3850 RETURN
3860 REM    PRINT HEADING OF PERCENTILE TIME IN QUEUE
3870 PRINT
3880 PRINT H$(6)
3890 PRINT H$(7)
3900 RETURN
3910 REM    I N I T I A L I Z A T I O N
3920 H$(1)="S T E A D Y     S T A T E     P R O B A B I L I T I E S"
3930 H$(2)="-------------------------------------------------------"
3940 H$(3)="    N          P(N)         P(K <= N)        P(K > N)"
3950 H$(4)="---------    ----------    --------------    -------------"
3960 H$(5)="   ###       #.####          #.####           #.####"
3970 H$(6)=" P E R C E N T I L E     T I M E     I N     Q U E U E"
3980 H$(7)=" ------------------------------------------------------"
3990 H$(8)=" ## PERCENT OF CUSTOMERS WAIT LESS THAN    ###.####"
4000 H$(9)="90TH PERCENTILE TIME IN THE SYSTEM ......  ####.####"
4010 H$(10)="95TH PERCENTILE TIME IN THE SYSTEM ......  ####.####"
4020 F$(1)="Q U E U E    P E R F O R M A N C E    S T A T I S T I C S"
4030 F$(2)="==========================================================="
4040 F$(3)="MEAN ARRIVAL RATE ......................  ####.####"
4050 F$(4)="MEAN SERVICE RATE PER SERVER ..·........  ####.####"
4060 F$(5)="MEAN INTERARRIVAL TIME .................  ####.####"
4070 F$(6)="MEAN SERVICE TIME PER SERVER ...........  ####.####"
4080 F$(7)="NUMBER OF SERVERS ......................  ####"
4090 F$(8)="TRAFFIC INTENSITY (ERLANGS).............  ####.####"
4100 F$(9)="SERVER UTILIZATION......................  ####.####  %"
4110 F$(10)="EXPECTED NUMBER IN THE SYSTEM...........  ####.####"
4120 F$(11)="EXPECTED NUMBER IN THE QUEUE ...........  ####.####"
4130 F$(12)="EXPECTED NUMBER BEING SERVED ...........  ####.####"
4140 F$(13)="EXPECTED TIME IN THE SYSTEM ............  ####.####"
4150 F$(14)="EXPECTED TIME IN THE QUEUE .............  ####.####"
4160 F$(15)="EXPECTED TIME IN SERVICE ...............  ####.####"
4170 F$(16)="QUEUE TIME FOR THOSE WHO MUST WAIT .....  ####.####"
4180 F$(17)="EFFECTIVE ARRIVAL RATE .................  ####.####"
4190 F$(18)="PROBABILITY THAT A CUSTOMER MUST WAIT ... ####.####"
4200 G$(1)="MM1   - SINGLE-SERVER, EXPONENTIAL SERVICE TIME"
4210 G$(2)="ME1   - SINGLE-SERVER, ERLANGIAN SERVICE TIME"
4220 G$(3)="MD1   - SINGLE-SERVER, CONSTANT SERVICE TIME"
4230 G$(4)="MG1   - SINGLE-SERVER, GENERAL SERVICE TIME"
4240 G$(5)="MM1K  - SINGLE-SERVER, EXPONENTIAL SERVICE, FINITE QUEUE"
4250 G$(6)="MM1KK - SINGLE-SERVER, EXP. SERVICE,FINITE QUEUE & SOURCE"
4260 G$(7)="MMC   - MULTI-SERVER , EXPONENTIAL SERVICE TIME"
4270 G$(8)="MMCK  - MULTI-SERVER , EXPONENTIAL SERVICE, FINITE QUEUE"
4280 G$(9)="MMCKK - MULTI-SERVER , EXP. SERVICE,FINITE QUEUE & SOURCE"
4290 RETURN
4300 END
```

V. Dimensional Specifications

    Program QUEUE

    Let m be the maximum system capacity for the case of finite capacity systems. It is currently equal to 50. In order to modify m, the following statements are affected:

    Line No.
    ----
    20 DIM P(m),P1(m),W(m)
    50 M1=m

    Note that the same vectors in line number 20 are used in the calculation of the steady state probabilities for the M/M/1 and M/M/c models whose system capacities are infinite. For these two models, if the server utilization ( ) is very close to one, a maximum of m+1 values of the cumulative probability is reported.

VI. References
    ----
 1. Allen, A.D., PROBABILITY, STATISTICS AND QUEUEING THEORY, Academic Press, New York, 1978.

 2. Cook, T.M., and R.A. Russell, INTRODUCTION TO MANAGEMENT SCIENCE, 2nd ed., Prentice-Hall, Englewood Cliffs, N.J., 1981.

 3. Dilworth, J.B., PRODUCTION AND OPERATIONS MANAGEMENT - MANUFACTURING AND NON-MANUFACTURING, Random House, New York, 1979.

 4. Gross, D., and C.M. Harris, FUNDAMENTALS OF QUEUEING THEORY, John Wiley, New York, 1974.

 5. Hillier, F.S., and G.J. Lieberman, INTRODUCTION TO OPERATIONS RESEARCH, Holden-Day, San Francisco, 1980.

 6. Hillier, F.S., and O.S. Yu, QUEUING TABLES AND GRAPHS, North Holland Publishing, New York, 1980.

 7. Martin, J., SYSTEMS ANALYSIS FOR DATA TRANSMISSION, Prentice Hall, Englewood Cliffs, N.J., 1972.

 8. McClain, J.O., and L.J. Thomas, OPERATIONS MANAGEMENT - PRODUCTION OF GOODS AND SERVICES, Prentice-Hall, Englewood Cliffs, N.J., 1980.

INDEX

A

Abramowitz, M., 200
Acceptance sampling, 359
Activity-on-arrow network, 160-61
Adaptive smoothing, 66
Aggregate planning, 122-25
Allen, A., 401
Antill, J., 160
AOQL, 359-61
Approximate inventory cost models, 287
Aquilano, N., 141
Archibald, R., 160
Assembly-line design, 233, 244-48
Assignment problem, 141
Attributes, inspection by, 359
Average outgoing quality limit (AOQL), 359-61

B

Backorder case of inventory, 287-89
Balance delay, 245
Balintfy, J., 176
Barnes, R., 231
Barr, R., 158
BASIC language, 1, 5
Besterfield, D., 399
Bowman, E., 141
Brown, R., 106, 269, 272-73
Burdick, D., 176

C

Capital budgeting, 10-40
Cash flows, 11
Caterer problem, 141
Chambers, J., 106
Chase, R., 141
Chu, K., 176
Clark, J., 40
Claycombe, W., 106
Computer programs for:
  aggregate planning, 126
  assembly line planning, 245
  capital budgeting, 13
  control charts, 322, 340
  critical path method (CPM and PERT), 163, 174
  curve fitting, 44
  exchange curve, 270
  forecasting, 68
  inventory cost models, 287
  lease-purchase analysis, 28
  linear programming, 109
  price breaks, 260
  MRP, 297
  quality control sampling plans, 362, 371, 381
  queueing models, 412
  transportation problems, 141
  time study analysis, 204
  work sampling, 218
Confidence level factor, 202, 217
Continuous review inventory system, 269
Control charts:

  $\bar{X}$, R, 322
  p, u, c, 340
Cook, T., 430
CPM (critical path method), 160-73
Criticality index, 174
Curve fitting, 42-43 (see also Forecasting)
Cycle stock, 270

D

Dantzig, G., 141
DeMatteis, J., 297
Dependent demand items, 258
Depreciation:
  straight-line, 11, 28
  sum-of-years-digits, 11, 28
  double-declining, 11, 28
Dilworth, J., 430
Dodge, H., 359-60, 397
Double-declining deprec., 11
Double exponential smoothing, 65
Duncan, A., 361

E

Economic life, 11
Economic order quantity (EOQ), 270
Elmaghraby, S., 200
Equipment replacement

problem, 141
Exchange rules in inventory, 259, 270-79, 287-89
Exponential smoothing, 65-68, 269
  single, 65
  double, 65
  triple, 66

F

Financing :
  debt, 26-29
  lease, 26-29
Forecasting, 41-106
  moving average, 63
  exponential smoothing, 65-68
  adaptive smoothing, 66
  regression models, 42, 64
  time-series analysis, 42, 63-71
Francis, R., 233

G

Glover, F., 158
Groover, M., 233
Gross, D., 270, 401
Gerson, G., 269, 272
Grant, E., 399
Grubbs, F., 399

H

Hadley, G., 142, 261, 270, 287, 298
Hanssmann, F., 122
Harris, C., 270, 401
Heiland, R., 231
Hess, S., 122
Hillier, F., 141, 401
Hindeland, T., 40
Holt's 2-parameters smoothing model, 67
Horngren, C., 40

I

Ince, R., 270
Incremental quantity discount, 260
Independent demand items, 258, 269

Internal rate of return, 11-12
Inventory analysis, 257-313
  backorder case, 287-89
  cycle stock, 270
  economic order quantity (EOQ), 270
  exchange rules, 259, 270-79
  independent demand items, 270
  Lagrange multipliers, 272
  lead time, 258, 270
  lot-sizing (MRP), 258, 297-99
  lost sales case, 287-89
  price breaks, 260-62
  quantity discount, 260-62
  safety factor, 272
  safety stock, 259, 270-72
  service level measures, 259, 270-79
  shortage cost, 259
Investment analysis, 11-14 (see also capital budgeting)

J

Johnson, L., 261, 298

K

Kannan, N., 122
Kendall notation, 403
Kerzner, H., 160
King, W., 200
Klingel, A., 200
Klingman, D., 158

L

Lagrange mutiplier, 272
Layout:
  of facilities, 232-56
  process, 233-43
  product, 244-55
Leach, D., 66
Lead time, 258, 270
Lease financing, 26
Leavenworth, R., 399
Lieberman, G., 125, 158, 401, 430
Linear:
  moving average, 64
  regression, 43, 64

Linear programming, 107-58
  aggregate planning by, 122
  for capacity planning, 122
  for personnel scheduling, 122
  for production scheduling, 122
  solution of, 109
Loomba, N., 158
Lost sales case of inventory, 287-89
Lot-by-lot single sampling, 359
Lot-size reorder-point (Q,r) model, 269-70
Lot-sizing (MRP), 258, 297-99

M

Makridakis, S., 43, 67, 69
Martin, J., 430
Material requirements planning (MRP), 297-311
Mathematical programming (see linear programming)
McClain, J., 430
Methods Time Measurement (MTM), 202
Military standard, 397
Moder, J., 160, 175
Montgomery, D., 106, 298, 261
Moore, J., 233
Moving average, 63-64
Multi-server queueing models, 408-10
Mundell, M., 231
Muther, R., 233

N

Nahmias, S., 270, 298
Naylor, T., 176
Neibel, B., 231
Net present value (NPV), 12, 27
Networks :
  project management (see CPM and PERT)
  simulation, 176
  transportation-type problems, 141-43

O

Orlicky, J., 297

P

Part-period algorithm, 297
Payback period, 12
Performance of:
  inventory lot-sizing techniques, 259
  waiting-line models, 403
PERT, 159-200
  project control, 159-63
  network diagram, 159-63
  uncertain activity times, 174-77
  scheduling with, 174-80
  simulation of, 176
Peterson, R., 297
Phillips, C., 175
Phillips, D., 176
Planning aggregate, 122
Precedence diagram, 161
Precedence relationships, 245
Predetermined motion time system, 202
Prenting, T., 244
Present value, 12, 27
Price breaks (see Inventory analysis)
Prime-400 computer, 8, 69
Pritchard, R., 40
Process:
  layout design, 233-43
  quality control, 316
Product:
  layout design, 244-55
  quality control, 359
Production and operations management, 2
Production scheduling, 122 (see also aggregate planning)
Profitability index, 12
Program:
  8CURVS, 44
  AGG, 126
  ASSMBLY, 245
  AQL.LTFD, 371
  CAPBUD, 13
  CPM, 163
  EXCHNGE, 270
  FORCST, 68

INVCOST, 287
LBL.AOQL, 362
LBL.SDV, 381
LEASE, 28
LOTSZ.MRP, 297
LP, 109
PBREAK, 260
PERT, 174
PLAYOUT, 235
PUC, 340
QUEUE, 412
TIME.STY, 204
TRANSP, 142
WORK.SMPLG, 218
XBARR, 322
Program structure, 4-5

Q

Quality Control, 314-99
  Acceptance sampling, 359
  AOQL, 359-61

  control charts ($\bar{X}, R$), 322
  control charts (p,u,c), 340
  lot-by-lot single sampling, 359
  process control, 316
  product control, 359
  rectifying inspection plan, 359
  single sampling by attributes, 359
Quantity discount (See also inventory analysis)
  All units, 260-62
  Incremental, 260-62
Queueing theory, 400-30 (See also waiting line)

R

Rate of return, 11
Rectifying inspection plan, 359
Regression coefficients, 43
Regression analysis, 42, 64 (see also forecasting)
  linear, 64
  models with one independent variable, 43
Richardson, W., 231
Roming, G., 359-60

S

Safety factor, 272
Safety stock, 259, 270-72
Scheduling:
  of personnel, 122
  of projects, 160
  of production, 122, 141
  using linear programming, 122, 141
  with PERT/CPM, 159-200
Schuermann, A., 122
Service facility, 401
Shannon, R., 176
Shortest route problem, 141
Silver, E., 297
Silver-Meal heuristic, 297
Simplex, 108-9, 126
Simulation of:
  inventory lot-sizing, 297
  networks, 176
Single exponential smoothing, 65
Single moving average, 63
Single-model assembly-line design, 44-55
Single sampling by attributes, 359 (see also quality control)
Single-server queueing models, 405-8
Srinivasan, V., 158
Standard time, 202-3, 217-8
Stegun, I., 200
Stevens, G., 40
Stopwatch technique, 202
Straight-line depreciation, 11
Sullivan, W., 106
Sum-of-years-digits depreciation, 11

T

Taylor, F., 202
Thomopoulos, N., 63-69, 244
Thompson, G., 158
Time-series analysis, 42, 63-71 (see also forecasting)
Time study, 202-7
Tippett, L., 217
Tompkins, J., 233
Traffic intensity, 404

Transportation-type problems, 141-43
　assignment, 141
　caterer, 141
　Bowman production scheduling, 141
　equipment replacement, 141
　shortest/longest route, 141
　transshipment, 141
　warehousing, 141
Transshipment, 141
Trigg, D., 66
Triple exponential smoothing, 66 (see also forecasting)
Turban, E., 158

V

Van Horne, J., 40
Vancil, R., 40
Van Slyke, R., 174-78
Variables sampling, 379-80

W

Wagner, H., 141, 270
Wagner-Whitin algorithm, 297
Waiting-line models, 400-30
　single-server, 405-8
　multiple-server, 408-11
　performance statistics of, 403-4
　notation of, 403
Warehousing problem, 141
White, J., 233
Whitin, T., 261, 270, 287, 297-8
Wild, R., 244
Winter's 3-parameters linear exponential smoothing, 67
Work element, 202-3, 244-45
Work factor, 202
Work measurement, 201-31
Work sampling, 202, 217-24
Work sheet, 219
Work stations (centers), 244-5